Data Security and Privacy Protection

A Comprehensive Guide

Data Security and Privacy Protection

A Comprehensive Guide

Anyu Wang

Cloud Security Alliance Great China Region

電子工業出版社
PUBLISHING HOUSE OF ELECTRONICS INDUSTRY
http://www.phei.com.cn

World Scientific

Published by

World Scientific Publishing Co. Pte. Ltd.
5 Toh Tuck Link, Singapore 596224
USA office: 27 Warren Street, Suite 401-402, Hackensack, NJ 07601
UK office: 57 Shelton Street, Covent Garden, London WC2H 9HE

Library of Congress Cataloging-in-Publication Data
Names: Wang, Anyu (Writer on data security), author.
Title: Data security and privacy protection : a comprehensive guide / Anyu Wang,
Cloud Security Alliance, Great China Region, China.
Description: New Jersey : World Scientific, [2025] | Includes bibliographical references and index.
Identifiers: LCCN 2024035243 | ISBN 9789819800278 (hardcover) |
ISBN 9789819800285 (ebook for institutions) | ISBN 9789819800292 (ebook for individuals)
Subjects: LCSH: Computer security--Textbooks. | Data privacy--Textbooks.
Classification: LCC QA76.9.A25 W25 2025 | DDC 005.8--dc23/eng/20241221
LC record available at https://lccn.loc.gov/2024035243

British Library Cataloguing-in-Publication Data
A catalogue record for this book is available from the British Library.

For any available supplementary material, please visit
https://www.worldscientific.com/worldscibooks/10.1142/14027#t=suppl

Desk Editors: Soundararajan Raghuraman/Julio Hong

Typeset by Stallion Press
Email: enquiries@stallionpress.com

Identify, detect and protect, for a brighter future and better world!

Foreword by Changjun Jiang

In recent years, the digital economy is in an unprecedented high-speed development process, covering finance, commerce, and trade, manufacturing, logistics, transportation, education, and other walks of life, which has brought extensive and profound impact to people's production and life. Especially, in the face of the sudden novel coronavirus pneumonia epidemic, the digital economy has hedged the negative impact of the epidemic on traditional industries and highlighted its multi-faceted value to society.

Data is the key production factor of the digital economy and the key strategic resource affecting global competition in the era of the digital economy. The world's largest technology companies by market capitalization, such as Facebook, Amazon, Google, Alibaba, and Tencent, are basically the first enterprises to use data and application data for services on a large scale. In the era of digital economy, the requirements for data collection, storage, analysis, and sharing capabilities are far greater than before, and once a large-scale data leak occurs at a key node, it may endanger users' privacy, property, and even personal safety for individuals. For enterprises and countries, if these data are obtained by competitors, it may have an impact on the development of enterprises and even the country's economy and security. Therefore, a high level of data security governance is the basic guarantee for the high-quality development of the digital economy.

Governance of data security has become the consensus of all countries, and more than 107 countries and regions around the world have formulated laws related to data security and privacy protection, such as the European Union, the United States, Russia, Japan, Singapore, Brazil, and India. This also indicates that the governance of data security is no longer just the responsibility of security engineers but also involves every participant in the production process of the digital economy.

Data security governance runs through the entire lifecycle of data collection, transmission, use, storage, and destruction, not only technically involving multiple security fields such as data classification, identity authentication, communication security, storage security, and operational security but also involving non-technical fields such as policies and regulations and industrial ecology, which is a complex and systematic system.

This book is a tool book for professionals, researchers, college students, and others interested in the field of data security in the fields of IT and digital construction. It is believed that the hierarchical classification system of data, data-centric security architecture, data security protection capabilities of the whole life-cycle, and various application scenarios described in this book can inspire readers. At the same time, this book also summarizes the cutting-edge research progress of AI data security, IoT data security and 5G/6G data security, which can be used as a reference for relevant practitioners. Written by industry experts with both industry and academic backgrounds, this book covers the core content of the field of data security and to a certain extent represents the research and industry practice of new technologies in the field of data security. It is believed that this book can play a positive role in the popularization of data security knowledge, the dissemination of data security thinking, and the popularization of data security technology.

Changjun Jiang
Academician of the Chinese Academy of Engineering

Foreword by Yale Li

In the current wave of global technological and industrial revolution, emerging technologies such as cloud computing, artificial intelligence, and big data are rapidly permeating various fields and integrating deeply with traditional industries. New technologies and business models are constantly emerging, marking the onset of a major era of transformation in the digital economy. This evolution is driving profound changes in the modes of production, social relationships, economic structures, and lifestyles of human society. Data has become the cornerstone of this new technological revolution, serving as the core resource of the digital economy and information society. Dubbed the "oil" and "diamond mine" of the 21st century, data is akin to the "land" of the agricultural era and the "capital" of the industrial age. Those who control data ownership possess resources and wealth, wield societal control, and hold the key to the future.

Data security, therefore, is a critical issue that impacts national security, public safety, individual privacy, international security, and social equity and justice. However, there is still no universally accepted or comprehensive definition of "data ownership". Data breaches are the most significant threat brought about by digital technology, and the flow and protection of data present a series of security challenges for enterprises. Data breaches impose substantial costs on businesses. According to a report by IBM, the average total cost of a data breach is $4.24 million, with the per-record cost rising by 10.3% from 2020 to 2021. In 2021, the average total cost of lost business opportunities was $1.59 million. The COVID-19 pandemic has led many companies to adopt remote work models, creating opportunities for cybercriminals and exacerbating the risk of data breaches. IBM's research indicates that when remote work contributes to a data breach, the average total cost rises by approximately $800,000, reaching $5.01 million. Organizations with more than 60% of employees working remotely bear higher average costs of data breaches compared to those with no remote staff.

The promulgation of the **Data Security Law of the People's Republic of China** on September 1, 2021, marked a groundbreaking milestone in the digital era. This law implements a comprehensive approach to national security, focusing on potential risks in data security, enhancing the coordination of national data security efforts, and establishing fundamental systems, such as data classification and hierarchical management, data security review, risk assessment, monitoring, early warning, and emergency response. At its core, data security aims to prevent misuse, abuse, and illegal use of data. Achieving this goal requires clarifying the rights, responsibilities, and interests of stakeholders, including data subjects, users, and regulators. It necessitates close collaboration among all parties and the establishment of rules and standards for data collection, storage, transmission, and usage. Enhancing privacy protection measures and regulating actions in access control, identity authentication, data encryption and masking, disaster recovery, and security auditing are also essential. Additionally, the **Personal Information Protection Law of the People's Republic of China**, which came into effect on November 1, 2021, further strengthens privacy protection. Together with the **Cybersecurity Law**, these three laws form a fundamental legal framework covering all aspects of the internet, personal information, and data activities. As such, it can be said, "Data is the lifeline of cybersecurity; without data security, there can be no cybersecurity".

The publication of this book reflects the dedication of its author, Anyu Wang, who has demonstrated exceptional leadership and professional expertise. It is my belief that this book will help readers gain a deep and comprehensive understanding of the field of data security, thereby contributing to the overall improvement of industry capabilities in data security and privacy protection, and supporting the sustainable development of the digital economy.

Yale Li
Chairman of CSA Greater China Region

Preface

At present, the world is in the midst of a new round of technological revolution and industrial revolution. The rapid development of new generation information technology such as big data, cloud computing, 5G, and smart terminals, and the rapid penetration of artificial intelligence technology and applications into various industries and fields are an important force to promote economic and social change.

Looking ahead, with the rapid rise and popularization of 5G, the era of the Internet of Everything has arrived. The Internet of Things expands the Internet to tens of billions of smart devices, artificial intelligence (AI) technology provides massive amounts of information, computing, and insights, augmented reality and virtual reality technology provides users with rich experiences, and the virtual world and the real world are further integrated.

Today, data is at the heart of artificial intelligence and cloud computing. Basically, all institutions, organizations, enterprises, and even everyone generate, process, transmit, and store all kinds of data to varying degrees.

The widespread popularity of AI technology and the Internet of Things has brought profound changes to production and life scenarios, making data show a spurt of growth, resulting in more and more data being collected and faster and faster data transmission. Human society is generating massive amounts of data all the time, which has sharply increased data security risks. At the same time, data, as the basis of artificial intelligence technologies such as machine learning, continues to promote the rapid development of intelligent technology and brings a wider range of artificial intelligence applications, which in turn will collect more data and form a larger data platform. In addition, with the development of artificial intelligence technology, the ability to analyze and mine data is also rapidly increasing, which will lead to risks to personal privacy, social security, and even national security. In particular, as more and more data are collected and utilized, data security and privacy protection have become serious security challenges faced by big data and intelligent systems in the development and application process.

The data security landscape is surrounded by environments and businesses that are constantly evolving. Whether it is the international political game, laws and regulations, international standards, national and industry standards, etc., all of them have put forward higher and higher requirements for data security. In addition, the continuous emergence of new scenarios, new services, and new formats has also brought huge risks and challenges to data security.

From the dimension of the overall framework of data security and data security model, this book researches and formulates data security architecture and design models according to the basic system of data security, the principles and cases of security design, and forward-looking analysis and elaboration of key technologies of data security and privacy protection in view of regulatory requirements, industry needs, and user demands, and faces future evolution, in order to provide management and professionals in information security, network security, cloud computing, and privacy protection with systematic cognition and comprehensive annotations in the field of data security. This book can also be used as a reference book for researchers related to data security, as well as others interested in consulting, management, and architecture design in the field of data security.

This book consists of six chapters.

Chapter 1 clarifies the scope of data security, defines the overall principles and implementations of data security, and provides an overview of the data security architecture.

Chapter 2 introduces the risks and challenges of data security from multiple perspectives and systematically explains the response mechanism of data security.

Chapter 3 introduces the framework for the design and implementation of data security architecture, analyzes the key technologies involved in data security, combines application scenarios, and provides some excellent cases.

Chapter 4 introduces the cornerstone of data security: cryptography, based on the key technologies of cryptography and applied encryption, combined with a wide range of industrial application practices, and looks forward to the new development of cryptography in the quantum era.

Chapter 5 introduces the regulatory requirements for privacy protection and data security compliance, the principles and systems of privacy protection, and the key technologies of privacy protection.

Chapter 6 describes the evolution of data security systems. This chapter elaborates on AI and data security, the system and algorithm behind blockchain, IoT data security, 5G, and data security and looks forward to data security in the 6G era.

In the process of writing this book, we have received strong support and assistance from the leaders and experts of the Cloud Security Alliance, and I would like to thank them here!

This book embodies the theoretical summary and practical experience of the two authors in cyberspace security research for many years, and in the process of compilation, they have extensively referred to the latest works and research reports at home and abroad, and I would like to express my heartfelt thanks to the relevant authors.

About the Author

Anyu Wang is a distinguished expert in data security and a Fellow of the Cloud Security Alliance Great China Region, where he also serves as Co-chair of the Data Security Working Group. With a bachelor's degree in Electronic Engineering from Shandong University and over two decades of experience in the IT industry, Mr. Wang has made significant contributions to advancing cybersecurity practices. His career includes pivotal roles, such as Chief Data Security Architect at a leading telecommunications company, where he developed cutting-edge security solutions for 5G technologies, and a tenure at a globally renowned smartphone manufacturer overseeing product security and compliance.

As an accomplished author, Mr. Wang has published several influential books on data security, widely respected within China's tech community. Beyond his professional life, he enjoys hiking, reading, and spending time with his family. He divides his time between Hong Kong and Shanghai, drawing inspiration from both cities for his innovative work in IT and cybersecurity.

Contents

Chapter 1

An Introduction to Data Security

Since the 20th century, the wave of scientific and technological progress has profoundly affected the global economy, changing all aspects of people's production and life. In the 1950s, the emergence of computers and digital technologies marked the beginning of the third industrial revolution. Computers were rapidly promoted and popularized in various fields, such as industry and agriculture, and triggered profound changes in industries, such as banking, energy, and communications. The rapid development and popularization of network technology, as well as the rise of the digital revolution, have brought the world closer together. Innovative technologies are on the rise and becoming mainstream.

The new generation of network technology and the transformation of information technology have affected human society with unprecedented strength and breadth. Based on an investigation of the global IoT market (Transforma Insights, 2020), the number of active IoT devices in 2019 was 7.6 billion. By 2030, there will be 24.1 billion connected devices worldwide. These networked devices, which include not only computers, servers, and mobile phones but also various smart terminal devices, such as smart wearable devices, Augmented Reality (AR) glasses, Virtual Reality (VR) headsets, and even holographic projection devices, will also be further popularized in people's lives. In addition, vehicles, drones, industrial robots, home appliances, and smart sensors will also account for an increasing proportion of network-connected devices.

The world today has entered the era of the fourth industrial revolution. The Fourth Industrial Revolution incorporates artificial intelligence (AI), robotics, the Internet of Things (IoT), 3D printing, genetic engineering, quantum computing, and other technologies that are the backers behind many of the products and services that are indispensable to modern life. For example, while driving, the intelligent navigation system can quickly calculate the fastest route to the destination. Voice assistants in smartphones provide rich human–computer interaction methods and can even replace the owner's call to reserve seats in restaurants; video sites can provide viewing and recommendation of movies based on personal interests; social apps can recognize faces in uploaded photos and tag friends associated with them.

As cyberspace and real space converge, the lines between the physical, digital, and biological worlds blur or converge. In this scenario, the scope of data has been greatly extended, the amount of data has expanded sharply, the importance and value of data have become higher and higher, the transmission speed of data has also been rapidly improved, and the risk of data has also increased rapidly.

In the era of convergence of all things, the volume, quantity, velocity, and quality of data have increased significantly, leading to a corresponding increase in data security risks. The evolution of the different dimensions of data is shown in Fig. 1.1. In the ever-changing and rapidly evolving digital world, systematic analysis and identification of data security risks, targeted formulation and implementation of various security control measures, design and implementation of security architecture, and continuous security operation are the key mechanisms for any organization to protect data security.

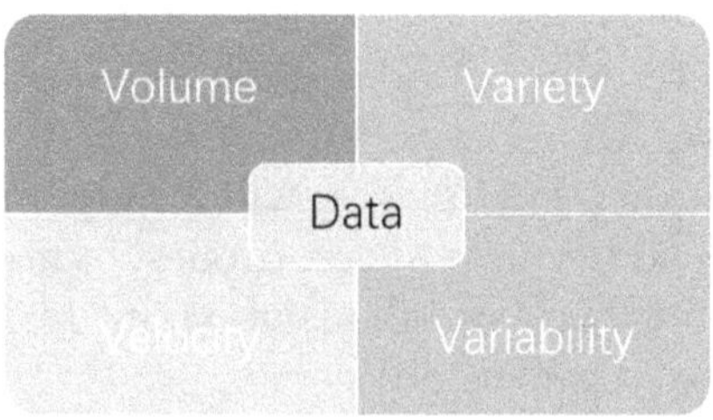

Fig. 1.1. The evolution of different dimensions of data.

1.1 What Is Data Security?

Data security encompasses a set of related standards, technologies, frameworks, and processes that focus on protecting data from unauthorized access and corruption throughout its lifecycle.

The primary purpose of data security is to protect data as it is collected, stored, created, received, or transmitted. Regulatory compliance is also an important consideration. Regardless of the device, technology, or process used to manage, store, or collect data, it must be protected.

Due to the rapid spread of digital transformation and its gradual integration into all aspects of life, almost all government agencies and businesses process data to some degree. Therefore, the data security landscape faces a variety of environments and enterprises. However, whether it is a banking system processing personal and financial data, basic cloud services provided by operators and Internet providers, or storing user photos on mobile phones, the three most important elements for data security are the definition of data, the data lifecycle, and the data security requirements, as shown in Fig. 1.2.

These three concepts are elaborated on in more detail in this chapter.

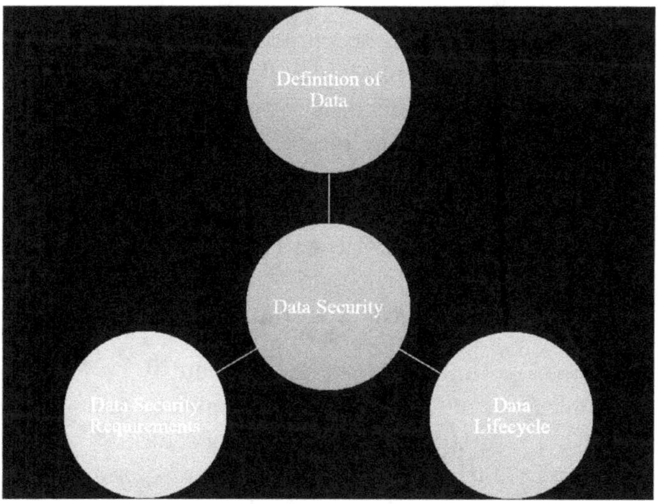

Fig. 1.2. Three elements of data security.

1.1.1 *Definition of data*

According to the basic definition, data is the result of facts or observations, a logical induction of objective things, and unprocessed raw materials used to represent objective things.

Data is the representation and carrier of information, which can be symbols, words, numbers, voice, images, videos, etc. Data and information are inseparable; data is the expression of information, and information is the connotation of data. Data itself has no meaning, and data becomes information only when it affects the behavior of the entity.

Formally, data can be a continuous value, such as sound or image, called analog data; it can also be discrete, such as symbols or words, called digital data. Strictly speaking, digital data is data represented by the binary systems 0 and 1 rather than in analog form.

In modern (post-1960) computer systems, data is represented in terms of binary information units $(0, 1)$. Unless otherwise noted, the term "data" in this book generally refers to narrow numerical data that computer systems can handle.

Of course, not all data has the same value. Therefore, data security is primarily concerned with the security of "data that needs to be protected". Two related concepts are of particular concern to organizations and individuals in general: personal data and sensitive data.

Personal data, also known as personal information, is any information that relates to an individual's identity. The scope of personal data is very broad, and its meaning varies slightly in different countries and regions, but the principles of personal data protection are similar. The two main types of personal information to be aware of are personally identifiable information (PII) and personal privacy information. The National Institute of Standards and Technology (NIST) defines personally identifiable information as "any information about an individual that is maintained by an agency, including:

(1) any information that can be used to identify or locate an individual, such as name, social security number, date and place of birth, mother's birth name, or biometric data;
(2) any other information that is linked to or accessible about individuals, such as medical, educational, financial, and employment information".

It is worth noting that according to this definition, an Internet user's IP address is "linked personal data". However, in the European Union (EU), an Internet user's IP address is personal data.

Sensitive personal data is a subset of personal data. Its scope is defined differently in different countries and regions. In general, personal financial data (e.g., bank cards), personal health data (e.g., genetics and medical records), and personal biometric data (e.g., facial features and fingerprints) are considered sensitive personal data. When handling sensitive personal data, it is important not only to implement the necessary security safeguards but also to comply with applicable legal, regulatory, and other compliance requirements.

1.1.2 *Data lifecycle*

With the rapid increase in data scale, the advent of the era of big data, and the continuous development and evolution of the Internet of Things, Data Lifecycle Management (DLM) has become increasingly important. More and more devices are generating huge amounts of data in every corner of the world, so it is important to identify the entire lifecycle of various data and properly secure it while maintaining its usability.

Data lifecycle, which is the sequence of lifecycle stages from the initial creation (or capture) of data through storage, use, sharing (transmission), and archiving, to its end of life and destruction, as shown in Fig. 1.3.

DLM is the process of managing the flow of data throughout its lifecycle. Accurately identifying and protecting data, especially sensitive data, throughout its lifecycle is a fundamental requirement for data security. Analysis based on the data lifecycle can also identify where security controls are applied and appropriate security mechanisms.

Different types of data have different lifecycle stages, and the security protection focus of each stage is different. This section provides only a basic explanation.

At the data creation stage, different types of data are created in very different ways. For example, digital thermometers collect indoor temperature data from offices, telecommunication equipment generates bills for each call a user makes, and mobile phones synchronize photos to cloud storage. However, from the perspective of the

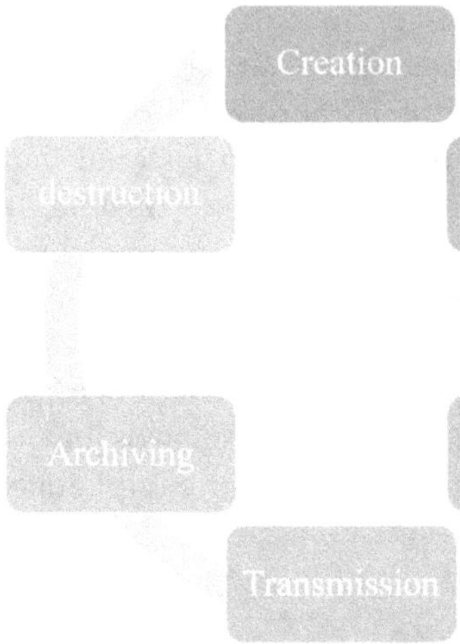

Fig. 1.3. Data lifecycle.

basic principles of data security, the data should be identified and appropriately classified at the data creation stage, and the protection requirements for such data should also be clarified.

At the data storage stage, similar to the data creation stage, the way the data is stored is also very different, which can be stored on a local storage device, or distributed storage, cloud storage. At this stage, the necessary data access controls should be implemented to ensure that the data is properly protected to reduce the risk of data leakage, tampering, and destruction.

At the data use stage, the key point of data security protection is to apply the security control mechanism to the data in use. In general, organizations should be able to monitor access to data, especially sensitive data, and apply various security controls to ensure the security of the data.

The data exchange stage is often the weakest link in traditional data security controls. For data as an information carrier, data can only create value after it is shared, but to some extent, data security and data value are in conflict, so it is important to apply the right security controls at the right time.

At the data archiving stage, it is necessary to consider the mandatory requirements of various laws and regulations on data retention, and decide on the archiving method, protection mechanism, and data isolation requirements according to organizational policies and user needs.

At the data destruction stage, the time limit for archiving the data has expired, or the data should be deleted based on legal or business requirements. At this point, the focus of data security is to implement an appropriate data destruction mechanism to ensure that the data cannot be recovered. Even if the data is not stored in a traditional file type, an appropriate destruction mechanism must be considered.

These are the six typical stages of the data lifecycle. It's important to note that different types of data don't have exactly the same lifecycle stages. For example, from the perspective of a digital thermometer, there may only be two stages for the temperature data of a digital thermometer: creation and sharing.

As data access interfaces become more complex, attack vectors increase, and attacker strategies become more systematic, all aspects of data security become very important to data security management — from secure data storage, use, and transmission to access control and effective key management, if there is a link that is vulnerable to attack, it will undermine the effectiveness of the entire security mechanism.

Addressing risks from these multiple dimensions requires a comprehensive, data-centric approach to security. This means focusing on protecting the data itself, not just the network, applications, or servers, at all points in the data lifecycle.

Data in the three states of storage (sometimes referred to as at rest), in use, and in motion (see Fig. 1.4) face different security threats and challenges, so different protection mechanisms should be employed for data in these three states.

Next, the three states in the data lifecycle are described in detail.

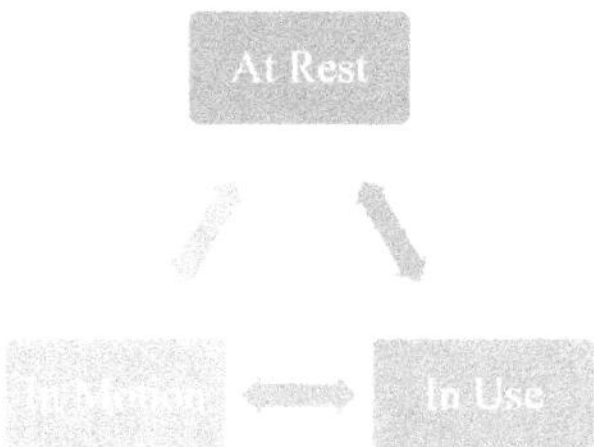

Fig. 1.4. Three states in the data lifecycle.

1.1.2.1 *Data at rest*

Data is static while it is stored on disk. In this relatively safe state, traditional layered defenses can help. First, the protection of the physical facilities where the storage is located (such as computer rooms and offices), second, the protection of the perimeter-based defenses (such as firewalls, and intrusion detection and protection systems), third, the endpoint-based defenses (such as antivirus software, endpoint policy protection, and patch management), and finally, the defenses of the data itself (including encryption and access control).

Encryption can be divided into two subtypes: disk encryption and file encryption. Hard drive encryption can prevent data security risks such as hard drive loss and theft, and can also have a positive impact on secure destruction scenarios. It is important to note that most disk encryption techniques have technical shortcomings, such as the fact that the contents of the disk are decrypted after power is turned on, at which point the files on the disk can be considered unencrypted. Therefore, application-level file encryption should be used for files that contain some sensitive data. For structured data storage types, many database engines also support features such as database encryption, table encryption, and even column-level encryption so that fine-grained protection for sensitive and personally identifiable information can be implemented.

Other storage security measures can also help protect data, such as storing different data classifications in different locations to reduce the likelihood that an attacker will have enough information to commit fraud or other crimes.

The above describes traditional host, mobile, and server storage. For cloud storage, there are other aspects of the requirement to focus on. For example, storage space must be geographically located where contracts, service level agreements, and regulations are allowed, and stored data must include all copies and backups to prevent loss due to data loss. For example, storing and using electronic health records, which are subject to the EU's General Data Protection Regulation (GDPR), can be a challenge for both data owners and cloud service providers. After storing data in the cloud, consider its reliability, confidentiality, and integrity, whether the supplier's security authority management measures are perfect, whether the stored data is securely managed, and whether the data storage format is reasonable. Therefore, additional integrity assurance, data encryption, and data isolation measures are required to ensure the security of data in this process.

In summary, data at rest refers to data that resides in locations such as hard drives, external storage, cloud storage, or tape drives, also known as static data. Data at rest is generally considered to be the most secure state of data. In this state, network and physical perimeter-based solutions can be considered the first line of defense. Depending on the purpose and sensitivity of the data itself, additional lines of defense may be added. Data storage must also comply with relevant laws and regulations.

1.1.2.2 *Data in motion*

While data at rest is easier to protect, there's no real value in data at rest. Data is being transferred to other consumers, allowing data to be shared and monetized. It is at this stage that data is most vulnerable.

Cloud data is transferred to other customers and virtual services for use over the network, process communications, etc., and due to the open nature of the network, data should not be transferred without security controls. The creator of the data needs to consider whether to set administrative permissions on the data, so to ensure the security of the transmitted data, it is necessary to use data encryption technology to maintain both confidentiality during data transmission and integrity after transmission.

Data in transit is more vulnerable than data at rest, whether it is transmitted over a public or private network or using other transport mechanisms. Figure 1.5 illustrates some typical data transport mechanisms.

The common practice for protecting data in transit is data encryption. Appropriate cryptographic algorithms and encryption practices should be used to leverage encryption in transit as an effective line of defense.

1.1.2.3 *Data in use*

Protecting data at rest is the easiest to solve and the focus of most attention. Protecting data in transit is the easiest to understand.

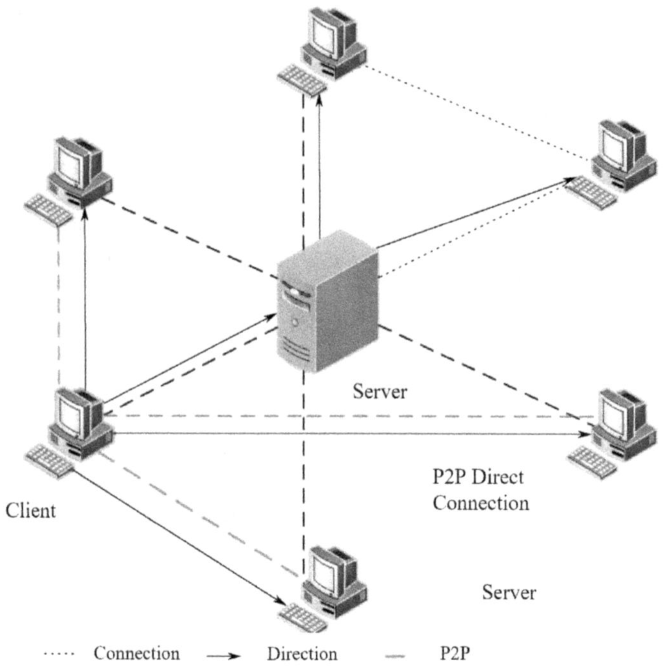

Fig. 1.5. Example of data transmission.

In contrast, the security of data in use is the easiest to overlook and the easiest for attackers to penetrate.

Data in use is more vulnerable than data at rest because, depending on its availability, it must be accessible to those who need it. Of course, the more people and devices that have access to data, the greater the risk that it will fall into the wrong hands. The key to securing data in use is to control access as tightly as possible and to combine one or more types of authentication to ensure that users are not using identities or credentials that have been stolen.

Organizations also need the ability to track and report on suspicious activity, diagnose potential threats, and proactively improve security. For example, an account that is locked out due to a certain number of failed login attempts may be a warning sign that the system is under attack.

Experience has shown that if there is data that is valuable to an attacker, it is important to identify and protect every entry point that can access that data. Limiting encryption to only one part of

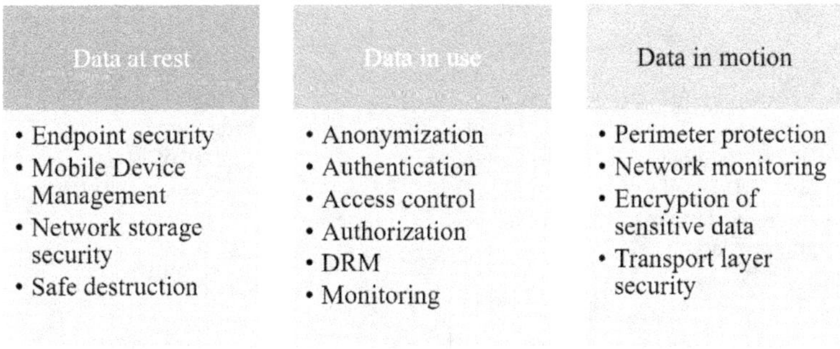

Fig. 1.6. Common protection mechanisms in various states of the data lifecycle.

the data lifecycle is dangerous. Protecting data at rest, in transit, and in use in all three states is critical.

In three different states of the data lifecycle, the challenges are different and the defenses are different. Figure 1.6 summarizes the common protection mechanisms at different stages of the data lifecycle. More is discussed in the following sections.

1.1.3 *Data security requirements*

When designing and implementing systems, products, and services that involve users' personal data, sensitive data, or trade secrets, the requirements for data security need to be considered.

Typical requirements for data security are as follows:

(1) Requirements for compliance with laws and regulations.
(2) Requirements for international and domestic standards.
(3) Requirements for industry standards, conventions, and best practices.
(4) The needs of the customers are served.
(5) Competitive needs.
(6) Demand for product internal quality improvement.
(7) Feedback Driven.

The above typical sources of data security requirements can be summarized into three categories: compliance requirements, external requirements, and internal requirements, as shown in Fig. 1.7.

Compliance requirements specifically refer to the laws of the corresponding country or region, the administrative regulations of

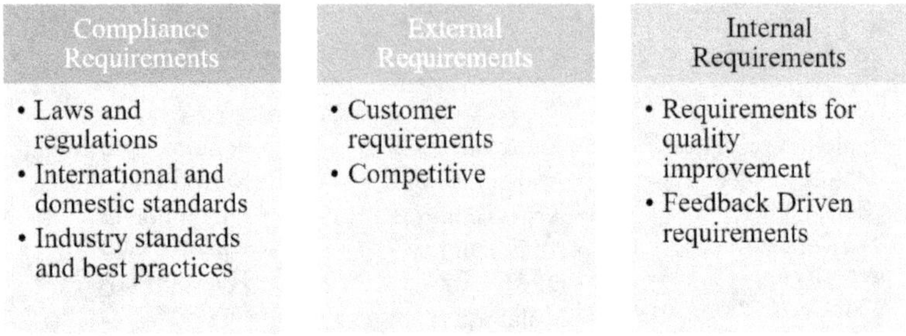

Fig. 1.7. Typical sources of data security requirements.

Fig. 1.8. Effectiveness of compliance requirements.

the region, international standards and national standards, industry standards, and best practices, and their effectiveness is shown in Fig. 1.8 from high to low.

For example, in Mainland China, when a retailer wants to establish a website for providing online retail services, it should first meet the legal requirements, especially the requirements of the "Cybersecurity Law of the People's Republic of China" (Qi *et al.*, 2018). Second, it is necessary to consider the administrative regulations of the region, such as the "Provisions on Protecting the Personal Information of Telecommunications and Internet Users" and "Provisions on the Cyber Protection of Children's Personal Information". Third, the requirements of relevant international standards, national standards, and industry standards should be analyzed, such as GB/T 35273-2020 "Information security technology — Personal information security specification" (Weber *et al.*, 2020).

It is worth noting that the "rules" of compliance requirements, i.e., laws and regulations and various standards, are not static.

For example, in 2021, both the Data Security Law of the People's Republic of China and the Personal Information Protection Law of the People's Republic of China have been implemented. Practitioners should pay attention to the dynamics of applicable laws and regulations and various standards, analyze their impact, and formulate or improve security control measures to achieve compliance.

At the implementation level, data security focuses on each stage of the data lifecycle to provide end-to-end security protection. Different stages of the data lifecycle face different threats and require different protection mechanisms and key controls. For example, at the data storage stage, the key to protection lies in mechanisms such as access control of files or databases and encryption key management of sensitive data. As for the data use stage, the key to protection lies in access control and system security mechanisms when the data is in use.

1.1.4 *Overall goal of data security*

The previous chapters pointed out that from an organization's perspective, the core of security technology is to protect data security. On the implementation path, according to the different stages of the data lifecycle, the technologies to focus on and adopt are also slightly different.

Historically, the technology that has received the most attention is network security technology. Network security technology refers to the isolation of an organization's internal network from the external network and the use of firewalls and gateways to ensure the security of the organization's internal network, thereby protecting the organization's information assets. At the same time, host security technology is also gradually being developed and popularized. For example, device and service identity authentication based on public key infrastructure (PKI), endpoint defense technology, data loss prevention (or data leak prevention, DLP), and so on. These technologies are constantly evolving, changing, and merging. Concepts and practices such as "defense in depth" security architecture and "data-centric security" architecture have also been proposed and are becoming more widely used. The increasing popularity of AI and big data brings new opportunities and challenges to an organization's security architecture. At the same time, the zero-trust security

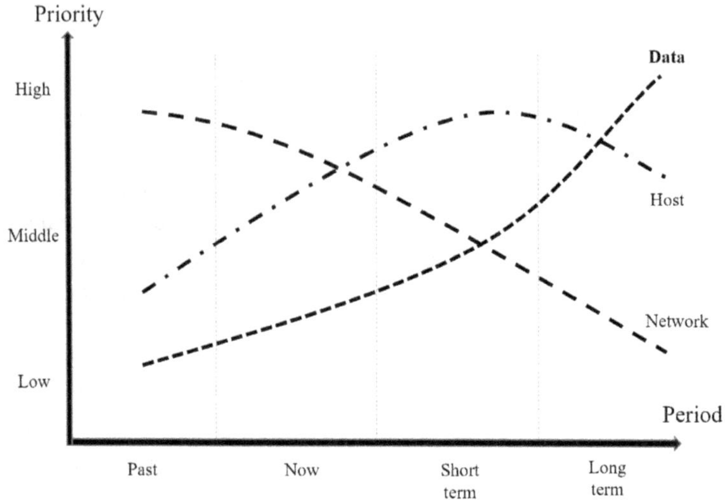

Fig. 1.9. The Evolution of information security technology.

architecture, as a new direction in recent years, has also received extensive attention.

In terms of the overall development trend of information security technology, the technology of network protection itself has matured, and the technology related to host security has also entered a mature stage, while the technology related to data security is developing rapidly and attracting more and more attention. The development of information security technology is shown in Fig. 1.9.

NIST Federal Information Processing Standard Publication 199 (FIPS 199) "Standards for Security Categorization of Federal Information and Information Systems" (Radack, 2004) proposed that the three core elements of the data security model are Confidentiality, Integrity and Availability, often referred to as the CIA Triad, as shown in Fig. 1.10.

Creating a data security model helps illustrate the relationship between the two areas of data security and information security. Data security and information security are inextricably linked, and in some cases, where the distinction is not strict, the two terms can be used interchangeably.

Information security generally refers to the protection of trade secrets, technical secrets, and other key information assets of a

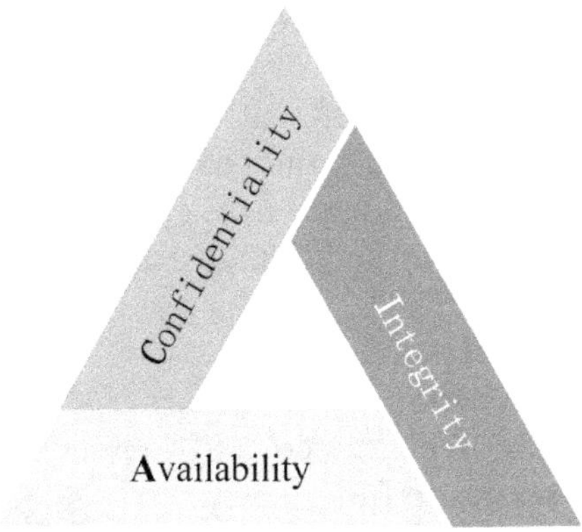

Fig. 1.10. Three core elements of a data security model.

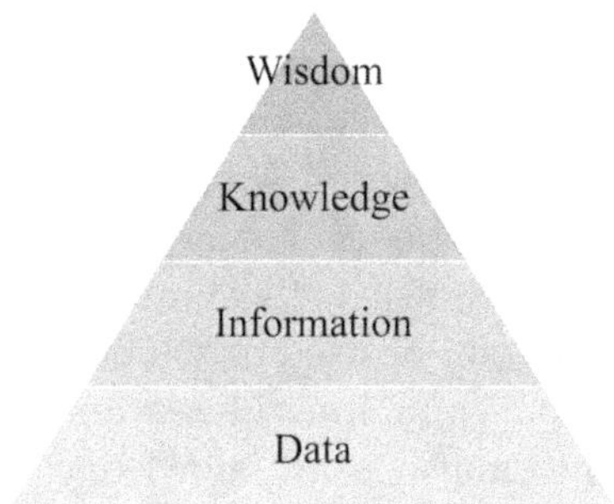

Fig. 1.11. DIKW pyramid model.

company or organization. In general, information security focuses more on confidentiality in the data security model.

The famous DIKW (Data, Information, Knowledge, and Wisdom) pyramid model (Rowley, 2007) clearly expounds the relationship between the four terms of data, information, knowledge, and wisdom, as shown in Fig. 1.11.

The bottom layer is data. Data is a collection of facts expressed in raw or unclassified form. The meaning of the data itself is not clear without association or context. For example, 12122020 is a string of numbers. It's easy to get the information when you add the context of the date: December 12, 2020. This brings us to the second concept: information.

Information is data that has been cleaned and processed for measurement, visualization, and analysis. Depending on the purpose, data processing may involve different operations, such as combining different datasets (aggregation) and ensuring that the data collected is relevant and accurate (validation), i.e., information is generated from the data associated with the context of the purpose. For example, 3371.96, 3373.28, and 3347.19 are a series of discrete data points. Knowing that these data points are data when the Shanghai Composite Index closes on certain days, a chart of the data points during that specific time period can be created to analyze the performance of the Shanghai Composite Index.

Information draws associations from data and can be used to answer the "4W" questions of who, what, when, and where. But once the question becomes "how", it moves to a higher level: knowledge.

How does the information derived from the collected data relate to our goals? How do "pieces" of that information relate to other pieces to add meaning and value? And, perhaps most importantly, "how" will the information be used to achieve goals?

Information becomes knowledge when it is not just a description of the facts collected but can be used to achieve goals. This knowledge often gives organizations and individuals an advantage over their competitors. Using information and knowledge to discover potential or implicit associations is called "insight".

Wisdom, the top of the DIKW pyramid, is achieved when an organization or individual uses the knowledge and insight gained from information to make proactive decisions. Wisdom is the application of knowledge to decision making. Wisdom can answer questions such as "why" in the first place, and it can also answer the question "which solution is best".

According to the conceptual analysis of information security and data security and according to the scope and connotation of information and data, data is more general and abstract. Data security focuses more on the business and technology of the domain, while information security focuses on the confidentiality of the organization's information and on confidentiality protection solutions and application scenarios from the organization's perspective.

The meaning of the three core elements of the data security model is explained in detail in the following.

1.1.4.1 *Confidentiality*

Confidentiality refers to ensuring that only those authorized to access information are able to obtain specified information. The subject of information access can be an actual user or a process, application, service, etc.

The definition of confidentiality is relatively abstract and can be understood intuitively from the opposite direction. Common cases of compromised data confidentiality include the following: loss of cell phones leads to leakage of stored photos; leakage of Internet account passwords leads to malicious login and operation of accounts; emails containing sensitive information such as quotes are mistakenly sent to others.

Common methods of ensuring confidentiality include data encryption and multi-factor authentication. Of course, extremely sensitive data may also require physical isolation, storage on non-networked devices, and other measures.

The purpose of taking protective measures in the process of maintaining information confidentiality is to prevent sensitive information from being leaked after being misrepresented or stolen during the transmission process. For this purpose, certain technical means must be used so that access to the real data in clear text is limited to those who have the right to see the data. General measures are implemented before the information is disclosed. There should also be plans for what to do if the data falls into the wrong hands. Once data has been obtained through unauthorized access, it is typically categorized according to the scope and nature of the breach, and appropriate remediation measures are implemented according to the category of these hypothetical situations.

In addition to technical countermeasures, protecting data confidentiality usually requires specialized training for personnel who handle critical data. This type of training can help those authorized to access critical information become familiar with potential risk factors and implement targeted preventive measures.

1.1.4.2 *Integrity*

Integrity refers to ensuring that data is not tampered with or illegally destroyed throughout its lifecycle to maintain its accuracy and integrity.

The term is very broad and has slightly different meanings in different contexts. In some scenarios, data integrity is synonymous with data quality (for example, no two records with opposite logic appear in the database), and in some scenarios, data integrity is the antonym of data corruption. In the realm of data security, data integrity aims to prevent data from being accidentally or intentionally altered.

Data integrity is a critical issue that must be considered in the design, implementation, and use of any system that stores, consumes, and transmits data. Integrity means that effective measures must be taken to ensure that data is not altered in transit or by unauthorized persons (e.g., when the confidentiality of information is compromised).

Any data that is inadvertently changed as a result of storage, retrieval, or use operations, including malicious tampering, unexpected hardware failure, and human error, is a data integrity failure scenario. If the change is caused by unauthorized access, it is also a data integrity failure scenario.

The purpose of data integrity techniques is the same: to ensure that data is recorded and later retrieved exactly as intended, making sure that the data is the same as when it was originally recorded.

Common data integrity measures include file access controls and user access controls. In addition, as part of day-to-day production, organizations must periodically take steps to detect changes in data caused by unintentional, non-human events, such as electromagnetic pulses or server crashes, including, but not limited to, checking the checksums of plaintext or encrypted data. In addition to regular testing, stable and secure backups must be in place to ensure that affected data can be restored to the correct state.

1.1.4.3 *Availability*

Availability refers to ensuring that legitimate users' access to and use of data can be guaranteed.

Any system used to store, use and transmit data should provide data in a timely and accurate manner when users need it. This means that the business functions of storage, use and transmission, the security controls used to protect the information and the communication channels used to access the information must function properly. High availability systems are designed to remain available at all times to prevent service interruptions due to power outages, hardware failures, and system upgrades. Ensuring availability also involves preventing Denial of Service Attacks (DoS attacks), such as situations in which a flood of incoming messages is sent to a target system, forcing that system to fail to serve normal users.

Rigorous maintenance of hardware equipment, regular hardware repairs, maintenance of the operating system environment, and keeping the system up-to-date with the latest security patches and updates are important guarantees for maintaining availability. Of course, it is also important to provide sufficient communication bandwidth and avoid performance bottlenecks. While appropriate preventative measures are taken, sufficient and stable redundancy, failover, Redundant Array of Independent Disks (RAID), and even high-availability clusters and other follow-on safeguards can effectively mitigate the effects of hardware problems. Rapid and adaptive disaster recovery is critical in the event of a disaster, and strong recovery capabilities rely on a comprehensive disaster recovery plan (DRP).

The DRP's risk assessment must include unpredictable events, such as disasters caused by force majeure. To prevent data loss from such events, backup copies should be kept as many times as possible and stored in geographically isolated locations. In addition to force majeure, in the case of possible malicious attacks such as denial-of-service attacks, the DRP requires the deployment of additional security devices or software (e.g., firewalls and proxy servers) to ensure that normal service can be restored quickly enough, even if the attack results in a short period of downtime.

As mentioned above, in the field of data security, usability is one of the three core elements, and it is also the key to ensuring that system users can use the system normally. At the same time, the three core elements of the CIA Triad also need to balance certain conflicts. For example, for enterprise IT systems, access to shared drives, configuration of network proxy servers, permissions for employees to send external email, and so on, all have confidentiality, integrity, and availability conflicts that require different teams (such as network operations, development, incident response, change management teams, and so on) to solve problems, address risks, and secure data across the enterprise.

1.2 Principles and Implementation of Data Classification

Data classification is the process of classifying data according to predefined, domain-specific criteria and continuously improving it. There is no doubt that accurately identifying and classifying the data processed in an organization's information systems is critical to properly selecting security controls and ensuring the confidentiality, integrity, and availability of the system and its data.

Data classification is the foundation of data security protection. Microsoft stated in the "Data Classification: Reclaiming Infosec's Redheaded Stepchild" at RSA 2016 (Preimesberger, 2016) that information security begins with data classification.

Data classification must cover the various stages of the data lifecycle. There are many ways to classify data, depending on the purpose and scenario of data use. In data security scenarios, data is generally labeled according to the type, sensitivity, and value of the data.

From a business perspective, data classification is easy to understand. In both production and everyday life, different types of data are often used. In an enterprise environment, different teams have access to different types of data. For example, sales and technical support teams maintain customer data, including mailing addresses, email addresses, and phone numbers. Financial and strategic analysts have access to operational and financial data. Human resources departments manage personnel data, such as employee salaries and performance. Software developers manage the source code library of software products, while network and security personnel monitor the company's network topology and account information. Cross-data access based on data classification should be limited by the principle of least privilege. For example, human resources should not have access to the source code library of software products, and network operations and maintenance personnel should not have access to customer data. The concept of data classification is derived from this simple scenario.

According to the 2019 Data Breach Investigations Report (Verizon Enterprise Solutions, 2019) data released by the US operator Verizon, based on an analysis of 41,686 incidents and 2013 confirmed data breaches in 2019, 34% of data breaches were triggered by factors within the organization and users from within the organization misused the data by virtue of the organization's trust in them.

The investigation of a total of 292 data breaches involving misuse showed that privilege abuse accounted for nearly 80% and data misuse accounted for 45%. Therefore, data classification and access control based on data classification are imperative in the practice of data security.

In reality, however, Microsoft's presentation at RSA 2016 noted that, based on its survey, 55% of IT professionals believe that data classification is too complex to plan, manage, and deploy. Sixty-three percent of IT professionals are unsure if their organization's classification policy aligns with how data is created, used, and shared. Even, 88% of IT professionals say they ignore or bypass an organization's data classification policies. The popularization and implementation of data classification, whether from the management level, or from the technical level and personnel awareness level, have a long way to go.

In the implementation dimension, the data classification model can be summarized as the identification method of data and the persistent labeling mechanism of data.

Based on the description of the data lifecycle in this book, data can be divided into structured data and unstructured data. The identification methods for these two types of data are also different. The data classification model is shown in Fig. 1.12.

When creating data, it is necessary to identify valuable data and relatively low-value data based on the business value of the data,

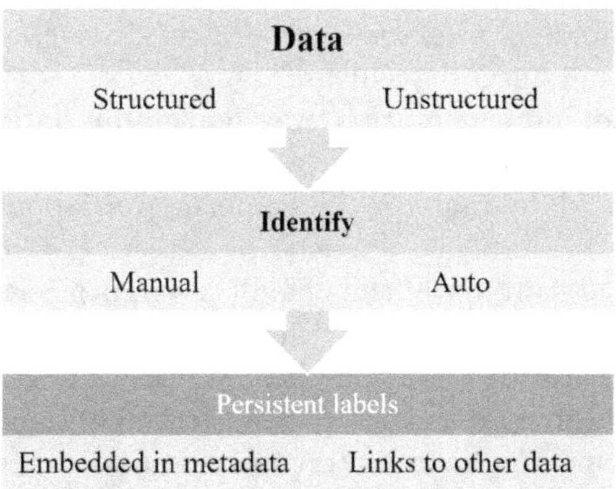

Fig. 1.12. Data classification model.

and define and implement appropriate access control measures for high-value data.

For structured data, its identification and classification are relatively simple; for unstructured data, its identification and classification are difficult.

There are three main types of data classification, which are content-based, context-based, and user-based. Among them, the first two can be automated to a certain extent. Automated data classification is the process of searching, identifying data content, and classifying data based on data content. It can be vividly compared to finding all the pages on the website that contain the words "data security". Of course, what is more important in practical scenarios is to identify and discover data assets that need to be protected in massive data storage, especially personal data, sensitive data, and intellectual property data.

For example, content-based data classification technology can read the content in the supported files through the file parser and then match the content with the target string or data. Context-based data classification techniques consider the location and application of the data. User-based data classification technology belongs to artificial identification, which depends on the knowledge and judgment of the user. Automated data identification and classification have limitations in various application scenarios, and manual identification is also required as a supplement. In fact, for small organizations and those that do not handle sensitive data, manual identification and classification may be the primary solution. In addition, manual identification and classification tend to be more accurate and help improve stakeholders' awareness of data sensitivity and to build an organization's data security culture.

Part of the implementation mechanism of data classification involves the processing of metadata. Metadata refers to "data about data", that is, "data that provides information about other data". Metadata can be implicit, explicit, or direct. For example, when receiving temperature data, the default assumption is that the data has a time reference of "current time", so the device logs the date, time, and temperature data together. When a data logger communicates temperatures, it must also report the date and time of each temperature, which is metadata in this scenario, where metadata is given implicitly. In the scenario of data classification, the classification of data is generally given explicitly in metadata.

In short, organizations of all types collect, generate, process, and transmit large amounts of data. Implementing effective data classification can help organizations do the following:

(1) **Organize relevant data effectively:** Storing unnecessary or duplicate data is not only expensive, but it can also overwhelm important data, lead to misjudgment, and even harm the business. Through data classification, potential correlations in the data can be discovered and outliers discarded.
(2) **Make data accessible:** Data classification ensures reliable and timely access to data by the right people. Additionally, labeling data helps the data to be discovered and increases productivity. With a clear data structure, each role in the organization can find what they need faster.
(3) **Ensure data security:** Classification is key to identifying the types of data an organization holds and properly protecting sensitive information. Data classification policies are used to authorize who has access to critical data. Securing data and limiting access to it provides organizations with a level of protection against cyberattacks and mitigates the impact of a data breach.
(4) **Compliance with regulatory requirements:** Commercial data is often tied to industry-specific regulations that require organizations of all types to protect sensitive data, such as personally identifiable information, credit card information, and health records. Data classification is critical to ensuring compliance standards and successfully passing audits.
(5) **Perform data analysis:** By classifying data, organizations can identify trends and gain insights into answering questions and make informed decisions. Through data analysis, organizations can understand the cause of a particular event, predict future outcomes, or measure the effectiveness of a particular action.

1.2.1 *Standards for data classification*

The ISO 27001:2022 standard "Information security, cybersecurity and privacy protection — Information security management systems — Requirements" (International Organization for Standardization, 2022) suggests that "classification of information" should be implemented to ensure that information is protected in a manner appropriate to its significance and states in the principles of information classification that "information shall be classified according to the information security needs of the organization based on confidentiality, integrity, availability and relevant interested party requirements". Correspondingly, the ISO 27002:2022 standard "Information security, cybersecurity and privacy protection — Information security controls" (International Organization for Standardization, 2022) provides clear recommendations on the principles of data categorization, in particular with regard to scenarios for the establishment and maintenance of information security management systems. The NIST Special Publication 800-60, Volumes 1 and 2, "Guide for Mapping Types of Information and Information Systems to Security Categories" (Stine *et al.*, 2008) provides similar advice.

ISO 27002 is intended to be used by organizations as a reference for implementing and managing information security controls. Many organizations use the ISO 27001 and ISO 27002 standards together as a framework to create laws and regulations. For example, the United States' "Sarbanes–Oxley Act" (Sarbanes, 2002) and the EU's "Data Protection Directive", the official name Directive 95/46/EC. The European Union's "Data Protection Directive" was promulgated in October 1995 and was replaced by GDPR in May 2018.

Section 5.12 of the ISO 27002:2022 standard states that the principles of data classification are as follows:

(1) legal requirements,
(2) data value,
(3) sensitivity of data leakage or tampering.

First of all, the classification of data and the formulation and implementation of related protection control measures should consider compliance with the legal requirements applicable to that type of data. Data types to focus on include financial data, financial data, health data, etc.

Second, something to consider is the value of the data. Taking a business organization as an example, the value of data can be seen as its importance and impact on business objectives. Typical confidential-level data may include intellectual property data (the design and formulation of a product), strategic planning, contract and customer information, investment plans, etc.

Finally, classification based on legal requirements and data value may be coarse-grained and do not cover all data types. Therefore, data classification is often implemented based on the sensitivity of the data to leakage or tampering.

The NIST data classification based on the degree of impact on the organization is shown in Table 1.1.

From the data classification methods of ISO and NIST, the principles of data classification are as follows: first, it is based on regulatory mandatory requirements, followed by classification based on the degree of data value (positive) or breach impact (negative).

When classifying data, it is important to note that data exists in one of three basic states: in storage (at rest), in use, and in transit middle. All three states require unique technical solutions for data classification and protection, but the principles for each data classification should be the same. Data classified as confidential needs to remain confidential while in storage, in use, and in transit.

Furthermore, data can also be structured or unstructured. The typical classification process for structured data in databases and spreadsheets is simpler and less time-consuming to manage than the classification process for unstructured data, such as documents, source codes, and emails. However, organizations typically have more unstructured data than structured data.

Data classification is not static, and there is a need for continuous refinement and continuous correction of deviations as organizations and businesses change.

Data classification can be implemented step by step by referring to the PDCA cycle in quality management. PDCA cycle is shown in Fig. 1.13:

(1) **Plan:** Identify the organization's data assets, designate responsible persons, initiate classification procedures, and develop protection configuration strategies.
(2) **Do:** Based on the data classification policy, deploy procedures and implement access restrictions for confidential data as needed.
(3) **Check:** Check and validate the report to ensure that the tools and methods used can be effectively processed based on the data classification strategy.

Table 1.1. Potential impact definitions for security objectives.

	Potential Impact		
Security Objectives	Low	Moderate	High
Confidentiality (Preserving authorized restrictions on information access and disclosure)	... Disclosures ... have a limited adverse effect on organizational operations, organizational assets, or individuals.	... Disclosures ... have a serious adverse effect on organizational operations, organizational assets, or individuals.	... Disclosures ... have a severe or catastrophic adverse effect on organizational operations, organizational assets, or individuals.
Integrity (Guarding against improper information modification or destruction)	The unauthorized modification or destruction of information could be expected to have a limited adverse effect. ...	The unauthorized modification or destruction of information could be expected to have a serious adverse effect. ...	The unauthorized modification or destruction of information could be expected to have a severe or catastrophic adverse effect. ...
Availability (Ensuring timely and reliable access to and use of information)	The disruption of access to or use of information or an information system could be expected to have a limited adverse effect. ...	The disruption of access to or use of information or an information system could be expected to have a serious adverse effect. ...	The disruption of access to or use of information or an information system could be expected to have a severe or catastrophic adverse effect. ...

(4) **Processing:** Check the status of data access, and review files and data that require revised classification, adopt changes, and respond to new risks.

Data classification requires ongoing maintenance. Periodic reassessment of data classification can confirm whether the original classification of the dataset is still applicable in light of changes in

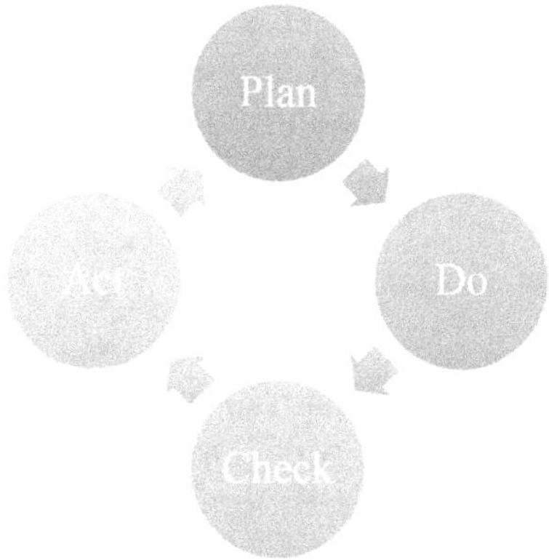

Fig. 1.13. PDCA cycle.

legal and contractual obligations, changes in data usage, or changes in the value of the data. This assessment should be performed by an appropriate data steward. Data stewards should determine the most appropriate frequency for reassessing data classification based on available resources, but it is recommended that reassessment occurs at least annually. When data stewards determine that the classification of a record has changed, an analysis of security controls should be performed to determine whether existing controls are consistent with the new classification. If gaps in existing security controls are identified, they should be remediated in a timely manner based on the level of risk posed by those gaps. There are also two scenarios for upgrading or downgrading data classification maintenance. For example, after the confidentiality period of a secret document expires, the scenario where the data classification of the document is downgraded from secret data to internal public data belongs to the downgrade scenario.

1.2.2 *Data classification framework*

In modern enterprise application systems, some systems support the use of information discovery technology and data classification templates to perform initial data classification. The results of data classification can be embedded in the data or associated with metadata, such as database field associations, field types, or file systems. Data classification schemes and their applications should span the entire data lifecycle to help organizations better store, use, and transfer data. Data classification is the foundation of data protection, which can effectively reduce data risk and improve the effectiveness of data leakage prevention, encryption, and other security measures.

1.2.2.1 *Data identification*

Data can be divided into structured and unstructured data. Structured data is data that conforms to a data model, has a well-defined structure, follows a consistent order, and can be easily accessed and used by humans or computer programs. Structured data is typically stored in a well-defined schema, such as traditional relational databases. Structured data is usually in tabular form, with rows and columns that have well-defined properties. SQL (Structured Query Language) is commonly used to manage structured data stored in databases.

In contrast, unstructured data is data that is not organized in a predefined way or has no predefined data model, making it unsuitable for mainstream relational databases. There are alternative platforms for storing and managing unstructured data. Unstructured data typically includes data in formats, such as text, voice, images, video, PDF documents, and media logs. It is becoming more prevalent in IT systems and can be used in various business intelligence and analytics applications.

An example of an organization's common data classification scheme is shown in Fig. 1.14.

Public	Internal	Confidential	Restricted
Information that is in the public domain	Data that should only be accessible to individuals authorized by the organization	Data that has special sensitivity and should require special clearance	Any strategic data that, if exposed, can cause major damage to the business
Marketing materials, information on official websites, sales representative contact information, etc.	Customer research reports, organization charts, employee training documents, etc.	Supplier Bill of Materials, Employee Performance Forms, R&D Design Documents, etc.	Trade secrets, intellectual properties, Protected health information

Fig. 1.14. Example of a data classification scheme.

1.2.2.2 *Persistent label*

The digital representation of the data classification results must persist with the data and be stored in the form of metadata, or be stored in a separate location and maintain a mapping of the data to its classification label. This persistent data classification result is commonly referred to as a "persistent label".

Consider the Microsoft Purview Information Protection solutions as an example. Microsoft's information protection solutions are implemented using "sensitivity labels". Users of Microsoft 365 client applications, such as Word and Excel, can apply sensitivity labels to documents or open protected documents to which they have access.

In Office applications, a sensitivity label appears to users like a tag on an email or document. Each content item can have a single sensitivity label applied to it. An item can have both a sensitivity label and a retention label.

Based on different protection requirements, such as privacy compliance requirements, the default sensitivity labels are personal, public, general, confidential, and highly confidential.

After the user applies a sensitivity label to content, the label is stored in the metadata of that email or document. The label and associated protection settings travel with the content and become the basis for policy application and enforcement.

Based on organizational needs and regulatory policies, administrators can revise the scope and operational requirements for sensitivity labels. For example:

(1) Highly confidential should apply to information whose unauthorized and reasonable disclosure could result in damages in excess of $1 million.
(2) Confidential should apply to information whose unauthorized and reasonable disclosure could result in more than 100,000 dollar loss.
(3) General should apply to information whose unauthorized disclosure should reasonably be expected not to cause measurable damage.
(4) Public should be applicable to information that is public and available to the outside world.
(5) Non-business should be applicable to direct or indirect information that has nothing to do with the company's business.

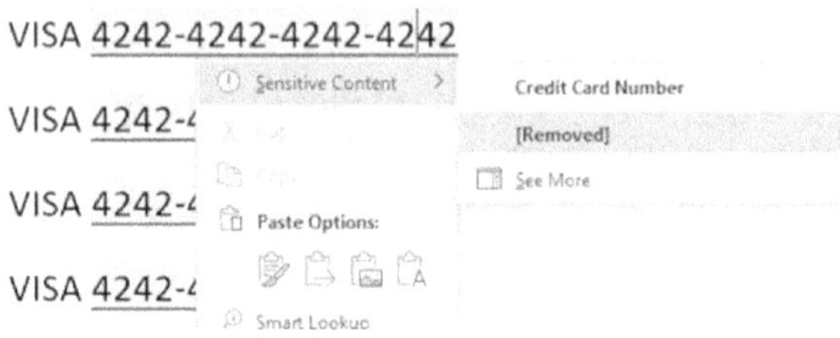

Fig. 1.15. Office 365 document labeling mechanism.

Each data classification describes the risk to the organization from unauthorized disclosure of information. After identifying the requirements for these classifications, the attributes of the data should be identified to help the data owner classify the data.

Microsoft's Office 365 suite supports manual and automatic sensitivity labeling. For example, when a user saves a Word document that contains a credit card number, the user may see a custom tooltip that advises the user to apply a sensitivity label to classify and protect the document. The Office 365 document labeling mechanism is shown in Fig. 1.15.

1.2.3 *Data classification implementation steps*

According to different data classification scenarios and requirements, the steps of data classification are slightly different. Good data classification work requires the support of efficient automated processing methods to cope with the large amount of data generated every day in information systems.

Usually, the implementation of data classification generally includes the following steps:

(1) Define the object to be classified, that is, the goal of data classification.
(2) Determine the data lifecycle state involved.
(3) Select the classification method: manual or automatic.
(4) Define and apply labels.

The implementation steps of data classification are shown in Fig. 1.16.

First, it is necessary to clarify the purpose of data classification. Note that not all data has the same value, and there is no need to classify all data, only to focus on data that has a business impact.

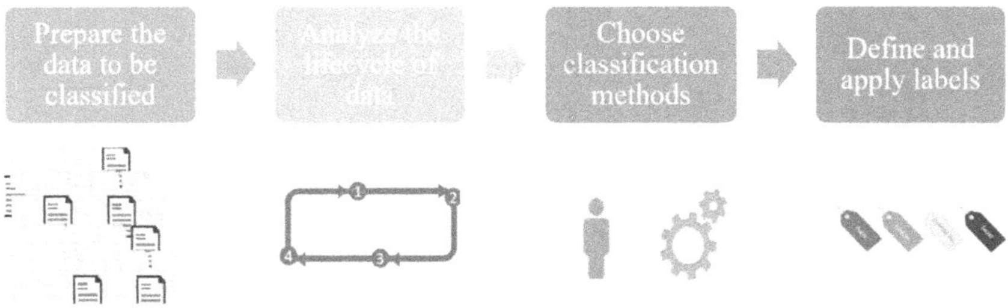

Fig. 1.16. Data classification implementation steps.

It's also worth noting that data classification is an ongoing, iterative process, and there is no one-size-fits-all data classification strategy. As the organization's business goals change or as time passes, the classification to which that data belongs will change. For example, before a new product is released, its exterior design, hardware configuration, and pricing strategy may be confidential information that needs to be fully protected. But after the product is released, the above information becomes public information.

Second, decide what data lifecycle states are involved. Determine the lifecycle state of the targets selected during the data classification phase. Most importantly, analyze when and where the sensitive data to be classified was created to recommend classification at the source of creation. Possible identified sources include applications, file servers, databases, code repositories, etc. Classification based on the context of the source would be more accurate. It is also more accurate and easier to audit based on the judgment of the creator. Identifying and classifying data as early as possible based on the data's lifecycle status is more conducive to covering the security of the entire data lifecycle.

Again, select the appropriate classification method. Available classification methods include automatic classification (content based, context based, etc.) and manual classification. The ideal classification method is one that minimizes disruption and inconvenience to users. Therefore, modern data classification solutions typically use a combination of automated pre-identification, automated suggestions, and manual judgment. Artificial intelligence is also playing an increasing role in data classification.

An ideal machine learning model for data classification is as follows: in the training phase, user information, document content, context of creation, and document classification results (labels)

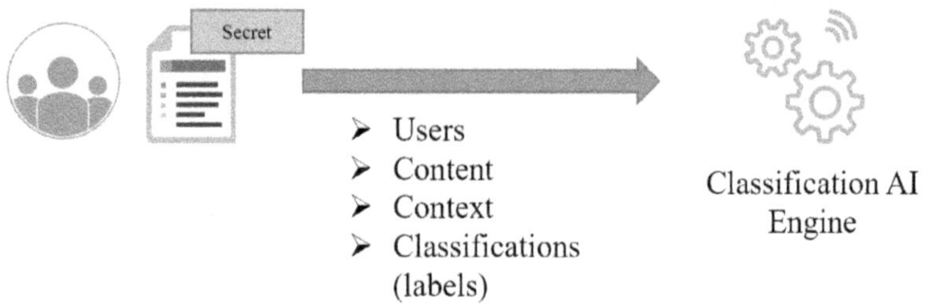

Fig. 1.17. Example of machine learning-based classification engine training.

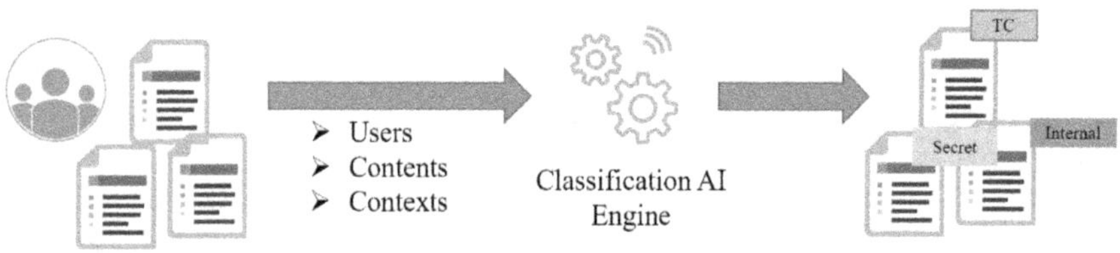

Fig. 1.18. Application example of classification engine based on machine learning.

are input. An example of machine learning-based classification engine training is shown in Fig. 1.17.

After that, the input documents are given classification suggestions using the machine learning model obtained from the training. An example application of machine learning based classification engine is shown in Fig. 1.18.

In the field of machine learning, "Classification" strategies usually fall under the umbrella of supervised learning. This book discusses in Section 6.1 in depth.

Finally, define and apply labels. The category of data should be able to carry the purpose of protection, and the label of data should also be self-explanatory. Whether it is Data Loss Prevention (DLP) or information security management based on departments and roles within the organization, appropriate data labels need to be applied.

Note that as mentioned earlier, a large number of different labels come with additional administrative costs. For data leakage prevention and general information security scenarios, Section 1.2.2 of this book suggests the protection based on 3 5 classification levels (such as high confidential, confidential, internal, and public) is more appropriate. After defining the labels, the protection policy matrix can be

Table 1.2. Data classification and protection strategies.

Class	Topics	Protection Strategies
Internal	Most documents	All employees
Confidential	Database of customer information	Account managers
Secret	Employee Information Sheet	HR
Top secret	Financial statements	Finance and executive management
Public	Company annual report	No additional protection required

defined based on the concept of access control matrix. For details about data classification and protection policies, see Table 1.2.

Today, there are many mature IT systems or solutions related to data classification. When selecting a solution, organizations can consider the following suggestions:

(1) It should support both manual and automatic data classification methods.
(2) It should support as many data forms and storage formats as possible, not just the document type of the Microsoft Office system.
(3) The classification mechanism can be integrated with existing DLP, archiving, and electronic discovery (or e-discovery) systems.
(4) Classification labels can be obtained throughout the data lifecycle (cross-platform).

1.2.4 *Regulatory-based data classification*

Based on the data classification framework and implementation steps introduced in the previous sections, personal data can be categorized as financial and fiscal data, social data, extrinsic or physiological data, intrinsic data, etc., in terms of the top-level classification of data according to EU standards.

On April 14, 2016, the European Parliament voted to adopt the General Data Protection Regulation (GDPR). The regulation includes 91 articles and a total of 204 pages. It officially came into effect on May 25, 2018. The predecessor of GDPR was the 1995 Data Protection Act in European Union.

With the adoption of this regulation, the protection and control of personal data in the EU has reached an unprecedented level. When comparing the personal data protection laws of different countries around the world, this regulation can be considered as one of the strictest data protection laws in the world.

The GDPR contains definitions of personal data and sensitive data. Among them, one or more of the following categories of personal data are classified as sensitive data:

(1) basic identity information, such as name, address, and ID card number,
(2) network data, such as location, IP address, cookie data, and RFID tags,
(3) healthcare and genetic data,
(4) biometric data, such as fingerprints and irises,
(5) racial or ethnic data,
(6) religious/philosophical beliefs,
(7) political views,
(8) data related to health, sex life, or sexual orientation.

The GDPR's definition of sensitive data mostly inherits categories from the earlier Dutch Personal Data Protection Act, with additions and amendments only in the dimensions of healthcare, genetic data, and biometric data.

The enactment and implementation of the GDPR have had a wide-ranging and profound impact on a global scale. The principles of its legislation, the definition and categorization of personal data, as well as the strict protection requirements have profoundly influenced subsequent personal data protection legislation in various countries and regions.

The GDPR definition focuses on principles and is relatively broad. Other personal data protection laws and regulations have similar or different requirements. From a general perspective, personal data can be categorized as follows.

1.2.4.1 *Financial data*

Based on the data classification framework and implementation steps presented in the above sections, personal data can be categorized

Table 1.3. Categories of financial and monetary data.

Categories	Description	Examples
Accounts	Data relating to financial accounts of individuals	Bank card number, bank account information
Possessions	Data related to the ownership of individuals	Cars, real estate, apartments
Transactions	Personal income, expenditure and all types of spending	Purchase records, lease records, tax records
Credits	Creditworthiness in relation to one's financial situation	Credit card repayment history, credit score, credit card credit limit

into financial and fiscal data, social data, extrinsic characteristics or physiological data, intrinsic characteristics data, etc., according to EU standards.

The categories and examples of financial and financial data are shown in Table 1.3.

1.2.4.2 *Social data*

Social data includes data of various dimensions as a "social person". Organizations providing social media services, communication services, and public services need to focus on this type of data. The legal constraints associated with such data are complex and require further identification of relevant legal requirements in the country and region.

The categories and examples of social data are shown in Table 1.4.

1.2.4.3 *Extrinsic characteristics or physiological data*

The External characteristics or physiological data category contains a large amount of data attributed to various dimensions of "natural persons". Companies or government agencies providing communications services, Internet services, medical and health services, and public services need to focus on this type of data. In addition, there are often more stringent regulatory requirements for medical and health data.

See Table 1.5 for categories and examples of extrinsic characteristics or physiological data.

Table 1.4. Categories of social data.

Categories	Description	Examples
Professional	Information about a person's educational or professional career	School attendance, employment history, employee files, evaluations, references, interviews, certifications
Criminal	Information about a person's criminal activity	Convictions, charges, pardons
Public life	Information about a person's public life	Character, general reputation, marital status, religion, political affiliations, communications metadata
Family	Information about a person's family and relationships	Family structure, siblings, marriages, divorces, relationships
Communication	Information communicated from or to a person	Email, telephone recordings, social messages

Table 1.5. Categories of external characteristics.

Categories	Description	Examples
Identifying	Information that identifies a specific person	Name, username, government-issued identification, biometric data
Physical characteristics	Information that describes a person's physical appearance	Height, weight, age, hair color, skin tone, tattoos, gender, piercings
Ethnicity	Information that describes a person's origins and lineage	Race, national or ethnic origin, language spoken, dialects
Behavioral	Actions or activities in the virtual or real world	Browser history, call logs, links visited, walking posture
Medical and health	Information that describes a person's health, medical conditions	Physical and mental health, drug test results, disabilities, health history, blood type, DNA

Table 1.6. Categories of intrinsic characteristics.

Categories	Description	Examples
Knowledge and belief	Information about what a person knows or believes	Religious beliefs, thoughts, what a person know and don't know
Authentication	Information used to authenticate, what you know	Passwords, PIN, mother's birth name
Preference	Information about personal preference	Favorite TV channels, favorite types of food, music

In addition, intimate information related to a person's sexual life also belongs to this category.

1.2.4.4 *Intrinsic feature data*

This category contains data specific to each individual. The categories and examples of intrinsic feature data are shown in Table 1.6.

In addition to the above categories, there are two other categories: historical data and tracking data. The former refers to things that have happened in an individual's life that affect that individual. Tracking data includes both data that can be traced to a specific individual, such as device identifiers (IP addresses, MAC addresses, and browser fingerprints), as well as information about the individual's location (latitude, longitude, and region), contact information (cell phone number and email address), and so on.

Using the above principles of personal data classification as a framework, and taking into account the data actually collected and controlled, an organization can define the data categories and data lists appropriate to the organization's situation and implement appropriate protection measures in compliance with regulatory requirements.

1.2.5 *Strategies for effective data classification*

Most organizations have their own data security policies that are made available on internal websites or otherwise, and new employees are even trained on them. But in reality, employees rarely know how to apply these policies in their day-to-day activities.

Security policies should be actionable. Data classification is a prerequisite for this, followed by the implementation of technical solutions to protect the data. Data classification also makes it more effective to implement an appropriate data protection scheme.

The method of data classification is to classify and label data according to sensitivity or value. These tags may be attached as visual markers (like the "Top Secret" stamp on a dossier) or embedded in the file's metadata. Numerous data protection schemes support labeling of file metadata, ensuring that data can only be accessed or used according to the rules corresponding to its label.

Data classification and labeling can be fully automated, but research and practice have shown that the best results are achieved using a combination of human input and software tools. This is called human-driven data classification.

When using a combination of human input and software tools, it is the user's responsibility to determine which label is appropriate and to apply the label to the data as it is created, edited, sent, or stored. The user's insight into the context of the data makes classification decisions more accurate than those made by computers.

A prerequisite for effectively implementing a data classification strategy is the discovery and location of data assets. On this basis, the appropriate data-type classification method is determined. In general, for most organizations, it is not recommended to use too many data types, 3 4 data types may be sufficient. Of course, regulated data types need to be labeled separately, such as user privacy and personal data, which are regulated by the EU GDPR.

Data classification requires the use of appropriate tools to complete the automation of data discovery and labeling, and to combine with existing business processes to provide a seamless experience for users. Of course, it is important to ensure that the tools work on all platforms and operating systems, databases, and document processing systems that an organization has.

A better solution for establishing appropriate data classification policies and implementation strategies is to first classify existing processed data, such as emails, various forms and documents required for day-to-day business. This not only verifies that the data classification strategy is appropriate and also ensures that after the data

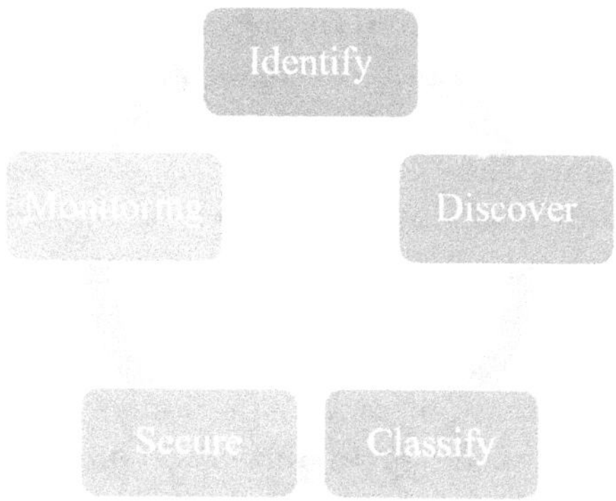

Fig. 1.19. Data classification lifecycle.

classification is implemented, the documents are "put in the vault" after a period of time. You can then continue the process of discovering, identifying, and protecting historical data.

Data classification is a dynamic process that cannot be accomplished overnight. Data classification follows the lifecycle process of "identify discover classify protect monitor", as shown in Fig. 1.19.

1.2.6 *Examples of data classification for organizations*

Data classification helps determine which security controls are appropriate to protect specific types of data. Organizations can classify data into categories such as highly confidential, confidential, internal, and public data.

For example, consider the data classification principles published by Carnegie Mellon University. The university's data is divided into three categories: restricted data, private data, and public data:

(1) **Restricted data:** Data should be classified as "restricted" if its unauthorized disclosure, alteration, or destruction could pose a significant risk to the university or its affiliates. Examples of restricted data include data protected by state or federal privacy statutes, and data protected by non-disclosure agreements. Restricted data should have the highest level of security controls.

(2) **Private data:** Data should be classified as "private" if its unauthorized disclosure, alteration, or destruction would pose a certain level of risk to the university or its affiliates. By default, all institutional data that is not explicitly classified as "restricted" or "public" data shall be considered "private data". An appropriate level of security controls should be applied to private data.
(3) **Public data:** Data should be classified as "public" when there is little risk to the university and its affiliates from unauthorized disclosure, modification, or destruction of the data. Examples of public data include press releases, course information, and research publications. While few controls are required to protect the confidentiality of public data, some level of control is required to prevent unauthorized modification or destruction of public data.

The categories included in the pre-defined restricted data are shown in Table 1.7.

Carnegie Mellon University is located in Pittsburgh, Pennsylvania, USA, and is governed by local laws. Therefore, the principles of the classification examples can be referred to, but the specific classification needs to be formulated based on scenarios and actual conditions.

1.2.7 *Example of cloud data classification*

With the advent of new business models such as cloud services and mobile office, the perimeter of enterprise network are no longer clear. The modern work environment has migrated to the clouds. In this transformation, how to protect corporate and customer data is the first problem that needs to be solved. As a result, enterprise security leaders need to define a data protection strategy that keeps pace with the times to update the data classification framework and implement IT systems for data protection.

Most cloud service providers support a robust data classification strategy. Some cloud service providers also support further customization based on the actual needs of the organization. Take the Azure cloud service used internally by Microsoft as an example, which provides the Azure Information Protection (AIP) framework.

Table 1.7. Types of restricted information.

No.	Data Type	Description
1	Authentication credentials	An authentication credential is confidential information held by an individual that proves that person is who they say they are. In some cases, authentication credentials can also be shared among a small group of people. Authentication credentials can also be used to prove the identity of a system or service. Examples are (1) passwords, (2) shared secrets, and (3) cryptographic private keys.
2	Financial information	Financial information governed by Gramm–Leach–Bliley Act (GLBA).
3	Electronic protected health information (EPHI)	EPHI is defined as any Protected Health Information (PHI) that is stored in or transmitted by electronic media.
4	Export controlled materials	Export Controlled Materials are defined as any information or materials that are subject to the United States export control regulations, including, but not limited to, the Export Administration Regulations (EAR) published by the US Department of Commerce and the International Traffic in Arms Regulations (ITAR) published by the US Department of State.
5	Federal tax information (FTI)	FTI is any return or return information received from the IRS or any secondary source which is protected by the confidentiality provisions of Internal Revenue Code section 6103.
6	Payment card information	Payment card information is defined as a credit card number (also referred to as a primary account number or PAN) in combination with one or more of the following data elements: cardholder name, service code, expiration date, CVC2, CVV2, or CID value, PIN or PIN block, contents of a credit card's magnetic stripe. Payment card information is also governed by PCI DSS.

(*Continued*)

Table 1.7. (*Continued*)

No.	Data Type	Description
7	Education records	Personally Identifiable Education Records are defined as any Education Records that contain one or more of the following personal identifiers: name of the student, name of the student's parent(s) or other family member(s), social security number, student number, a list of personal characteristics that would make the student's identity easily traceable, or any other information or identifier that would make the student's identity easily traceable.
8	Personally identifiable information (PII)	PII is defined as a person's first name or first initial and last name in combination with one or more of the following data elements: social security number, state-issued driver's license number, state-issued identification card number, financial account number in combination with a security code, access code, or password that would permit access to the account, and medical and/or health insurance information.
9	Protected health information (PHI)	PHI is defined as individually identifiable health information transmitted by electronic media, maintained in electronic media, or transmitted or maintained in any other form. PHI is considered individually identifiable if it contains one or more of the following identifiers: name, address, all elements of dates (except year) related to an individual, including birth date, admissions date, discharge date, and date of death, telephone numbers, fax numbers, electronic mail addresses, social security numbers, medical record numbers, health plan beneficiary numbers, account numbers, certificate/license numbers, vehicle identifiers and serial numbers, including license plate number, device identifiers and serial numbers, Universal Resource Locators (URLs), Internet protocol (IP) addresses, biometric identifiers, including finger and voice prints, full-face photographic images and any comparable images, and any other unique identifying number, characteristic, or code that could identify an individual.

Table 1.7. (*Continued*)

No.	Data Type	Description
10	Controlled technical information (CTI)	Controlled Technical Information means technical information with military or space applications that is subject to controls on the access, use, reproduction, modification, performance, display, release, disclosure, or dissemination.
11	For official use only (FOUO)	Documents and data labeled or marked For Official Use Only are a pre-cursor of Controlled Unclassified Information (CUI) as defined by the National Archives and Records Administration (NARA).
12	Personal data from European Union (EU)	The EU's GDPR defines personal data as any information that can identify a natural person, directly or indirectly, by reference to an identifier, including: l Name l An identification number l Location data l An online identifier l One or more factors specific to the physical, physiological, genetic, mental, economic, cultural, or social identity of that natural person.

Microsoft's data classification framework is jointly developed and updated by the security and compliance teams. New frames and labels are more intuitive and enable employees to make better data handling decisions. Data protection standards can be more clearly aligned and easier for employees to understand and follow. Microsoft Corporation's classification framework includes the following five labels:

(1) **Non-business:** Non-business data that is created using Microsoft's resources but is unrelated to Microsoft's business or to Microsoft Corporation's customers. This data is characterized by the fact that it does not require encryption and cannot be traced or revoked.
(2) **Public:** Preparing and releasing Microsoft business information for public use. The data is not protected by Azure Rights Management Services (RMS), and the owner cannot use RMS to track or revoke content authorization.
(3) **General:** Business information for non-public use; however, it may be shared with employees, temporary workers, business

customers, and external partners. The data is not protected by RMS, and the owner cannot use RMS to track or revoke the content. If the employee does not select a classification label, the AIP defaults to using this label but provides label suggestions based on what is entered in the document.

(4) **Confidential:** Sensitive strategic business information that could cause harm if not properly disclosed. Personal information, whether identifiable or not, also falls into this category. Data is characterized with RMS protection, the owner can track and revoke content, and recipients are trusted and have full delegated rights, including the ability to remove RMS protection.
(5) **High confidential:** Highly sensitive, critical, and high-risk data that requires the strictest protection. Such data includes data subject to laws and regulations and sensitive personally identifiable information. The data is characterized by the use of RMS protection; the owner can track and revoke the content; the recipient has no delegated rights and cannot modify or remove RMS protection.

For confidential and highly confidential, Azure cloud services also provide sublabels that identify visibility scopes.

(1) **Microsoft management:** Only visible to Microsoft senior management.
(2) **Microsoft FTE:** All full-time employees (FTE) can see it.
(3) **Microsoft extension:** Visible to both full-time and temporary employees.

After data classification is implemented, Microsoft Azure Information Protection protects data throughout its lifecycle through the protection mechanism integrated into DLP, as shown in Fig. 1.20.

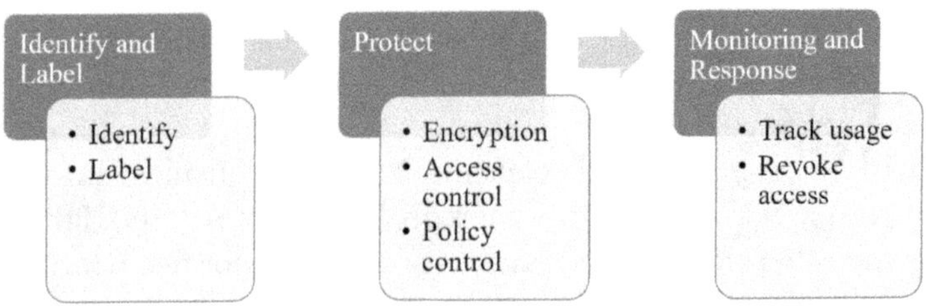

Fig. 1.20. Microsoft Azure information protection.

1.2.7.1 *Identify and label*

Azure Information Protection classifies, tags, and protects approximately 15 million documents and emails within Microsoft each month. With Azure Information Protection, employees can classify the content of information under their control or use Azure Information Protection scanners to automatically classify documents to scan for sensitive content such as credit card numbers and social security numbers.

The right combination of people, processes, and technology can effectively protect an organization's data. Microsoft employees play a critical role in protecting the data of the companies they create, collaborate with, and manage. In the absence of tools and automation, Microsoft still expects employees to think about the proper way to classify and handle confidential data. Microsoft has chosen to have employees classify data and then use tools to scan content and suggest classifications. Whether inside or outside the organization, employees who create, edit, store, and share data on corporate and personal devices need the tools and information to easily and correctly classify and label data.

Azure Information Protection software is an add-in to Office 365 productivity applications (including Outlook email, Word, Excel, and PowerPoint). When Azure Information Protection software is installed, the Protect button provides employees with classification prompts and an automated way to classify files they open or create in Office 365 productivity applications.

Microsoft Corporation requires employees to classify their own files and provides prompts and suggestions similar to those shown in Fig. 1.21 to help employees choose the appropriate classification and encourage them to continue to participate in the classification.

By default, uncategorized documents open with the "General" category, but Azure Information Protection makes suggestions as content and data are added. The suggestions are based on keyword scans, or numerical pattern matching based on financial or personal information.

1.2.7.2 *Protect*

Labeling and classification use encryption, identity, and authorization policies for automated protection. Azure RMS integrates with

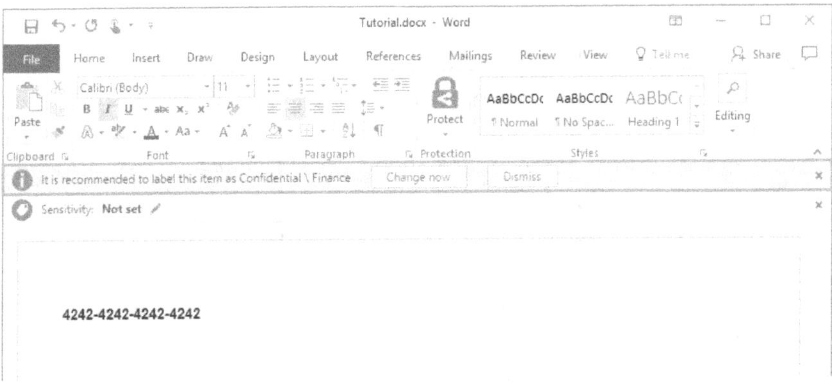

Fig. 1.21. Microsoft Office 365 document sensitivity label.

cloud services and applications, including Office 365, Azure Active Directory, and Windows Information Protection. Protection travels with documents and email, whether the files are located inside or outside the corporate network, file servers, and applications:

(1) **Encryption**

Azure RMS data encryption in Azure Storage helps protect data at rest and in transit. Azure Disk Encryption can be used to encrypt operating systems and data disks used by virtual machines. Data is protected in transit between the application and Azure so it remains secure at all times.

The usage of stored keys can be recorded, monitored, and audited through Azure Logs, and logs can be imported into Azure HDInsight or a Security Information and Event Management system (SIEM) for additional analysis and threat detection.

(2) **Access control**

Azure cloud services can manage user identities and credentials to control access to data in several ways.

Azure AD ensures that only authorized users have access to computing environments, data, and applications on Microsoft systems. Azure role-based access control helps manage access to Azure services that contain personal data. Azure AD Privileged Identity Management reduces the risks associated with administrative privileges through access control, governance, and reporting.

(3) **Policy control**

Classification and labeling will also inform policy controls, which can restrict access to certain document features based on the classification level. Once employees apply classifications, labels, and scopes, Microsoft's corporate policies specify which recipients and recipient actions are allowed. For example:

By applying a non-commercial label, recipients can view, forward, print, and save the content without RMS protection.

By applying the "Confidential/Recipients Only" label, recipients can view, reply, print, and save content. However, they cannot forward content or remove RMS protection.

By applying the "Confidential/Microsoft Executives" label, only executives and members of the appropriate user group can view, forward, reply, print, and save the content. Document owners can use the Azure Information Protection application or utilize the console to view traces and logs for their confidential documents.

1.2.7.3 *Monitoring and response*

Using Azure Information Protection, employees can securely collaborate with people inside and outside the organization, track usage, and even revoke access remotely. With the Azure Information Protection application, employees can see who has access to their content and where that content is located, even if it extends beyond the traditional perimeter of the organization.

1.2.8 *Example of Microsoft data classification*

For organizations, all information of business value should be protected. Whether it's emailing or sharing documents online, keeping organizations' information secure is every employee's responsibility.

Obviously, the greater the value of the information, the more controls should be imposed.

Microsoft divides the data within the enterprise into three levels according to the business impact: high, medium, and low. It is represented by the abbreviations of HBI, MBI, and LBI, see Table 1.8.

In typical scenarios, the classification of various data types is shown in Table 1.9. Of course, it needs to be combined with the

Table 1.8. Microsoft data classification.

Class	Name	Description
HBI	High Business Impact	Unauthorized disclosure of HBI could cause severe or catastrophic material loss that includes (without limitation) theft of financial instruments or property, operational disruption, identity theft, brand misappropriation, damage to Microsoft's reputation, or significant legal and regulatory liability.
MBI	Medium Business Impact	Unauthorized disclosure may cause serious material loss due to identity or brand damage, operational disruption, damage to Microsoft's reputation, or legal or regulatory liability.
LBI	Low Business Impact	Unauthorized disclosure could cause limited to no material loss.

Table 1.9. Microsoft data classification examples.

Data containing the following information	HBI	MBI	LBI
Email		Yes	
SSN or ID card information	Yes		
Business process documents		Yes	
Private keys and passwords	Yes		
User ID or user name	Yes		
Publicly accessible information			Yes
Company trading secrets		Yes	
Financial information	Yes		
Customer's information	Yes		
Postal code of employee's address			Yes

specific circumstances of the organization to determine the final class of various data.

It is the responsibility of the organization's employees to judge the class to which the data belongs. When a computer is sent for repair, the HBI and MBI data needs to be cleared.

Microsoft provides corresponding tools in its Office suite to implement protection. It mainly includes the following tools:

(1) **Information Rights Management (IRM):** A part of the Office suite. This feature can set permissions on Office documents

Table 1.10. Data security controls of Microsoft.

	IRM	S/MIME	EFS	BitLocker
Send via internal mail	Suggested	Acceptable	Not applicable	Not applicable
Send via external mail	Should only be shared with organizations that support RMS	Suggested	Not applicable	Not applicable
Online document sharing	Suggested	Not applicable	Not applicable	Not applicable
Stored in computer	With BitLocker	Not applicable	With BitLocker	Needed
Storage on removable storage device	Acceptable	Not applicable	Acceptable	Suggested

to prevent unauthorized forwarding, printing, and copying and even to set an expiration date.

(2) **Secure/Multipurpose internet mail extensions (S/MIME):** Can encrypt and sign emails.

(3) **BitLocker disk encryption:** The disk encryption feature provided by Windows 7 and above. Can effectively protect data on hard drives in the event of laptop theft.

(4) **Encrypted file system (EFS):** Enabled when the system does not support BitLocker. Can encrypt some files or directories.

For High Business Impact and Medium Business Impact data (HBI and MBI), recommended controls for transmission, sharing, and storage are shown in Table 1.10.

Data management measures are shown in Table 1.11.

1.2.9 *Smartphone data classification example*

According to the "Number of Smartphone Users Worldwide from 2013 to 2028" report (Statista Inc, 2023) published by Statista (a provider of market research and business data), in 2013, 2.6 billion people worldwide were using smartphones, and by the end of 2022, the number had reached 4.7 billion. Smartphones have changed the way people work and live. On smartphones, various applications

Table 1.11. Securing business information from Microsoft.

Action	HBI	MBI	LBI
Send data	Approval for forwarding, exporting, and copying by the information owner is required. Both internal and external sending are required for encryption. IRM or S/MIME is required.	When leaving the organization, encryption is required. IRM or S/MIME is required when leaving the organization.	No additional measures required
Share data	IRM is required to restrict forwarding, copying, and printing. Restrict permissions based on the list specified by the information owner. Transmission to third parties requires authorization agreements and legal approvals.	Restrict permissions to use only in legitimate business scenarios. Transmission to third parties requires authorization agreements and legal approvals.	No additional measures required
Store data	BitLocker encryption is required. Storage is only allowed in scenarios where portable smart devices support strong encryption and authentication security controls.	The information owner determines whether encryption is required.	No additional measures required
Backup	It is recommended that backups be performed only by authorized personnel and at locations designated by the information security department. Encrypt storage media.	Stored in a physically secure location, the backup process is logged, and access is controlled and monitored.	No additional measures required
Destroy	Paper materials need to be shredded or incinerated. Physical damage is required to destroy a tape or disk. Be aware of the destruction requirements for decommissioned hardware and storage media.	Paper materials need to be shredded or incinerated. Destroy tapes or disks. Before reusing or retiring disks, clean up the data. If the disk is inoperable, physical destruction is required.	No additional measures required

such as socializing, shopping, payment, traveling, and entertainment enable users to enjoy the convenience of the Internet era anytime, anywhere.

At the same time, terminals, especially smart terminals, are increasingly storing various personal and sensitive data, even biometric information (fingerprints, facial features, etc.) and personal health information (heart rate, blood pressure, etc.). Whether these data are fully protected and whether the confidentiality, integrity, and availability of data can be ensured in the complex usage scenarios of smartphones are common concerns from users.

Fortunately, modern operating systems for smartphones implement different data classifications and implement complex protection mechanisms based on data classification. In the following, iOS (the operating system used by Apple's iPhone) is used as an example.

The files stored on the iPhone are divided into four classes: A, B, C, and D (Apple Inc., 2022). The usage scenarios of each class are different, and the strength of the corresponding protection mechanism is also different, see Table 1.12.

Table 1.12. iPhone data protection classes.

Type	Name	Description	Examples
Class A	Complete protection	Shortly after the user locks a device, all data in this class is inaccessible.	Apple Pay, Keychain data
Class B	Protected unless open	Some files may need to be written while the device is locked.	A mail attachment downloading in the background
Class C	Protected until first user authentication	The protection in this class has similar properties to desktop full-volume encryption and protects data from attacks that involve a reboot.	The default class for all third-party app data
Class D	No protection	This class key is protected only with the UID.	Alarms, wallpapers, Medical IDs

The above four classes are classifications of data files. For passwords, keys, etc., iOS also provides keychain data protection and uses secure enclave processor hardware to participate in data protection. Keychain items are encrypted using two different keys of the symmetric AES 256-GCM encryption algorithm: the table key (metadata) and the row-unique key (secret key). Keychain metadata (all properties except kSecValue) are encrypted with the metadata key for fast searching, while secret values (kSecValueData) are encrypted with the secret key. Metadata keys are protected by the Secure Enclave processor but are cached in the application processor for fast keychain lookups. Private keys must always be handled within the Secure Enclave processor.

1.3 Identity Authentication and Data Access Control

Access control consists of two important components: authorization and authentication. Access control relies on accurate authentication. Without authentication, there is no access control and data security.

In a narrow sense, identity authentication refers to the authentication of a user, i.e., the verification that the user is indeed who he or she claims to be. In a broader sense, authentication includes the authentication of the entity that initiates a resource access or operation. The entity that initiates a resource access can be a person, a physical device, or a process/software service, so accordingly authentication also includes authentication to a person, authentication to a service, and authentication to a device, as shown in Fig. 1.22.

Fig. 1.22. Different authentication objects.

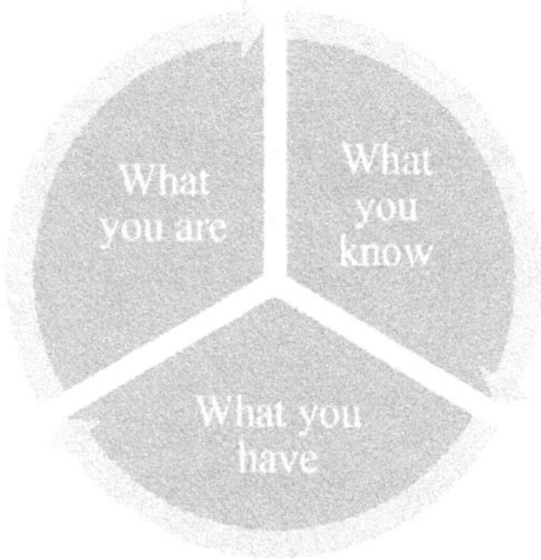

Fig. 1.23. Three dimensions of human authentication.

For different identity authentication objects and scenarios, different methods need to be adopted to ensure the accuracy of authentication.

1.3.1 *User authentication*

The authentication of people can be summarized into three authentication dimensions, as shown in Fig. 1.23:

(1) **Something you know:** Secrets only known to you.
(2) **Something you have:** The authentication credentials you have, which are generally physical entities.
(3) **Something you are:** Your biometric information.

1.3.1.1 *Something you know*

The well-known password-based authentication is an example of the "something you know" dimension of authentication. The user must enter a secret known only to the user to verify his or her identity.

Of course, it is also well known that password-based authentication is not secure enough. This vulnerability can be caused by password reuse (e.g., using the same password for multiple web applications), weak passwords (e.g., "123456" or "qwerty"), inappropriate password sharing (e.g., server administrators writing root's password on a sticky note and posting it on the server), attackers cracking or guessing passwords, etc.

Fig. 1.24. FIDO Alliance for two-factor authentication USB security key.

1.3.1.2 *Something you have*

"Something you have" can include various access control cards, USB smart cards, hardware tokens (One Time Pad token, OTP token), etc. The user proves his identity by proving to the authenticator that he owns the physical entity in some way.

There are also industry efforts to standardize hardware-based authentication protocols, which have received broad support from various Internet vendors and hardware manufacturers. For example, the FIDO Alliance, whose authentication protocol supports account authentication based on hardware keys, uses USB security keys for two-factor authentication (Srinivas *et al.*, 2015), as shown in Fig. 1.24.

In recent years, the rise and frequent use of mobile phone SMS (short message) authentication code and QR code (quick-response code, a kind of matrix barcode) scanning login also belongs to the "you have" authentication dimension. In general, the "what you have" authentication dimension is rarely used alone in identity authentication scenarios that require access to sensitive data.

1.3.1.3 *Something you are*

"Something you are" refers to the user's own biometric information, such as fingerprints, facial features, and iris lines. Common biometric usage scenarios in production and life are as follows:

(1) **Face recognition:** Measures the unique pattern of a human face by comparing and analyzing facial contours. It can be used not only for security authentication and law enforcement but also for authenticating and unlocking devices, such as smartphones and laptops.
(2) **Fingerprint scanning:** Captures unique patterns on fingers. Many smartphones and some laptops use this technology as a form of credential for unlocking.

(3) **Iris recognition:** Identifies the unique pattern of a person's iris. The iris is the colored part of the human eye and a component of the uvea. It is used in high-security scenarios, such as the military, government, and banking, but is not widely used in the consumer market.

(4) **Voice recognition:** When you speak to a device, it measures the unique sound waves in your voice. Banks can use voice recognition to verify a customer's identity when they call, and smart speakers like Amazon Alexa can also use voice recognition to determine a user's identity. Other use cases include Windows support for Windows Hello voice recognition for login.

Biometrics can identify and authenticate individuals based on an identifiable and verifiable set of data that is unique and specific to them.

Biometric authentication is the process of comparing a person's characteristics to a biometric "model" of that person to determine similarity. The reference model is first stored in a database or on a secure portable device, such as a smart card, and then the stored data is compared with the biometric data of the person being authenticated to verify that the user is who he or she claims to be. In this mode, the question is "Are you Mr. Bob?"

Another similar concept is biometrics identification. The first is to collect a person's biometric data, which may be a facial photo, a fingerprint image, or a voice recording, and then match the biometric data with the pre-stored biometric data of some people. In this mode, the question is "Who are you?".

Identity authentication based on biometric information is widely used on smartphones and is gradually being extended to products in the form of tablet computers and portable computers.

When designing and implementing a biometric authentication solution, privacy concerns need to be taken into consideration, especially the biometric data privacy protection. The main privacy risks are as follows:

(1) Any data collection can eventually be hacked. Biometric data can be a particularly attractive target for hackers. As biometric applications become more common, biometric data may be used in more places that may not be stored with the same level of security.

(2) Data stored in biometric databases may be more vulnerable to attack than any other type of data. Users can change passwords, but they cannot change fingerprints or irises. This means that once biometric data is compromised, it may be impossible to control.
(3) Some physical authentication credentials can be copied. For example, criminals can capture high-resolution images of ears from distant photographs or copy fingerprints from cups left in coffee shops. Such information could be used to hack into devices or accounts.

The three authentication dimensions mentioned above can also be used in combination in scenarios that require high authentication strength, known as multi-factor authentication (MFA). MFA is the process by which a user or device provides two or more different types of proofs of control associated with a particular digital identity in order to gain access to associated permissions, privileges, and memberships. Two-Factor Authentication (2FA) means that only two credentials are required for successful authentication and is a subset of MFA.

Today, many Internet sites, services, and applications, including those from Google, Microsoft, Facebook, and Amazon, offer more secure MFA options in addition to traditional password-based authentication solutions.

Google Inc. had published an article in July 2018 (Krebs, 2018) stating that it had great success in addressing phishing attacks against its employees by having its more than 85,000 employees to adopt a physical security key instead of the traditional username and password login method in early 2017. MFA solutions are supported by default in most popular operating systems. In addition, hundreds of third-party vendors offer MFA solutions for a variety of authentication scenarios. Several Open Authentication Standards, such as those promoted by the FIDO Alliance, have been widely adopted.

In addition, with the enrichment of identity authentication scenarios and the continuous evolution of technologies, some new authentication methods, technologies, and solutions have emerged and been promoted, and become popular. Notable among them are behavior-based authentication and passwordless authentication.

Most users are familiar with the various forms of authentication required to log on to a computer, access an e-mail account, or open a shared server in the organization where they work. But in most cases, the user is authenticated only once, leaving these systems vulnerable to security breaches for the remainder of the session. Behavioral biometric authentication is a new capability that uses a person's behavior to continuously verify their identity throughout a session, not just at the initial login.

Behavioral biometrics is the analysis of patterns of human activity (such as keystrokes, speech, gait, signature, and cognition) that can be used to identify individuals with a high degree of accuracy. The obvious advantage of behavioral biometrics is that it is a passive authentication method that requires no time or technical knowledge on the part of the user. In addition, the technology continues to protect someone even if their authentication credentials are compromised, such as a compromised password.

Take smartphones as an example. Behavioral biometrics are currently being developed to analyze gait (the way you walk) and typing style (speed, keystroke pressure, finger position, etc.). Voice recognition technology is also sometimes classified as a form of behavioral biometric authentication. The difference with other biometric authentication is that other biometric authentication (such as face and fingerprint) depends on a specific part of the body, while behavioral biometric authentication relies on each person's unique way of interacting.

In terms of implementation, continuous behavioral authentication typically calculates an "identity score" on a rolling basis that reflects the likelihood of who the user claims to be. In high-stakes situations (such as online financial transactions), behavioral authentication is gaining more and more popularity. In these cases, the authentication score calculation takes into account how the user behaves within the application, their location, and any personal information they enter. Continuous behavioral authentication takes traditional geolocation-based authentication to a new level by incorporating biometrics into algorithms that provide a more complete understanding of the user.

As a new form of identity authentication, passwordless identity authentication offers outstanding security and ease-of-use benefits. Mobile tokens such as Microsoft Authenticator and Google

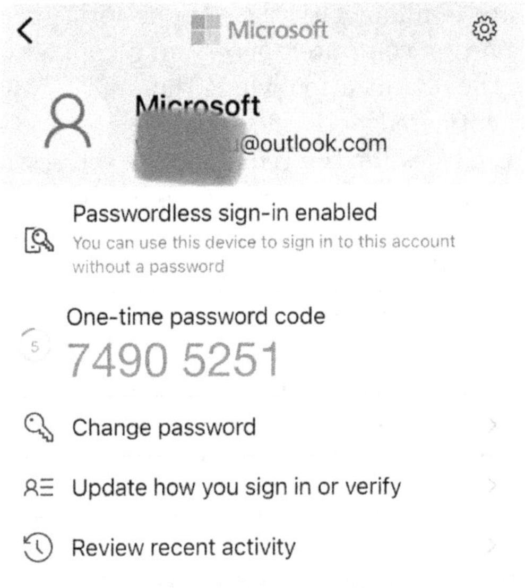

Fig. 1.25. Microsoft authenticator.

Authenticator are one type of passwordless authentication. Microsoft Authenticator is shown in Fig. 1.25.

Other passwordless identity authentication methods also include email-based authentication and SMS-based authentication. The basic idea is to authenticate users based on a previously authenticated identity. For example, based on email authentication, the user has been proven to be a user of an email address, i.e., has some sort of digital identity on the Internet.

More broadly, biometric-based authentication, such as fingerprint authentication to unlock a mobile phone, also falls under the umbrella of passwordless authentication when used alone.

The Microsoft white paper "Password-less Protection" (Microsoft Corporation, 2020) points out that different authentication methods have different levels of security and convenience. Passwordless identity authentication is much higher than password authentication in terms of security and much higher than "password + two-factor" authentication in terms of convenience. Figure 1.26 shows the comparison of different authentication methods.

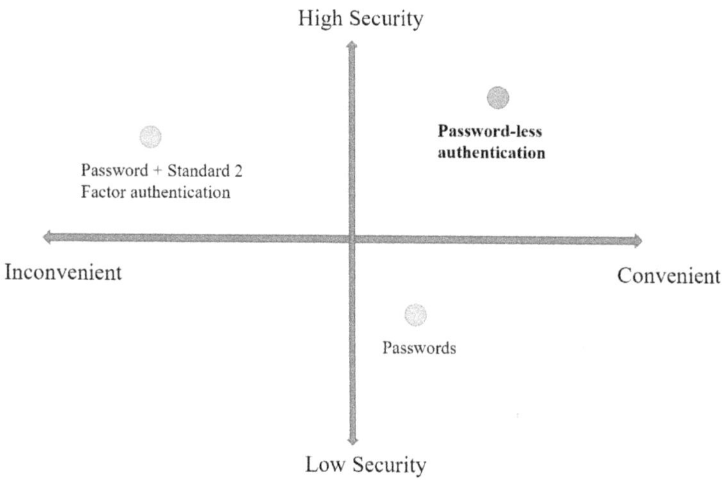

Fig. 1.26. Comparison of different authentication methods.

1.3.2 *Service authentication*

Services here generally refer to software entities that provide services to the outside world through interfaces. For services, the user authentication dimension discussed above is hardly useful. Therefore, two types of measures are generally adopted to realize the identity authentication of the service. One is identity authentication based on digital certificates, and the other is digital signatures to secure services to trusted publishers/developers. The implementation of both types of measures involves the basic concept of digital certificates.

A digital certificate binds an entity (such as a person, organization, or system) to a specific pair of public and private keys. Think of a digital certificate as an electronic credential that verifies the identity of an individual, system, or organization.

Different types of digital certificates can be used for different purposes, such as the following digital certificates:

(1) Secure Multipurpose Internet Mail Extensions (S/MIME) digital certificates for signed email.
(2) Secure Sockets Layer (SSL) and Internet Protocol Security (IPSec) digital certificates for verifying network connections.

(3) A smart card digital certificate for logging on to a personal computer.

Most digital certificates use the X.509 standard. This standard was developed by the Internet Engineering Task Force (IETF).

1.3.3 *Device authentication*

In scenarios such as device software version upgrade and near-field device interconnection, you need to authenticate the identity of the device. Device authentication is generally implemented by issuing a digital certificate to the device.

The Apple iPhone uses an additional implementation architecture beyond regular device certificates. The iOS security architecture (Apple Inc., 2022) is shown in Fig. 1.27.

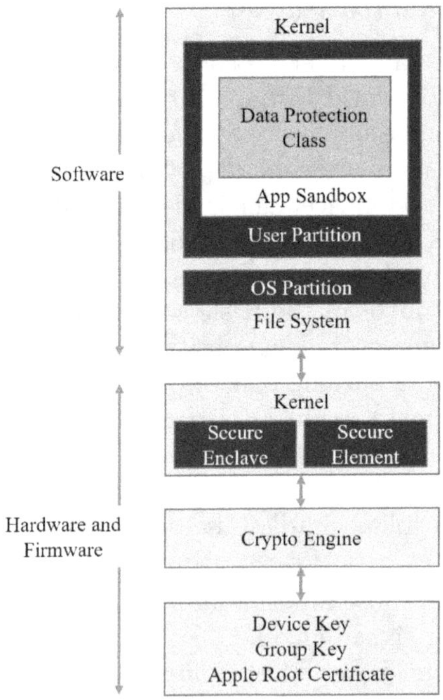

Fig. 1.27. iOS security architecture.

The device's Unique ID (UID) and the device's Group ID (GID) are 256-bit AES keys that are hardwired or compiled into the application processor and the secure enclosure during the manufacturing process. These two IDs are protected by hardware and cannot be read directly at the software and firmware level. These two IDs serve as the encryption and decryption keys, and the encryption or decryption operations are performed by a dedicated AES engine on the chip.

For example, the GID is common to the same type of device using the same model of processor and does not have a unique value for each device. For iOS firmware upgrade scenarios, Apple releases firmware images that are decrypted using the GID key and in this way ensures the integrity of the firmware image and also ensures that the firmware image is installed on a specific version of the Apple device to achieve some sense of device authentication.

1.3.4 *Data access control system*

Access control is the foundation of data security, relying on identity authentication, authorization, and access policies to determine whether to allow access to various data and resources. Simply put, access control is about ensuring that the right people or things are using the right devices, in the right place, at the right time, and have legitimate reasons to access the right resources.

At a high level of abstraction, access control is a method of ensuring that users are who they say they are and that they have the appropriate access rights to resources. Broadly speaking, there are two types of access controls: physical access control and logical access control. Physical access controls limit access to buildings, offices, and IT physical facilities. Logical access controls limit access to computer networks, system files, and data. This section focuses on logical access control.

Access control consists of two main components: authentication and authorization. Authentication is a technology used to verify whether the subject of data access control is the identity he claims to be. Authentication alone cannot guarantee data security, so authorization is also required. That is, you need to determine whether a user should access a particular data or perform a specific action.

At the implementation level, resource access control is typically used by server administrators to control user access to file resources and network resources. Its basic goals are to prevent unauthorized access to data resources, allow authorized subjects to access specific objects, and deny services to unauthorized subjects (illegal users or illegal use of system resources by legitimate users).

The goal of access control is usually described from the perspective of protecting system resources from inappropriate access, while from a business perspective, good access control measures should not hinder or even encourage reasonable and effective data sharing. Access control is the cornerstone of system confidentiality, integrity, and availability, and it is also one of the most important data security measures.

The access control system includes the following basic concepts:

(1) **Subject:** This is the initiator of specific activities to access resources. This activity may result in the flow of information or a change in the state of the system. The subject may be a specific user, or it may be a process or service started by the user, a device owned by the user, etc.
(2) **Object:** This refers to an entity that contains or receives information. Access to an object can be abstracted as access to the information it contains. Any information, resource, or item that can be accessed and manipulated can be an object. For example, objects can be files or directories on the server, tables or fields in the database, or even various hardware, such as printers, hardware servers, and mobile communication terminals.
(3) **Operation:** This is the process of activities initiated by a subject. For example, when a user inserts a bank card into an ATM and enters a PIN, the verification of the user's identity is an active process. After the verification of the user's identity is completed, the user can also initiate other operations, such as depositing and withdrawing money, and transferring money.
(4) **Access policy (attribution):** This is a set of relevant access rules or policies for the subject to the object. The access policy embodies an authorization behavior and is also the default of certain operation behaviors (such as reading, writing, executing, or denying access) performed by the object on the subject.

Figure 1.28 shows the relationship between the three access control elements: subject, object, and access policy. In this case, the subject is the initiator of the arrow, the object is the pointing party of the

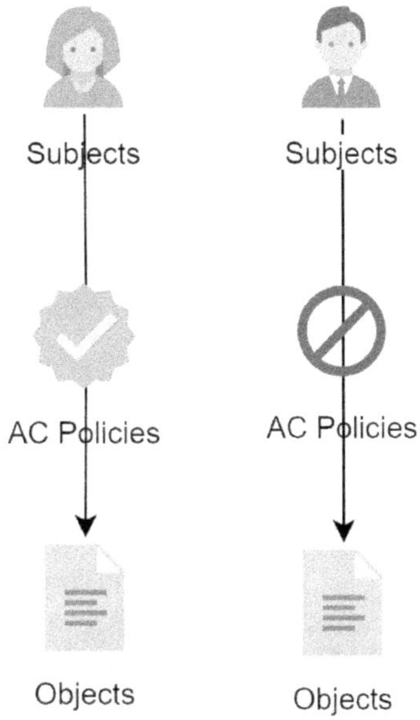

Fig. 1.28. The relationship between the three elements of access control.

arrow, and the arrow represents the access policy. The subject can be the person, process, or other logical entity that initiates the access, and the objects may be applications, data, services, devices, and so on. An access policy both ensures that the right subject accesses the right object and blocks access by unauthorized subjects.

1.4 Summary of Data Security System

The core goal of data security is to protect the confidentiality, integrity, and availability of the data to be protected during the entire lifecycle stages including storage, use, and transmission. It is important to keep this goal in mind when designing and analyzing systems that contain data.

To achieve the core goal of data security, it is necessary to combine several methods and technologies such as data lifecycle definition, data classification framework and implementation, and data access control, and requires the support of hardware root of trust, cryptographic algorithms, and key management systems. The basic data security system is shown in Fig. 1.29.

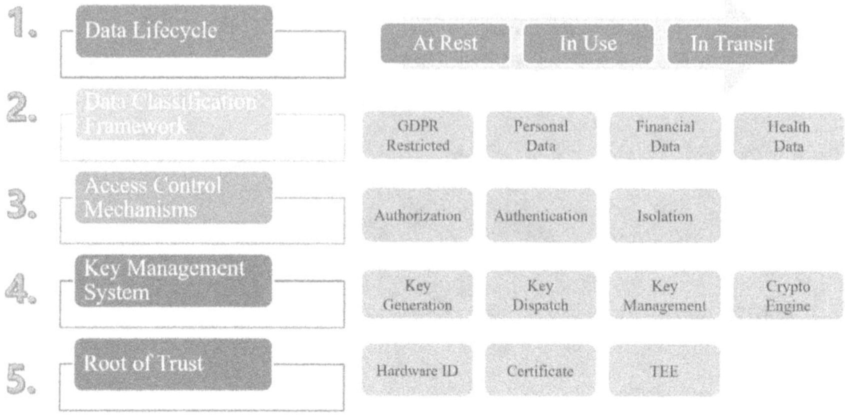

Fig. 1.29. Basic data security system.

Whether it's designing a new online shopping site, developing a fitness app, or even designing a new IoT smart terminal, one or more of the methods and techniques in the data security solutions described above will be involved.

To support the core goal of data security, it is first necessary to organize the lifecycle of the data to be protected from the dimension of the system or product, and analyze the storage, use, and transmission scenarios. Second, an accurate data classification framework and implementation strategy are required to identify the data that really needs to be protected. Finally, based on the identification of protected data, and based on its lifecycle process, build an appropriate access control mechanism. Protecting sensitive data usually requires a combination of two protection mechanisms: access control and encryption. Therefore, it is necessary to consider designing or using an existing key management system. Key management is a major difficulty, and its lifecycle process often requires the participation of the root of trust and trusted environment.

The five levels in the data security infrastructure system support and cooperate with each other to build a complete data security framework together.

Chapter 2

Data Security Risk Challenges and Responses

This chapter analyzes and summarizes the risks and challenges of data security in different eras, scenarios, and fields, and systematically analyzes and summarizes the major threats to data security. On this basis, based on the threat analysis, the principles, processes, and methods of data security responses are given by comprehensively utilizing systems and methodologies such as the Cyber Security Framework (CSF), Risk Management Framework (RMF), and threat modeling analysis.

2.1 Risks and Challenges of Data Security

Review the four eras of industrial revolutions in human history. The first industrial revolution, which began in the 1860s, was characterized by steam-powered mechanical manufacturing, typified by the invention and application of the steam engine. The second industrial revolution, which began in the late 1860s, was characterized by mass production powered by electricity, typified by the invention of the first automated assembly line. The third industrial revolution began in the 1940s and 1950s and is characterized by the use of atomic energy, the use of computers, the invention of the Internet, and controller-based automation, with the invention of the Internet as its typical symbol.

There is much debate about when the Fourth Industrial Revolution began and what exactly it means. However, it is widely accepted that the concept of Industry 4.0 was first proposed by the German government in a government memorandum in 2013. The German government's high-tech strategy document outlined plans to make manufacturing almost entirely computerized, eliminating the need for humans. German Chancellor Angela Merkel promoted the concept at the World Economic Forum in Davos in January 2015, calling "Industrie 4.0" (Hermann *et al.*, 2015) the fusion of the online world and the world of industrial production.

Upon review, it can be easily found that data security, as an important part of cyberspace security, originated in the era of the third industrial revolution and has become increasingly important in the production and life of society with the rapid development and popularization of the Internet. In the era of the Fourth Industrial Revolution, with the continuous emergence of new scenes, new businesses, and new ecologies, data security will again face new risks and challenges.

2.1.1 *Data security in the Internet era*

The Internet was born in the 1950s. Since then, it has grown far beyond the expectations of its inventors. With the advent of the digital revolution, people have become more connected — from offline to online, from telephone to smartphone, from local to cloud, and from private to shared. Along with technological innovation, new production and lifestyles are gradually emerging.

With the digital revolution came hackers. They break into systems, steal or destroy data, inject malware, and block or disrupt networks for personal interest, for profit, or even for political reasons. Since the 1990s, the rapid development of the Internet has provided hackers with endless "gold mines".

There is no doubt that data leakage is the biggest data security risk in the Internet world.

2.1.2 *Data security in the era of convergence of everything*

In the 18th century, the advent of the steam engine sparked the first industrial revolution, mechanizing industrial production for the first time and driving urbanization and social change.

In the second industrial revolution, electricity and other scientific advances made industrial mass production possible. The third industrial revolution, which began in the 1950s, saw the advent of computers and digital technologies, which led to profound changes in the automation of manufacturing and in industries, such as banking, energy, and communications. Computer and digital technologies are rapidly promoted and popularized in various fields such as industry and agriculture.

The Fourth Industrial Revolution, which includes artificial intelligence (AI), robotics, the Internet of Things (IoT), 3D printing, genetic engineering, quantum computing, and other technologies, is behind many of the modern and even intelligent products and services that are commonplace in modern life.

Figure 2.1 shows the most critical inventions in the four industrial revolutions, which significantly promoted the progress of society.

The concept of the Fourth Industrial Revolution was proposed by Klaus Schwab, the founder and executive chairman of the World Economic Forum and the author of the book "The Fourth Industrial Revolution" (Schwab, 2017). Schwab wrote in a 2016 article (Schwab, 2016), "The Fourth Industrial Revolution: what it means, how to respond" that "like the revolutions that preceded it, the Fourth Industrial Revolution has the potential to raise global income levels and improve the quality of life for populations around the world".

According to him, "In the future, technological innovation will also lead to a supply-side miracle, with long-term gains in efficiency and productivity. Transportation and communication costs will fall, logistics and global supply chains will become more efficient, and the trade costs will diminish, all of which will open new markets and drive economic growth".

Not all of the Fourth Industrial Revolution is good news, however. Schwab added that the revolution could lead to greater inequality

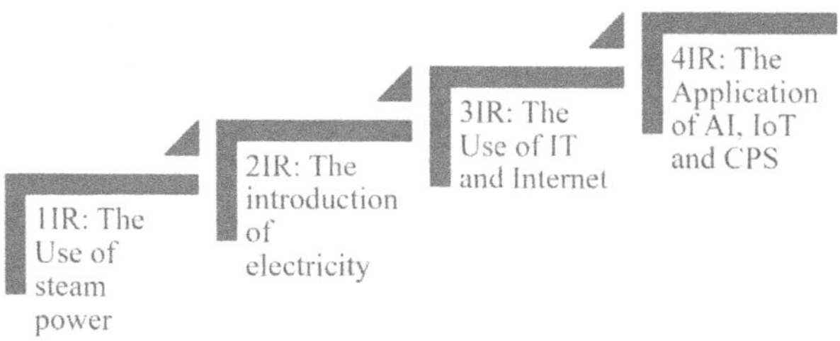

Fig. 2.1. The most critical inventions in the four industrial revolutions.

and further disruption in the labor market. In addition, the labor market could become increasingly divided into "low-skill/low-wage" and "high-skill/high-wage" roles, which could exacerbate social tensions.

According to Schwab, "The changes are so profound that, from the perspective of human history, there has never been a time of greater promise or potential peril" (Schwab, 2017).

The 10 key technologies driving the Fourth Industrial Revolution are as follows:

(1) **Artificial Intelligence (AI)** is dedicated to building machines that can "think" like humans — recognizing patterns, acquiring information, drawing insights, and making recommendations for decision-making from vast amounts of data. AI is being used in a variety of applications, from discovering patterns in large amounts of unstructured data to providing automated email responses on mobile phones.
(2) **Blockchain** is a secure, decentralized, and transparent way to record and share data without relying on third parties as intermediaries. Digital currencies are the best-known applications of blockchain. But the technology can be used in other ways, including enabling supply chain traceability, anonymizing sensitive medical data, and combating election fraud.
(3) **New computing technologies** are making computers smarter and faster. The advent of the cloud has allowed businesses to securely store and access their data anytime, anywhere with an Internet connection. Quantum computing technologies currently under development could eventually increase the power of computers by millions of times. These quantum computers will have the potential to improve AI, create highly complex data models in seconds, and could have a greater impact on materials engineering and medicine.
(4) **Virtual reality (VR)** provides an immersive digital experience that simulates the real world (using a VR headset), while augmented reality (AR) merges the digital and physical worlds. For example, L'Oréal makeup app lets users digitally try on makeup products before buying, and the Google Translate app lets users scan and instantly translate street signs (see Fig. 2.2), menus, and other texts.

Fig. 2.2. Street sign translation (Google Translate application).

(5) **Biotechnology** uses cellular and biomolecular processes to develop new technologies and products for various purposes, including the development of new drugs and materials, more efficient industrial manufacturing processes, and cleaner, more efficient energy sources.
(6) **Robotics** refers to the design, manufacture, and use of robots for personal and commercial use. While robotic assistants aren't yet in every home, technological advances are making robots increasingly sophisticated. They are widely used in industrial manufacturing, healthcare and emergency rescue, and other fields.
(7) **3D printing technology** allows manufacturing companies to print their own parts with fewer tools, at lower cost, and faster speed than traditional processes. In addition, companies can also use 3D printing technology to customize designs to ensure that products and designs fit perfectly.
(8) **Innovative materials technologies**, including new plastics, metal alloys, and biomaterials, are expected to drive the development of manufacturing, renewable energy, construction, and healthcare.
(9) **The Internet of Things (IoT)** is gradually permeating production and people's lives, including in various industries. From medical wearables that monitor a user's health to tracking devices for courier packages, more and more things are connected to the Internet and can be connected to each other. Companies can collect customer data from continuously

connected products, allowing them to better understand how customers use their products and develop marketing campaigns accordingly. There are also many industrial and agricultural applications, such as farmers placing agricultural IoT sensors in their fields to monitor soil conditions and be informed about when to apply fertilizer.

(10) **New energy technologies:** Energy harvesting, storage, and transmission is another growing market segment, driven by the declining cost of renewable energy technologies and increasing battery storage capacity.

Among the 10 key technologies driving the fourth industrial revolution, artificial intelligence, blockchain, cloud computing, virtual reality, robotics and IoT technology are directly related to data security, and some other technologies are more or less related.

With the rapid development of the fourth industrial revolution around the world, in addition to the above-mentioned new technology risks, data security risks such as new scenario risks, new business risks, and new regulatory risks are also worthy of attention. There is a significant correlation between these risks, as shown in Fig. 2.3.

For example, the development of quantum computing technology may pose a direct risk to existing public key infrastructure and

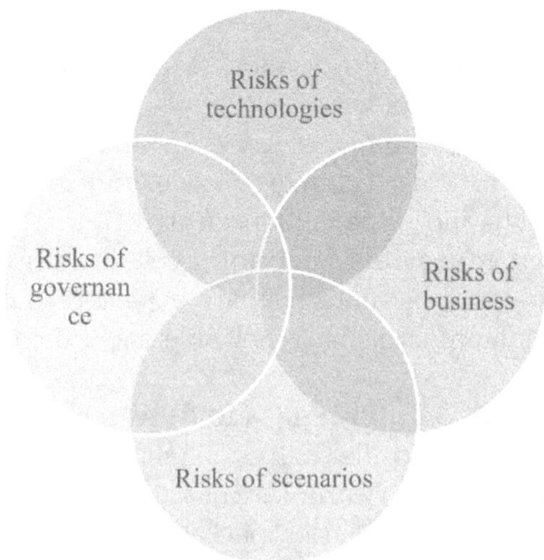

Fig. 2.3. New risks in the era of integration of everything.

encryption schemes, thereby affecting the confidentiality of sensitive data, which can be categorized as a new technology risk; the use of artificial intelligence technology to enhance the auto-response function of email by suggesting appropriate responses based on the content of the email, which may involve the collection and processing of confidential data, which can be categorized as a new scenario risk; the use of massive data from wearable electronic devices to conduct health research based on different populations (e.g., the relationship between average heart rate and body weight of healthy people of different ages), which can be categorized as a new business risk; and whether the business model will be able to comply with increasingly detailed and stringent or even regulatory requirements with different geographic characteristics and value considerations as various laws and regulations continue to improve and deepen. These can be categorized as new regulatory risks.

In addition, the above four types of risks have relatively many overlaps, creating new fusion risks. For example, the use of AI's deep learning technology to conduct genetic disease and gene defect screening based on gene sequence big data will involve risks in the above multiple data and privacy dimensions.

According to the research report of the Cloud Security Alliance (CSA) Big Data Working Group (Gartner, 2014), the characteristics of big data can be referred to as "3V": Velocity, Volume, Variety. These characteristics amplify the concerns of data security and privacy concerns. Traditional security mechanisms designed to protect small amounts of data at rest (as opposed to streaming) are often inadequate in this scenario.

CSA lists 10 big data security challenges (Big Data Working Group, 2013a), as shown in Fig. 2.4.

The solutions and measures to address these challenges are discussed in the following.

Of course, data security risks in IT systems still exist in traditional enterprise environments, and with the emergence of various new office scenarios, such as employee bring-your-own-device (BYOD), data security risks are gradually increasing.

Data, especially sensitive data in enterprises, has brought many difficulties to data security protection due to its characteristics of multiple forms, many access personnel, and rapid dissemination. In today's Internet and mobile office environment, factors such as

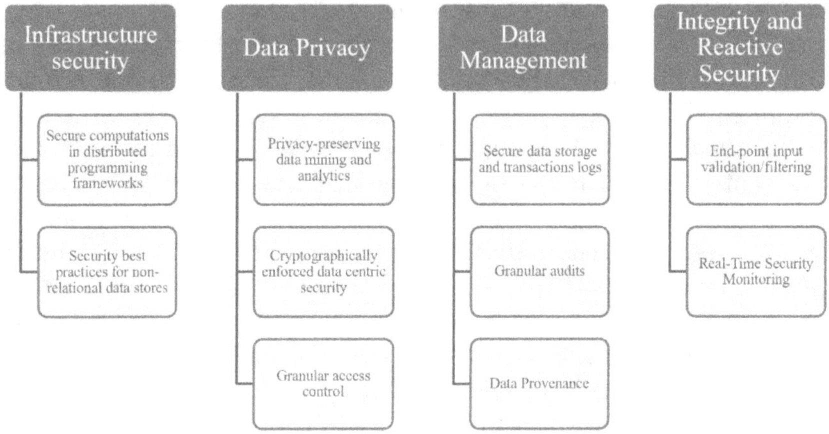

Fig. 2.4. Ten challenges of big data security.

multiple data storage media, enterprise development and testing requirements, internal staff active leaks, more and more outsourcing, and third-party cooperation have brought risks and challenges to data security protection.

Data risks are everywhere, and data security should be implemented throughout the entire data lifecycle. The full lifecycle of enterprise data includes data creation (classification, authorization, etc.), storage (access control, encryption, rights management, content distribution, etc.), transmission (channel security, encryption, rights management, etc.), use (behavior monitoring, rights management, logical access control, application security, etc.), and destruction (information fragmentation, secure deletion, content discovery, etc.). There are corresponding security risks in the entire lifecycle of enterprise data, and corresponding coping mechanisms need to be designed and implemented.

2.1.3 *Frequent data leakage incidents*

With the advent of the digital age, a large amount of sensitive information such as personal data, property data, and health data is stored and transmitted on the Internet, and the importance of data security protection is self-evident. However, in recent years, large-scale data leakage incidents have continued to occur, and people

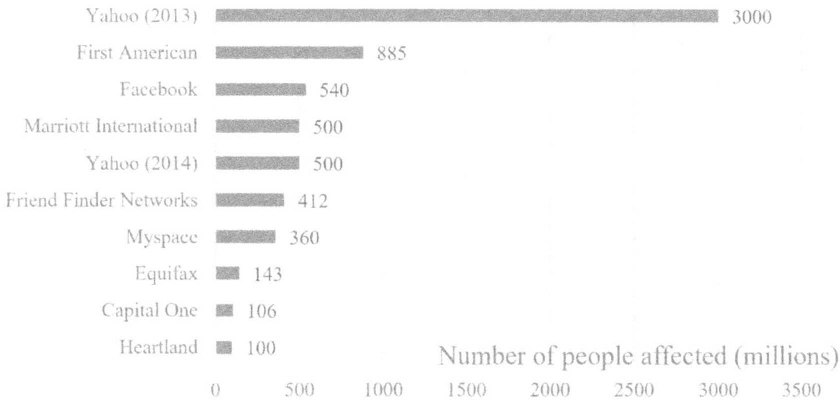

Fig. 2.5. Summary statistics of data leakage incidents.

from all walks of life have become increasingly concerned about data security and privacy protection.

According to a study (Purdue University Global, 2019) published by Purdue University in October 2019, in the 10 years since 2008, the 10 most serious data breaches have affected a total of about 6.5 billion people, and the data involved is diverse. There is individual personal data, such as email addresses, and there is very sensitive private data, such as bank account information. The summary statistics of data leakage incidents are shown in Fig. 2.5.

A few of the worst data breaches of the 21st century are summarized in the following based on the amount of data compromised, the level of confidentiality of the data, and the level of damage or risk to users.

1. **Yahoo**

Founded in 1994, Yahoo once held a dominant position among Internet information providers, with a market capitalization of more than $100 billion at its peak.

In September 2016, Yahoo disclosed a massive breach of account data. Due to the breadth of its impact, it has been called the largest Internet data breach in history. The investigation revealed that the real names, email addresses, birth dates, and phone numbers of up to 500 million users were compromised in a "possibly government-sponsored" attack that began in 2014. According to the company, the "vast majority" of the passwords involved were hashed using the robust bcrypt algorithm.

In December 2016, the company revealed that another group of hackers had gained access to potentially as many as 1 billion accounts in a 2013 cyberattack. In addition to names, birthdates, and email addresses, passwords and security questions and answers were also compromised. In October 2017, Yahoo revised this estimate, saying that all 3 billion user accounts may actually have been compromised.

These data breaches have embroiled Yahoo in privacy and compliance lawsuits, making it difficult for Yahoo to expand its business and even sustain its original business. US telecom operator Verizon eventually acquired Yahoo's core Internet business for $4.48 billion. The acquisition agreement requires the two companies to share liability for breach of contract. After the sale, Yahoo Corporation changed its name to Altaba Inc.

2. **Marriott International**

In November 2018, Marriott International announced that hackers had stolen the data of approximately 500 million customers from its website. The series of data breaches actually began in 2014, when Starwood-branded systems were compromised. The attackers remained in the system until September 2018, after Marriott International acquired Starwood in 2016, and were not discovered until then.

Some hotel customers' names and contact information were compromised. The attackers may have even used a combination of contact information, passport numbers, Starwood membership card numbers, travel information, and other personal information to facilitate their further social engineering attacks. Marriott International believes that more than 100 million customers had their credit card numbers and expiration dates stolen, although the company is unsure if the attackers were able to decrypt the credit card numbers.

3. **eBay**

eBay is the leader in online auctions. The company reported a cyberattack in May 2014 that compromised the names, addresses, dates of birth, and encrypted passwords of its 145 million users. The company said the hackers used the credentials of three company employees to break into the company's network and had full internal access for 229 days, during which time they were able to break into user databases.

The company urged customers to change their passwords and said that financial information (such as credit card numbers) is stored separately and was not affected by the breach. The company was criticized by the public for its lack of effective communication with users and poor implementation of the password update process.

4. **Equifax**

Equifax, one of the largest credit reporting agencies in the United States, announced on September 7, 2017, that an application vulnerability on one of its affiliated websites resulted in a data breach that exposed the personal information of 143 million consumers (including Social Security numbers, birthdates, addresses, and, in some cases, driver's license numbers). The credit card information of 209,000 consumers was also compromised. The breach was discovered on July 29, 2017, but the company said the attack and data breach may have begun in mid-May.

5. **Heartland payment system**

In 2008, credit card processor Heartland Corporation processed 100 million payment card transactions per month for 175,000 merchants, mostly small and mid-sized retailers, using its Heartland Payment System. In March 2008, hackers used Structured Query Language (SQL) injection to install spyware on the Heartland payment system's data systems, stealing information from 134 million credit cards. Heartland didn't learn of the serious data breach until January 2009, after Visa and MasterCard notified Heartland of suspicious transactions on accounts it processed.

As a direct result of the incident, Heartland's payment system was deemed non-compliant with the Payment Card Industry Data Security Standard (PCI DSS), and Heartland's core business was suspended until May 2009, when it was permitted to process transactions from major credit card issuers. The company also paid approximately $145 million to settle fraudulent payments.

Subsequent criminal and civil liability investigations revealed that a Cuban-American named Albert Gonzalez orchestrated an "international coordination" to steal credit and debit card accounts. A federal grand jury indicted Gonzalez and two unnamed Russian co-conspirators in 2009. Gonzalez was sentenced to 20 years in prison in March 2010.

The SQL injection vulnerability is well known, and security analysts have repeatedly warned that many payment systems could be affected. However, persistent vulnerabilities in many web-facing applications make SQL injection the most common form of attack against websites today.

6. **Target**

In December 2013, retail giant Target, the eighth largest retailer in the U.S., announced that hackers had accessed Target's POS payment card readers through a third-party HVAC vendor and collected approximately 40 million credit and debit card numbers. The breach actually began before Thanksgiving, but wasn't discovered until a few weeks later.

In January 2014, the company revised that estimate upward and reported that the personally identifiable information (PII) of 70 million of its customers had been compromised, including their names, addresses, emails, and phone numbers. The final estimate was that the breach could affect as many as 110 million customers.

Target's CIO announced his resignation in March 2014 and its CEO resigned in May. The company estimated the damage from the data breach at $162 million.

Since then, the company has made significant improvements in security. But a leading security expert declared that "the company is still using yesterday's security paradigm" because the company's security improvements focused on keeping attackers off the network and did not target improvements in incident response.

7. **Uber**

In late 2016, Uber learned that two hackers had obtained the names, email addresses, and phone numbers of 57 million Uber app users. The duo also obtained the driver's license numbers of 600,000 Uber drivers. This is a very serious breach of personal information. The scope of Uber's data breach is very large, and the company did not come up with a clear emergency response strategy in time.

Uber claims that no other information, such as credit card or social security numbers, was stolen. Hackers were able to access Uber's GitHub account, where they found username and password credentials for Uber's AWS (Amazon Web Services) account. These credentials should never be on GitHub, a publicly accessible code repository.

Unfortunately, it took Uber about a year to disclose the data breach. To make matters worse, they paid the hackers about $100,000 to try to destroy the data, claiming it was a "bug bounty" payment, but were unable to verify that the hackers did so. Uber blamed the breach on its chief security officer and fired him.

The breach cost Uber a lot of money and reputation. At the time of the breach announcement, the company was in talks to sell part of its stake to SoftBank. Initially, Uber was valued at $68 billion. By the time the deal closed in December, its valuation had fallen to $48 billion. The damage to goodwill caused by a serious data breach and a poor response is not necessarily the main reason for the losses, but business analysts believe it is an important factor.

8. **JPMorgan Chase**

JPMorgan Chase Group is the largest bank in the United States. In the summer of 2014, the bank was hacked and exposed data affecting more than half of all US households (76 million) and 7 million small businesses. According to its filings with the U.S. Securities and Exchange Commission (SEC), this data includes customers' names, addresses, phone numbers and emails, and even other private information about users.

The financial institution says that there is no evidence that account numbers, passwords, user IDs, dates of birth, or Social Security numbers were left vulnerable, and to date, JPMorgan has not seen any "unusual customer fraud related to this incident".

Despite this, hackers were reportedly able to gain "root" privileges on more than 90 of the bank's servers, meaning that hackers could take a variety of malicious actions, including even transferring funds and closing accounts. JPMorgan Chase spends $250 million a year on security, according to the SANS Institute.

In November 2015, the U.S. government charged four men with hacking JPMorgan Chase and other financial institutions. The hackers face 23 charges including unauthorized computer access, identity theft, securities and wire fraud, and money laundering, which allowed them to steal about $100 million.

9. **Sony Mobile PlayStation Network**

On April 20, 2011, 77 million PlayStation Network accounts were hacked. The site was down for a month with an estimated loss of $171 million. Of the more than 77 million accounts affected, 12 million were unencrypted credit card numbers. Hackers can obtain information such as names, passwords, emails, home addresses, purchase history, credit card numbers, and PSN logins and passwords. The Sony incident is a wake-up call for large multi-national corporations that also store vast amounts of user data.

In 2014, Sony agreed to an initial $15 million settlement in a class action lawsuit over the breach. This is considered the worst data breach of the gaming community ever.

10. **Anthem Insurance Company**

On March 13, 2015, Anthem filed a breach report with the U.S. Department of Health and Human Services Office of Civil Rights (HHS OCR), detailing that on January 29, 2015, it discovered that cyber attackers had gained access to the company's IT systems through an apparently undetected Advanced Persistent Threat (APT) and obtained data beyond their authorization. After filing a breach report, Anthem discovered that cyber attackers had infiltrated its systems through at least one employee who responded to malicious phishing emails. An HHS OCR investigation found that between December 2, 2014, and January 27, 2015, cyber attackers stole the personal health information of nearly 79 million individuals, including names, Social Security numbers, medical identification numbers, addresses, dates of birth, email addresses, and employment information.

The incident has been called "the largest health data breach in U.S. history". In October 2018, Anthem agreed to pay a $16 million settlement to HHS OCR and take significant corrective actions to address privacy and security concerns that may have violated the Health Insurance Portability and Accountability Act (HIPAA).

A post-mortem investigation found that Anthem failed to take adequate controls to detect hackers who could gain access to its systems to obtain passwords and steal individuals' private information, and failed to implement a robust password policy.

The 10 major data breaches mentioned above involved a wide range of industries, including finance, insurance, healthcare, hospitality, and gaming. The data breaches at these large companies, which touch the lives of consumers, have had far-reaching and serious consequences.

According to a report (IBM Security, 2019) published by the IBM Security Team (IBM Security, 2021) and the Ponemon Institute, "Data Breach Costs Growth, Average $4 Million Per Incident", based on in-depth interviews with more than 500 companies around the world that experienced data breaches between July 2018 and April 2019, and taking into account hundreds of cost factors ranging from legal and regulatory violations to loss of brand equity, customer churn, and reduced employee productivity and hundreds of other cost factors, each data breach involved an average of 25,575 data records, resulting in approximately $3.92 million in financial losses, equivalent to about $150 per record. Data breaches in the healthcare industry result in greater economic losses than in other industries.

It's not just large organizations or multi-national corporations that are affected by data breaches. Verizon's 2019 Data Breach Investigation Report (DBIR) (Verizon Enterprise Solutions, 2019), based on an analysis and survey of 41,686 security incidents, of which 2013 were confirmed compromises, found that 43% of data breaches involved small businesses. Figure 2.6 shows the percentage of organizations involved in data breaches in 2019.

The 2019 Data Breach Investigation Report also pointed out some concerns, and its findings are shown in Fig. 2.7. Among them, 56% of the data breaches took a month or more to be discovered; 71% of the data breaches were driven by commercial interests; 25% of the hacking attacks involved strategy and competition. In terms of attack methods, 29% of attacks involved the use of stolen login credentials; 32% used phishing attacks.

The staggering number of users affected by data breaches demonstrates the seriousness of the current data security situation.

Compared with developed countries such as Europe and the United States, which have more stringent legal requirements and higher breach costs, China is still in the development stage in the field of data security and privacy protection. It is not uncommon for organizations to collect and use personal data beyond their authority for

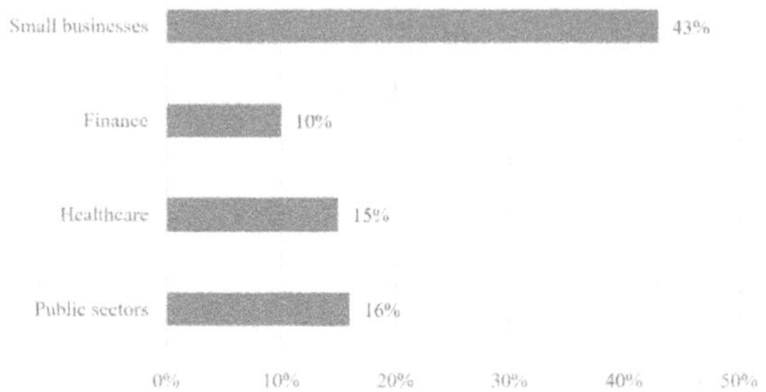

Fig. 2.6. Proportion of organizations involved in data breaches in 2019.

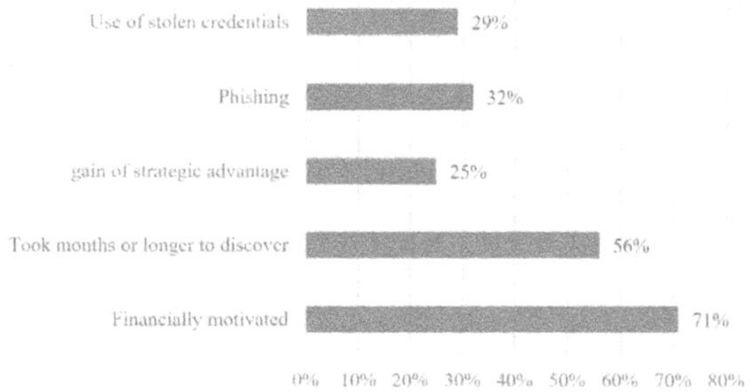

Fig. 2.7. Investigation conclusions of data leakage incidents.

various business purposes. As for companies responsible for storing, processing, and transmitting personal data, their data breach cases have repeatedly made headlines in various media, attracting great attention from regulators, industries, and ordinary users.

According to ZDNet (Catalin, 2019), in the first quarter of 2019, there were several resume information leaks involving 590 million resumes from Chinese companies. The main cause of the resume leak incidents was the lack of security measures in place for MongoDB (a document-based database) and Elasticsearch search servers. For example, default passwords or even blank passwords were used for the databases, or the database servers used only for intranet access were exposed to the public Internet due to misconfiguration of firewalls.

On March 24, 2020, Sina Weibo was exposed to a malicious call to the user query interface, which led to data leakage, and the Cybersecurity Administration of the Ministry of Industry and Information Technology (MIIT) questioned and interrogated the relevant person in charge of Sina Weibo, requiring it to take further effective measures to eliminate data security risks. On the Dark Web, hackers claimed to have obtained 500 million mobile phone numbers of Weibo users (Paganini, 2020). Post-analysis revealed that the hackers may have used a combination of social engineering and "credential stuffing" (i.e., obtaining login credentials, such as account passwords, from one website and attempting to log in from another website) to achieve their goal through the abuse of query interfaces.

The scope of these attacks demonstrates that even the largest, most well-protected organizations are vulnerable to attack and are prone to error. And that means it's getting harder and harder for organizations of all kinds to earn users' trust in data security and privacy.

2.1.4 *Growing concerns about data security*

On October 15, 2018, Peter Firstbrook, vice president and senior analyst at Gartner, discussed the six major security trends in 2018 and 2019 at the Gartner Symposium/ITxpo summit in Orlando (Firstbrook *et al.*, 2018). His speech focuses on the trend and evolution of the security industry, which will help enterprises realize their network security strategies.

Among them, the first trend he proposed is that business executives pay more attention to security than before, and security personnel need to master the ability to communicate with executives using business language. Regardless of the regulatory level, business level or technical level, the potential economic losses caused by security and privacy protection risks have reached the level that executives and corporate boards need to pay attention to. For example, the official implementation of GDPR and the strict enforcement by various countries have significantly increased the cost of corporate breaches and created a series of uncertainties. What's more, the outbreak of various ransomware viruses such as WannaCry has caused a lot of economic losses to the IT infrastructure and even production lines of companies.

According to Gartner's research report, almost all large commercial organizations will require the board of directors to report on the status, risks, and measures of network security and data protection. This requires security personnel to not only express and present from a technical perspective but also add a business perspective and summarize and report in a language that board members can understand. The data security governance system and framework, security risks, and mitigation measures should also partially address the organization's business competitiveness or business risk objectives.

According to a survey report published by Gartner in 2019: in terms of additional technology investments by Chief Information Officers (CIOs) of enterprises and government agencies, data analytics and cybersecurity are crowding out Cloud took the top position (McMillan and Proctor, 2018). In the area of network security, 95% of CIOs believe that "the threat to security will increase", 4% of CIOs believe that "the threat will remain the same", and only 1% of CIOs believe that "the threat to security will decrease". There are two main reasons for this: one is that with the increase in communication between organizations and the further blurring of the concept of internal and external networks, more and more external users will access the organization's information, making security management more challenging; the other is the booming development of the Internet of Things. Due to the limited computing power of IoT devices, there is a certain trade-off for security. For example, it is possible to attack an online camera and then use that device as a jumping off point to move laterally through the network in search of additional attack targets.

The majority of business strategies will explicitly identify data as a core asset of the organization and data analytics as an essential business capability. Most of the Chief Digital Officers (CDOs) and Chief Financial Officers (CFOs) will formally assess the value of their organization's data assets in order to improve the management of and benefits from data, and more large enterprises will use financial risk assessments of their data assets to prioritize investment decisions for IT, analytics, cybersecurity, and privacy.

2.1.5 *Data security has become a hot topic in the industry*

The RSA Conference is the world's leading information security forum and exhibition. Held each spring in San Francisco, USA, the RSA Conference attracts the attention of the global security community, the business community, and even the general public. During

[illegible]	[illegible]	[illegible]	[illegible]	[illegible]	[illegible]	[illegible]	[illegible]	[illegible]	[illegible]
Data	**Data**	**Data**	**Data**	**Data**	**Data**	**Data**	**Data**	**Data**	**Data**
Information	Information	Information	Information	Information	Information	Information	Information	Information	Time
Based	Based	Based	Organizations	Based	Based	Based	Time	Time	Information
Organizations	Organizations	Risk	Risk	Time	Organizations	Time	Organizations	Organizations	Organizations
Time	Risk	Organizations	Attack	Organizations	Time	Organizations	Based	Risk	Based
Risk	Time	Access	Threat	Risk	Attack	Risk	Real	Based	Threat
Access	Access	Time	Applications	Real	Risk	Real	Risk	Cyber	Risk
Applications	Applications	Compliance	Including	Including	Real	Threat	Cyber	Threat	Cyber
Including	Including	Key	Time	Key	Including	Threat	Threat	Attack	Real
Compliance	Tech	Attack	Access	Attack	Threat	Attack	Attack	Real	Attack

Fig. 2.8. RSA conference keyword ranking.

the week-long conference, the world's top cybersecurity professionals and business leaders come together to discuss emerging trends and develop the best strategies to address current and future threats.

The keyword rankings of the RSA Conference over the years can reflect the hot spots in the security industry. Figure 2.8 summarizes the keyword rankings for the decade from 2009 to 2018. This ranking may reflect that the keyword "data" always occupies the top spot. The keyword "information", which is related to "data", is also in the top three.

In addition, it is worth noting that the keyword "Cyber(space)" has appeared in recent years, and the frequency of the occurrence of "Threat" has increased overall, reflecting the current general trend in cyberspace security.

Data security is not only receiving more and more attention from the industry but has also become a hot topic in various fields, such as politics, law, and social life.

Under the theme of "Now Matters", the 2018 RSA Conference enabled enterprises from around the world to have in-depth discussions on the development trend of cybersecurity in the digital era, so as to jointly explore hot issues in cybersecurity, to pay attention to the progress of cybersecurity regulations, and to examine the progress of cutting-edge technologies. The conference explored the following two topics related to data security.

First, personal privacy protection. How to protect personal privacy from regulatory, supervisory, and technical perspectives is one of the important topics of the conference, including exploring the legislation on the use of AI data, the establishment of identity mechanisms, and privacy and security in machine learning and the IoT.

The second is the implementation of privacy laws and regulations. In May 2018, the European Union's General Data Protection Regulation (GDPR) went into effect. As the most stringent user data security and privacy regulations in history, any company operating in the European market will face GDPR compliance issues. Therefore, the conference organized a workshop with GDPR as the core topic, focusing on the main challenges of GDPR, how companies can meet compliance requirements and how individuals can use GDPR for privacy protection, and the place of "security" in GDPR compliance. Topics also include legal compliance for companies to take proactive defensive measures, cybersecurity and international trade, employee oversight in the digital age, and corporate data breaches.

At the 2019 RSA Conference, there was a wide variety of topics related to data security.

Data protection continued to be a very hot topic, both from a consumer and industry perspective. Whether it was GDPR compliance or multiple data breaches affecting a large number of consumers, it sparked discussions about compliance with privacy legislation, restrictions on the use of private data, data ownership between large internet companies and consumers, and defining responsibility and penalties for data breaches.

From the perspective of Internet users, beyond the dimension of data security, the authenticity of information has become a hot topic of cyberspace security. Under the conditions of highly popularized Internet and extremely fast information dissemination, there is also room for speculation and dissemination of fake news published on the basis of certain political, military, or commercial purposes. In particular, the development and misuse of new technologies, such as the use of artificial intelligence to process videos and images to create "deep fake" videos (commonly known as "face-swapping"), may have unforeseen effects on politics, economics, culture, and even religion.

From the perspective of security professionals and security vendors, there are many challenges in the area of data security. Attacks are becoming more sophisticated, attackers are investing more, and attack tools are becoming more intelligent and automated. The complexity of security products and solutions is also increasing with the evolution of attack and defense confrontations, and there are difficulties in deploying them and operating them effectively in the enterprise. In addition, the irreversible trend of enterprises moving their business to the cloud poses the challenge of how to effectively integrate the security capabilities offered by cloud providers with those already deployed by enterprises.

At the RSA Conference 2020, the data security and privacy landscape took another somewhat welcome turn. Driven by the implementation of privacy compliance frameworks, the use of privacy compliance automation tools, and other factors, security leaders and experts are less concerned and more understanding about "making privacy actionable" in products, services, and organizations. This understanding is driving the development of some data security technologies, such as homomorphic encryption. Another notable change is that many organizations that previously understood privacy as "legal compliance" and "social responsibility" are now focusing on its more positive business implications, i.e., gaining insight into and protecting user intent to provide a differentiated and positive user experience. In this dimension, "security" and "privacy" are working together in a more positive way in organizations. However, insights from the RSA Conference 2020 sessions also point to a world in which the privacy conversation is changing frequently. Some of the themes highlight the challenges and unintended consequences of privacy compliance and enforcement, particularly the rapidly exploding scenario of regional, national, and global privacy regulations (some of which even conflict with each other), the need to explore ethical considerations related to privacy and data security, and the overarching point that "we can and must do more and better!" overall perspective.

2.1.6 *Where is the sensitive data located?*

Trustwave Corporation pointed out in its "Trustwave Global Security Report" (Holdings, 2014) released in 2014 that most organizations do not have mature methods for controlling and tracking sensitive data.

When it comes to data security, organizations are generally aware that it can lead to significant risk and even legal liability, but there is no proven method for controlling risk by tracking sensitive data. Based on interviews with a wide range of IT and security professionals, 63% of organizations don't have a mature method for controlling and tracking sensitive data. "This means that many organizations don't know where their sensitive data is, who has access to it, and whether it flows". Phillip J. Smith, senior vice president at Trustwave, said, "This information is the first step in building a security strategy".

Indeed, data is critical to both organizations and users. One of the first tasks of a system designer is to identify sensitive data and determine how to properly protect it. Taking an inventory of all the data that needs to be protected is a critical step in data security. By maintaining an up-to-date inventory of all systems used to store and process sensitive data, organizations can manage their data security goals. Classifying data by sensitivity helps data management teams determine where to focus their security efforts.

Before counting data, the following questions should be considered first:

(1) Does the organization have an up-to-date inventory of all internal systems, computer equipment, software, and data files?
(2) Does the organization have a detailed and up-to-date inventory of all data that should be classified as sensitive data (both commercial and personal data that could be harmed by unauthorized or accidental disclosure)?
(3) Has the organization categorized data records based on the level of risk that disclosure of the data would pose to the organization?
(4) Does the organization have a written data inventory policy, and does the policy outline what should be included in the inventory and how and by whom the data inventory should be updated?

From a business perspective, there should be a comprehensive inventory of the data sources owned by the organization. Common data sources are listed as follows:

(1) An organization's production and business systems. Examples include financial analysis, sales, manufacturing, inventory, and warehouse management systems. Note that systems that use cloud storage and cloud hosting should also be included.
(2) Third-party systems that interface with the organization's internal systems, such as data sources from customers or suppliers.
(3) Data sources from mobile devices or the Internet of Things (IoT). For example, information obtained from devices such as cameras, sensors or thermostats.

In the case of telecommunications companies, for example, the main places where sensitive information is stored include support systems, business platforms, and communication systems.

The support systems mainly store users' personal information, which involves the protection of users' privacy, and mainly includes business support system, operation analysis system, customer service system, network management system, customer support system, signaling monitoring system, network security control platform, customer relationship management system, data service monitoring system, industrial customer service integrated operation platform, value-added service management platform, etc.

Business platforms mainly include SMSC, MMSC, industrial gateway, SMS gateway, WAP gateway, MMS gateway, industrial customer relationship system, spam monitoring platform, MISC/DSMP, video platform, e-commerce platform, payment platform, and so on.

Taking the traditional GPRS network as an example, the communication systems mainly include packet switched core network (including SGSN and GGSN), circuit switched core network (MSC and MGW), UMTS/3G access network (RNC and NodeB), GSM/2G access network (BSC and BTS), and so on.

These are the key network elements of sensitive data storage and circulation processing, and enterprises should start from these parts to protect sensitive data in order to prevent data leakage and other security accidents from bringing losses to the interests and reputation of enterprises.

2.1.7 *Behind the data security issues*

Ginni Rometty, Chair, CEO, and President of IBM, once stated: "We believe that data is the phenomenon of our time. It is the world's new natural resource. It is the new basis of competitive advantage, and it is transforming every profession and industry. If all of this is true — even inevitable — then cyber crime is, by definition, the greatest threat to every profession, every industry, every company in the world" (Lenzner, 2013). In this sense, data security cannot be overstated.

As far as the typical enterprise environment is concerned, data security faces a complex environment, and the risk of data security

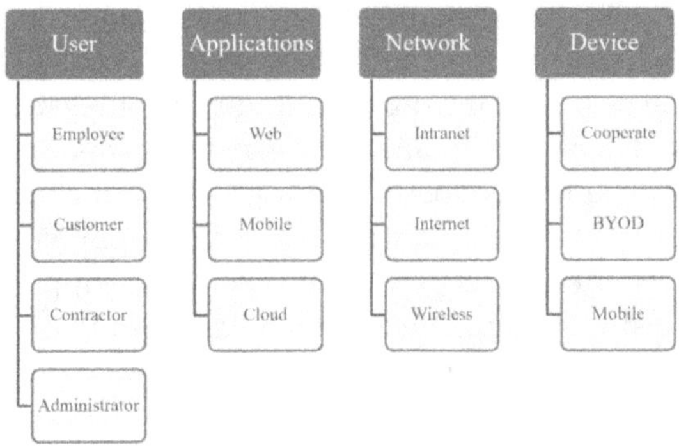

Fig. 2.9. Challenges to data security in complex scenarios.

is very high, especially the increasing complexity of people, applications, networks, and devices, which poses huge risks and challenges to data. Challenges to data security in complex scenarios are as shown in Fig. 2.9.

According to the survey, more than 60% of employees use their own computers and other devices to get work done, primarily for creativity, convenience, and ease of use. A deeper dive revealed that 50% of organizations rely on employees to manage their own personal devices, 70% of employees never receive operational guidance from their organization on how to deal with the risks of bringing their own devices, and 65% of employees download specific business applications on their own to make it easier to work with data.

Clearly, untrusted networks, devices, users, and applications will inevitably lead to data risks. Figure 2.10 shows the major sources of data risk.

To effectively address with these risks, it is necessary to build and implement a unified security policy system based on the full lifecycle process of "Identify-Protect-Detect-Respond-Recover" in combination with the data security governance framework introduced in Section 3.3.

Based on annual reports and industry trend analyses in the security field such as Verizon's "2019 Data Breach Investigation Report"(Verizon Enterprise Solutions, 2019) and FireEye's "M Trends: Frontline Report" (FireEye Inc., 2019), the root cause of more than 80% of data breaches or security incidents can still be

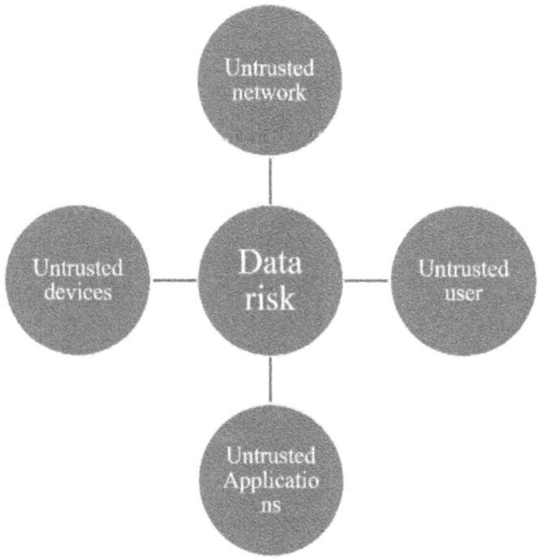

Fig. 2.10. Main sources of data risk.

attributed to basic security issues. The most basic security configuration and security awareness still need to be addressed.

2.1.8 *Evolution of data security attack surface*

With the continuous application of new technologies such as the Internet of Things, cloud computing, and artificial intelligence, the network perimeter in the traditional sense continues to collapse, the organization's attack surface is constantly expanding, and the organization's data security is facing new challenges. At the same time, cyberattacks also show a trend of complex means, clear purpose, and variable sources. Complex network changes and diverse network attacks have greatly expanded the original data security attack surface in depth and breadth. The evolution of the attack surface over time is shown in Fig. 2.11.

Before the 1990s, the Internet was relatively rare, and the Internet was probably built more by researchers and professionals, with less access by the general public. At that time, security was not as important, and the limited attack vectors were mainly physical access attacks. However, as more and more sensitive information is placed on the network, the importance of data and network security is gaining more and more attention.

In the 1970s and 1980s, researchers with access to the "Internet" used it to poke fun at each other. The jokes were harmless but still

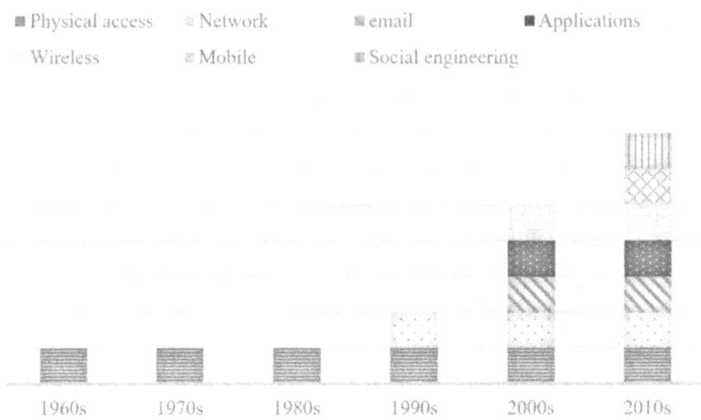

Fig. 2.11. Evolution of the attack surface over time.

exposed flaws in the security of ARPANET (Advanced Research Projects Agency Network, commonly known as "ARPANET"). The scope of network applications at this time is small, and many users know each other, so the risks and threats posed by network attacks are not so serious.

There were also several high-profile security incidents during this time. The movie "WarGames" popularized the term "computer hacking", but outside of narrow circles among politicians, the legal profession, and security professionals, little serious cybersecurity approach is taken to the risk of hacking.

In the late 1980s, the use of the Web began to grow rapidly. With the establishment of network connections between universities, government, and military facilities, the need for security increased. The first automated worm appeared on the ARPANET in 1988. The "Morris Worm", developed by the students at Cornell University in the United States, could exploit flaws in services such as finger, rsh/rexec, and sendmail on the Unix system to infiltrate another computer and use the vulnerabilities to replicate itself and then copy itself sent to the new computer. The self-replicating "Morris Worm" exposes vulnerabilities in networked computers. It consumes many resources and renders the infected computer inoperable, allowing the virus to spread quickly throughout the network. At this time, influential leaders in the network decided to develop a network security method to deal with network threats, which directly led to the birth of the first professional network security organization: the Computer Emergency Response Team (CERT).

Based on this, it can be said that since the 1990s, the "network" has officially begun to serve as an emerging attack surface. In the following years, the number of network-based attack methods increased, and automated and manual attacks became more frequent and mainstream.

Around the year 2000, there were many new attacks that used email to propagate and exploited vulnerabilities in instant messaging protocols.

The Melissa virus was created by David Smith in 1999. It spreads to computers by sending itself to the first 50 people in a recipient's address book, infecting many computers in the process. The Melissa virus disrupted government and commercial networks, causing a total of approximately $80 million in damage.

The ILOVEYOU virus, released in 2000 and believed to have been created by Onel de Guzman, infects computers through email and IRC clients. The ILOVEYOU virus spreads widely around the world, causing an estimated $1 billion in damage.

The Nimda virus, released in 2001, spreads via email, network shares, and other methods, infecting thousands of computers. The Nimda virus attacked Internet servers, causing the entire Internet to experience a sharp drop in performance or even to stop responding.

The year 2000 can be considered the beginning of the "mail" era as a new type of attack. Statistics show that in 2000, there were already 360 million Internet users worldwide. With the rapid development of the Internet, various new attack methods also emerge, such as attacks based on vulnerabilities in operating systems and application software, wireless attacks, and so on.

Among them, the "Blaster Worm" virus, which appeared in August 2003 is worth mentioning. The virus exploits a known vulnerability in Microsoft's DCOM RPC service, which is described in Microsoft Security Bulletin MS03-026. When run, the virus attempts to retrieve a copy of the file msblast.exe from the compromised host. Once this file is retrieved, it is executed and the infected system starts scanning for vulnerable systems, causing damage in the same way. In addition, the virus installs a remote command line backdoor on the infected computer that listens on TCP port 4444, allowing attackers to issue remote commands to the infected system.

Around 2000, attacks against devices on wireless LANs and even GSM/GPRS networks began to appear. Against the widely used

Wi-Fi networks, various attacks such as sniffing attacks, man-in-the-middle attacks, and attacks that exploit the vulnerability of the WEP (Wired Equivalent Privacy, the encryption technology used in the early days of wireless networks, which has essentially been abandoned) protocol to crack Wi-Fi passwords have emerged.

As we enter the new Internet era, new attack methods continue to emerge. It is worth mentioning social engineering attacks and attacks on mobile devices.

Social engineering attacks mainly refer to the use of deception to gain access to information systems. The medium of communication is usually telephone, SMS, instant message, or email. Attackers typically pose as executives or managers of a company to pressure IT support teams to provide server passwords, copy sensitive data, or perform some other unauthorized action. Unlike previous attacks based on technical vulnerabilities, social engineering attacks put the human factor at risk in cyber attacks. The human element is often cited as the weakest link in cybersecurity. The theme of the RSA Conference in San Francisco in February 2020 was "Human Element", highlighting the security industry's consistent consideration and concern about this issue.

In fact, based on "Cyber Security Threats 2018: Trends and Forecasts" published by Positive Technologies (2019) (a security company based in Russia), the top two cyber attack methods are the use of malware and social engineering attacks. Social engineering attacks accounted for 31%, and there is a clear annual growth trend. The "2019 Data Breach Investigation Report" (Verizon Enterprise Solutions, 2019) published by Verizon also shows that compared to 2018, social engineering attacks have increased very significantly in 2019.

Another attack is on mobile devices, especially smartphones. With the increase in the average download rate of 4G users and the decrease in mobile traffic tariffs, mobile applications are becoming more and more abundant, bringing convenience to daily life, but at the same time, a large number of malicious programs have appeared on the mobile Internet, jeopardizing the security of the public's personal information and property security. The White Paper on Personal Information Protection in the Intelligent Terminal Industry (2018) (CAICT, 2019), jointly released by the China Academy of Information and Communications Technology (CAICT), the Internet

Society of China, and the Telecommunications Terminal Industry Association, pointed out that of the more than 27.22 million intelligent terminal applications detected from October 2017 to October 2018, there were 2.99 million applications that violated users' rights and interests, such as tariff depletion, nuisance, and privacy theft, of which the proportion of applications that stole confidential information was 11.86%. In 2017, the number of malicious programs on the mobile Internet obtained through the National Computer Network Emergency Response Technical Team (CNCERT) and the number of malicious programs on the mobile Internet obtained through autonomous capture and vendor exchange amounted to more than 2.53 million, a year-on-year increase of 23.4%.

2.1.9 *Typical attack scenarios and types*

Based on the reports published by various network security research institutions in the industry, in terms of attack targets, attacks against infrastructure, websites and services, and specific users account for a very high proportion: attacks against infrastructure account for nearly half of all attack incidents; attacks against websites and services account for 26%; in addition, attacks against specific users, terminal devices, IoT devices, and financial terminals (such as POS machines and ATM machines) also account for a certain proportion. Figure 2.12 shows the proportion of attack targets.

From the identity dimension of the attackers, they can be broadly categorized as follows:

(1) Internal employee (disgruntled or inadvertently affiliated with the organization).
(2) Hacker individual/organization.
(3) Competitors.
(4) Cybercrime groups.
(5) Terrorists.
(6) Government/intelligence.

Taking the infrastructure attack as an example, the typical attack method is shown in Fig. 2.13.

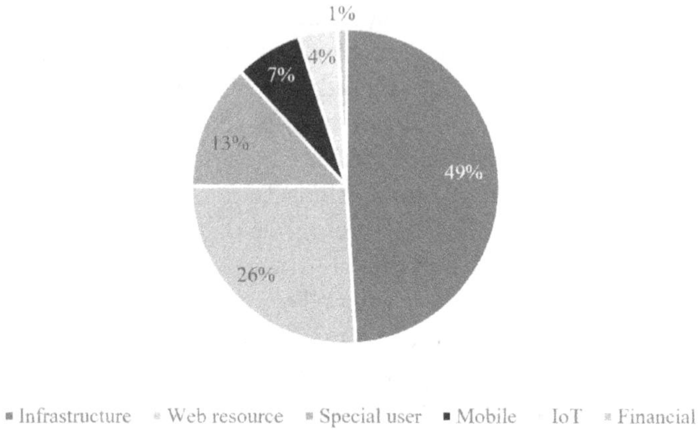

Fig. 2.12. Proportion of attack targets.

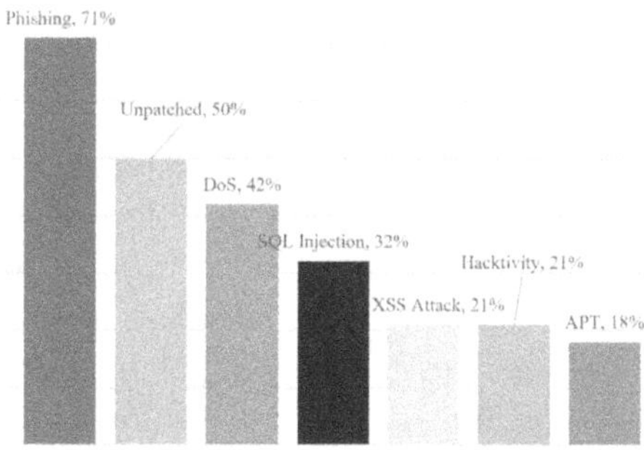

Fig. 2.13. Typical attack methods.

The top attack methods are phishing attacks, exploitation of unpatched vulnerabilities, denial of service attacks, SQL injection, cross-site scripting attacks, etc.

2.1.9.1 *Phishing attack*

Phishing is classified as a social engineering attack in which the attacker poses as a legitimate entity that the victim trusts and interacts with in order to trick the victim into revealing some confidential information. The most common confidential information is authentication credentials (account numbers, passwords, etc.)

and may also include financial information, etc., which can be maliciously exploited by the attacker. Phishing is essentially a form of information gathering.

When a victim opens a malicious email, instant message, or text message, if they are tricked into clicking on a malicious link, it can trigger various types of malicious actions, such as installing malware, downloading ransomware, or disclosing sensitive information.

Phishing attacks are common and can be devastating. While the impact of phishing attacks on individuals can include damage to reputation or property, the impact on organizations is more serious and can be used as a prelude to a large-scale attack, such as an Advanced Persistent Threat (APT) event, to bypass security perimeters, spread malware on local networks, or gain privileged access to sensitive data.

Organizations that fall victims to phishing attacks can see their reputation and consumer confidence suffer, even losing market share and ultimately suffering severe financial losses. Depending on the severity, phishing attacks can escalate into security incidents.

2.1.9.2 *Unpatched vulnerabilities*

In addition to phishing attacks, the percentage of scenarios where systems are not patched in a timely manner, exposing them to attacks using known vulnerabilities, is also high. According to the study, more than 50% of organizations that have suffered a data breach believe the culprit was a known vulnerability that they hadn't patched, and 34% said they knew their systems were vulnerable before they were attacked.

It may seem like a no-brainer for organizations to patch security vulnerabilities by installing software updates, but that's not the case. On the one hand, adequate testing must be done before applying security updates or patches to avoid "business disruption"; on the other hand, there are considerations of downtime, legacy systems, compatibility with existing software and operating systems, and the need to "race the hackers". Organizations should apply updates or patches to all of their systems in a timely manner before vulnerabilities are exploited.

Establishing and improving an effective vulnerability response process is necessary for unpatched vulnerability scenarios. Vulnerability response teams should work with IT and security personnel in a timely manner to prioritize patches based on the risk of vulnerability exploitation, automate the vulnerability response process as much as possible, and improve response efficiency.

2.1.9.3 *Denial of service attack*

A denial of service (DoS) attack is a malicious attempt to deny legitimate users access to a target system, such as a website or application. Typically, an attacker generates a flood of packets or access requests that eventually overwhelms the target system. A Distributed Denial of Service (DDoS) attack occurs when an attacker uses multiple compromised or controlled sources to generate an attack on a target system.

Following the layers of the Open Systems Interconnection (OSI) model, DDoS attacks typically occur at the network, transport, and application layers.

Broadly speaking, DoS and DDoS attacks can be divided into the following three types:

(1) **Traffic-based attacks** include UDP flooding, ICMP flooding, and other deceptive packet flooding. The purpose of the attack is to occupy the bandwidth of the attacked website so that the website cannot provide services to legitimate users. The unit of measurement for traffic attack strength is bits per second (BPS).
(2) **Protocol-based attacks** include SYN flooding and fragmented packet attacks. The purpose of this type of attack is to consume the resources of the website server or the resources of communication devices, such as firewalls and load balancing devices, so that the website cannot provide services to legitimate users. The unit to measure the strength of protocol attack is the number of packets per second (PPS).
(3) **Application layer attacks** include low-speed and slow-speed attacks, GET/POST flooding, Apache server vulnerability attacks, and Windows or Linux operating system vulnerability attacks. These attacks consist of seemingly legitimate requests with the goal of crashing a web server so that it cannot serve legitimate users. The unit of measurement for application layer attack severity is the number of requests per second (RPS).

2.1.9.4 *SQL injection*

SQL is a query language designed to manage data stored in relational databases. SQL can be used to read, modify, and delete data, and some database systems even support the use of SQL commands to execute operating system commands.

SQL injection attacks are a type of injection attack. Attackers insert malicious code into SQL statements and pass them to the database server to be parsed and executed for malicious purposes. A prerequisite for a successful SQL injection attack is that the attacker first finds vulnerable user input in the front-end web page or Web application. An SQL injection vulnerability is input that allows an attacker to manipulate the contents of an SQL query that is executed in the background using this type of attack. This input is often referred to as the malicious payload and is the key to the attack. After the attacker sends this content, malicious SQL commands are executed in the database.

In a successful SQL injection exploit scenario, it is possible to read sensitive data from the database, modify database data (insert/delete/update), and perform administrative operations on the database (such as shutting down the database instance). In some cases, OS commands can also be sent to the server operating system.

In addition to the "traditional" method of directly manipulating user input and splicing SQL query statements, some new types of SQL injection attacks are emerging. For example, manipulating local browser cookies attempts to inject SQL into the server database while the web application is processing the cookie information. Client-generated parameters in the network protocol, such as HTTP headers, can also be used as vectors for SQL injection attacks. If the Web application doesn't sanitize these inputs, the spoofed header file containing SQL injections could inject this code into the database.

2.1.9.5 *Cross-site scripting attack*

Similar to SQL injection attacks, Cross-Site Scripting (XSS) attacks are a type of client code injection attack, but the implementation mechanisms of the two are quite different. The attacker's goal in cross-site scripting attacks is to execute malicious scripts in the victim's web browser by adding malicious code to legitimate web pages or web applications. The actual attack occurs when a victim visits a web page or web application that contains malicious code. web pages that allow user input, such as message boards, forums, or web comment areas, are common vectors for cross-site scripting attacks.

In a successful exploit scenario, the attacker uses XSS to send malicious scripts to end users who are unaware of the attack. The end user's browser has no way of knowing that the script is untrusted, so it runs it normally. Since the browser considers the script to be from a trusted source, it allows the script to access cached information about the web site in the browser, such as cookies, session tokens, or other sensitive information. Malicious XSS scripts can even rewrite the content of HTML pages.

Cross-site scripting attacks occur for a variety of reasons, but the underlying principles are similar. If a web application uses user input in the output it generates without validation or encoding, the end user's browser, when parsing that output, will parse the malicious code embedded in it and be affected by this vulnerability. Malicious code content typically takes the form of JavaScript code embedded in HTML but can also take the form of HTML, Flash, VBScript, ActiveX, etc.

2.1.10 *Major threats to data security*

According to "Cyber Security Threats 2018: Trends and Forecasts" (Positive Technologies, 2019) published by Positive Technologies, the number of attacks aimed at stealing data is on the rise. Attackers' interest is focused on personal data (30%), credentials (24%), and payment card information (14%).

The report also shows that among the cyberattacks that occurred in 2018, the two most direct motivations are the acquisition of information (especially sensitive information) and interest-driven, accounting for 42% and 41%, respectively (see Fig. 2.14). At the same time, the sensitive information obtained can also be used for financial gain, such as stealing property, extorting, or more directly selling on the dark web. As a result, it is estimated that more than 80% of cyber attacks are targeted at data.

Insights into attack target industries, the most vulnerable industries, are government agencies, healthcare, and finance, as

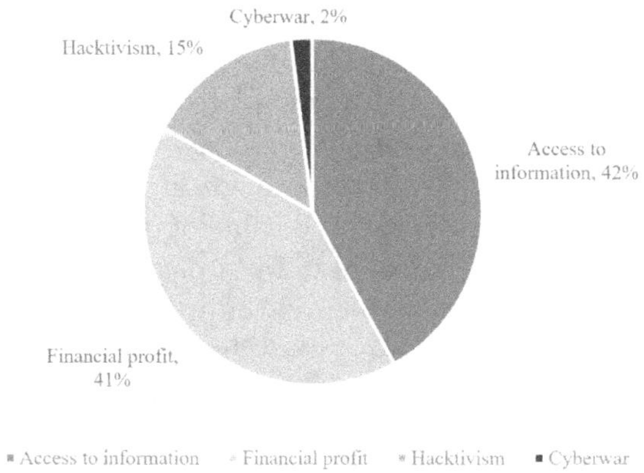

Fig. 2.14. Attacker motivation statistics.

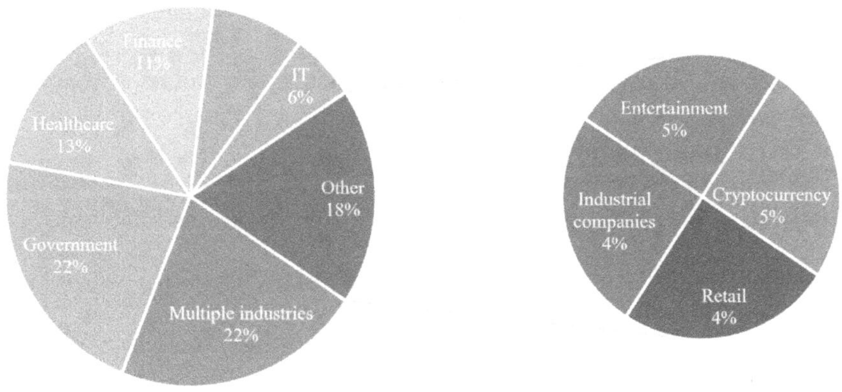

Fig. 2.15. Statistics of attackers' target industries.

shown in Fig. 2.15. This also confirms the principles based on data regulatory requirements, data value and sensitivity to leaks, and the degree of impact of leaks discussed in Section 1.2 "Principles and Implementation of Data Classification" from actual cases.

There is a famous saying in the field of data security, "All people are created equal, but data is not". Data, since its creation, has different values and sensitivities and should have its appropriate level of protection based on that.

There is an imprecise but vivid metaphor for the value of data: "See how much it can sell". According to research presented by the CBI organization at the RSA 2017 conference in San Francisco (Wolfgang Goerlich, 2017), on the dark web, a valid set of credit card information is worth \$10, a complete set of personal information (name, address, phone number, etc.) is worth \$20, and a set of an individual's health data (such as medical records) is worth \$60. All of this graphically illustrates the value of data in a general sense: health data > full set of personal data > partial financial data.

The reason for this is that behind the value of data, there is also the attribute of whether it is easy to change. For example, after it is confirmed that the credit card is lost or the online payment data is leaked, the credit card number can be changed or invalidated, and its value will be reduced accordingly. A person's home address and telephone number are relatively difficult to change or invalidate, but their value is relatively high. Personal medical records, health status, and biometrics (such as fingerprints) cannot be changed once they are compromised, so their value is very high.

From the data security model dimension, the major threats to data security can be summarized as threats to data confidentiality, integrity, and availability.

A threat to data confidentiality is unauthorized access or disclosure. The main sources of possible threats are as follows:

(1) Internal employees accessing confidential data in violation of business needs and the principle of least privilege.
(2) Poor access control allows unauthorized users to access protected data.
(3) Storage in an insecure area can lead to data leakage.
(4) Unauthorized disclosure of known confidential information.
(5) Disclosure of confidential information to persons outside the company.

A threat to data integrity refers to the unauthorized modification of data. Integrity refers to the accuracy of information, including its authenticity and trustworthiness.

Data integrity threats include all integrity threats to specific types of data. For example, the following are common types of data that require integrity protection:

(1) **Application code:** Must be accurate to ensure that the application functions completely and correctly. It should not be tampered with, contain malicious code, or contain backdoors.

(2) **System logs:** Must be accurate and unaltered to ensure proper detection of intrusions and system events.

Note that data integrity is not limited to storage. Depending on the stage of the data lifecycle, integrity must also be considered during transmission and use. To give a vivid example, when a transfer is sent over the network, the original transfer of $100 is altered to $1000 during the transmission process because the data packet is modified.

Data availability refers to the timeliness and reliability of data access and use. It also includes data accessibility. Data is only valuable if the right people can access it at the right time. The threat to data availability is primarily the inability of legitimate users to access data when they need it, which can lead to business losses or other negative impacts.

Common data where availability is a concern include website data and files, and critical databases such as financial databases and human resources databases. In addition, there are availability threats that don't target specific types of data; common examples include ransomware such as WannaCry.

Most enterprises in developed countries such as the United States, especially in the infrastructure industry and leading Internet companies, have done a relatively good job of data security and infrastructure protection, but still frequently encounter cyber attacks and data breaches. The data security of most enterprises in China is even less optimistic. A large number of leaks involving personal data and sensitive data have constantly touched the sensitive nerves of the public.

On August 28, 2018, a large amount of user data, suspected to originate from a well-known hotel group (Wang and Xu, 2021), was leaked on the Internet and offered for public sale. According to the content released by the seller, the data includes the guest information of more than 10 brand hotels under the hotel group. The leaked information includes the identity information of the hotel check-in registration and the hotel room opening records, such as the guest's name, mobile phone number, email address, ID number, login account number, and login password. Due to the very high market share of the hotels under the hotel group, some studies believe that the identity information of about 130 million people has been leaked.

Judging from the leaked content, the attacker should have used the method of "dragging database" (or data enumeration attack), that is, directly downloading the database file to obtain valuable information. From the analysis of the amount, value, and sensitivity of the leaked data, basic data security protection mechanisms, such as hash protection of passwords and separate encryption of important fields, have not been implemented, which is the main cause of the data leakage.

2.2 Data Security Response Mechanism

The above content focuses on the risks and challenges of data security. In short, there are more and more data categories, more and more data sizes, faster and faster data transmission speeds, and higher and higher data values, all of which pose great challenges to data security.

To cope with the increasingly complex environment, especially the openness and complexity of the network environment, and the increasing specialization and organization of attackers, it is necessary to adopt a data-centric security governance framework, which is the framework described in Section 3.3 "Data Security Governance Framework", to achieve complete end-to-end protection of sensitive and critical data based on its lifecycle process according to the "Identify-Protect-Detect-Respond-Recover" methodology.

Note that the goal of Data Security Response should not be to keep an enterprise or organization's network 100% safe from attackers, nor is it a simple idea of "guarding the network perimeter", but focus on the protection mechanism, solution, and implementation of critical data.

The first step in the data security response mechanism is the data security risk assessment.

2.2.1 *Risk assessment method*

Risk assessment in the area of data security refers to identifying, assessing, and addressing the risks that the use of various types of information systems may pose to an organization's data or critical information assets.

Risk assessment is more of a business concept. The first thing to consider is which data assets will affect the profitability of the organization, or which data assets, if compromised, will cause huge

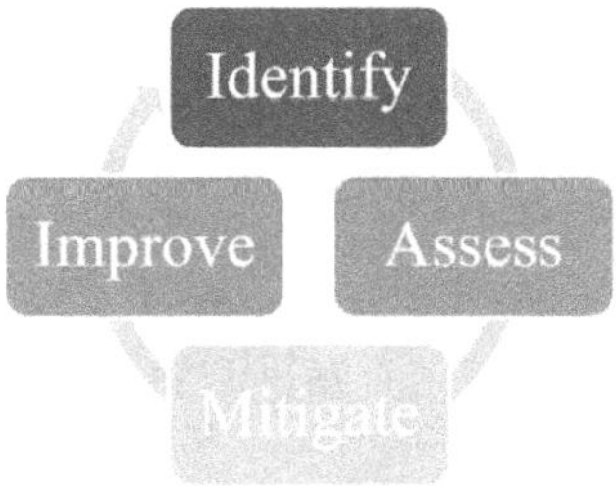

Fig. 2.16. Data risk assessment model.

economic losses to the organization. Then you look at the risks and threats to those data assets.

Data security risk assessment can be based on the four steps of "Identification, Assessment, Mitigation and Prevention" (see Fig. 2.16) of the Data Risk Assessment Model to formulate plans and carry out work.

During the identification phase, the organization's asset list and data are identified and classified in the manner described in Section 1.2 "Principles and Implementation of Data Classification". Note that the asset list should include not only static data but also data that is created and transferred.

During the assessment phase, each key data asset should have its security risks identified and an assessment made of how resources should be invested to mitigate or eliminate the risks.

During the mitigation phase, appropriate security control measures are designed and implemented for each security risk.

During the prevention phase, it is necessary to implement business and process improvements based on the risks and vulnerabilities identified in the previous phases to minimize the impact of threats.

The assessment phase is the main content described in this section.

A basic data security risk assessment includes three factors: the criticality of the data asset, the severity of the threat, and the vulnerability of the system to that threat. Using these factors, it is possible to assess the data security risk, which is the likelihood of business loss to the organization. Risk assessment is difficult to quantify but can be expressed intuitively with the following formula:

$$Risk = Asset * Threat * Vulnerability$$

There is another intuitive understanding of the above formula, the result of multiplying any number by 0 is 0. If the threat is high and the system is vulnerable, but the asset value is 0, then no additional protection is needed.

From the data risk assessment model, it can be concluded that the likelihood of security incidents is related to the vulnerability of assets and the frequency of threats, while the losses caused by security incidents are related to the value of assets and the vulnerability of assets.

Further evaluation can be done based on the risk value evaluation matrix, which describes the possibility of security incidents and the loss of security incidents, as shown in Fig. 2.17.

In Fig. 2.17, a score greater than or equal to 6 means high risk, greater than or equal to 3 means medium risk, and other means low risk. Organizations should give different levels of risk different levels of attention based on their risk management process requirements.

2.2.2 *Analysis of business impact*

The international standard ISO/IEC 27005–2018 (International Organization for Standardization, 2018) points out that the risk level should be based on the possibility of occurrence and the assessment of business impact. This section mainly discusses the business impact assessment method.

Business impact assessment is divided into four main steps:

(1) Identify the organization's key business activities.
(2) Identify key personnel, technology, and facilities.
(3) Determine the value to the organization of the normal operation of the identified business.
(4) Estimate the negative value of disruption to the organization.

	Likelyhood	Very Unlikely	Unlikely	Possible	Likely	Frequent
Business impact	Very Low	0	1	2	3	4
	Low	1	2	3	4	5
	Medium	2	3	4	5	6
	High	3	4	5	6	7
	Very High	4	5	6	7	8

Fig. 2.17. Risk assessment matrix.

The first step is to identify the key business activities of the organization. The purpose of these key business activities is to achieve the organization's objectives. Organizations typically perform different tasks and engage in different types of business activities. This is especially true of large and complex organizations. Each of its components typically focuses on one or two main business activities.

While all of these organizational components and their associated missions/businesses may be important and functioning, in reality, they are not equally important. The greater the importance of the organization's mission and business, the greater the need to ensure that risks are managed appropriately.

The second step is to identify the key people, technologies, and facilities, that is, the key people, technologies, and facilities that support the organization's key objectives and key business activities.

The third step is to determine the value to the organization of normal business operations. For example, in a typical business organization, the ordering system, the mail system, and the internal employee training system have different values.

The fourth step is to estimate the negative value to the organization due to the impact of the disrupted business. Note that this estimate should include not only direct economic losses (if any) but also all tangible and intangible losses.

Note that broad business impact analysis falls under the umbrella of business continuity management and disaster recovery planning. This section only discusses its use in risk assessment.

2.2.3 *Data flow analysis*

After conducting a comprehensive risk level assessment, the organization has qualitatively or even quantitatively analyzed the key data it owns and the data security risks it faces, and begins to address those risks. At this point, the first step is to perform a data flow analysis.

For organizations with healthcare and bank card data, the HIPAA Act and PCI DSS standards require organizations to achieve protection goals for data storage, use, and transmission scenarios by analyzing data flows and implementing appropriate encryption, network segregation, and access control mechanisms.

The PCI DSS standard (Payment Card Industry, 2010) states that organizations should comply with "Requirement 1.1.2 A current network diagram that identifies all connections between the cardholder data environment and other networks, including any wireless networks" and "Requirement 1.1.3 Current diagram that shows all cardholder data flows across systems and networks". Therefore, a complete and timely updated data flow analysis is not only helpful in protecting the organization's data security but also a basic requirement for compliance with laws and regulations.

Data flow analysis consists of the following steps:

(1) Identify key data and data flows according to the organization's business objectives and major business activities.
(2) Drawing information system diagrams and data flow diagrams.
(3) Identify the targets (systems, networks, and data) to be protected in order to achieve the organization's business objectives.

These steps are not strictly sequential. The sequence can be determined based on the organization's actual situation.

2.2.3.1 *Identify key data and data flows*

The first step is to identify key data based on the organization's business objectives and key business activities. As described in Section 1.2, key data that needs to be protected can be identified based on dimensions such as regulatory requirements, data value, data criticality, and sensitivity to data loss or tampering.

Once critical data has been identified, all flows of that data, both in and out of the organization, should be analyzed. There are four main parts to consider when analyzing the direction of flow:

(1) Where key data enters the organization, or where it is generated.
(2) What key data is being transferred and processed within the organization.
(3) Where key data flows out of the organization.
(4) What the potential leakage risk points are.

In the entire data lifecycle, the initial data phase is very important. Knowing where all sensitive data is created or collected is essential to determining where data security should begin. Note that the data scope should include both newly collected data and existing data in the organization, data that is generated or imported in electronic form, as well as data that is recorded and collected from physical media.

For examplc, medical data for a hospital may come from the following sources:

(1) A form completed by the patient online (web or email).
(2) Information provided by patients using the mobile application.
(3) Patient information actively entered by physicians or nurses.
(4) Test sheets, lab reports, imaging reports, etc. obtained through medical devices.
(5) Information from partner institutions (other hospitals, clinics, and insurance companies).

Knowing how critical data enters an organization is clearly not enough. The second step is to understand where critical data flows within the organization. Identifying data flows requires consideration of all the hardware, software, devices, and systems involved. In typical scenarios, databases, servers, workstations, mobile devices, email systems, laptops, encryption systems, etc. are all target objects that need to be considered.

After identifying the systems involved in storing, transmitting, and processing critical data, it is necessary to further clarify whether the existing data protection mechanism in the system is at risk of data leakage or tampering, and whether there is a more appropriate protection mechanism.

The movement of data outside an organization is also an issue that cannot be ignored. Organizations must take responsibility for the critical data that leaves them. When critical data is transferred out, organizations must ensure that it is transferred or destroyed in a secure manner.

When transferring data, organizations must consider the following:

(1) Whether the transfer is necessary for business purposes.
(2) What data protection mechanisms are in place at by the other party.
(3) Whether the transmission channel is encrypted.
(4) Whether the identities of both parties to the transfer have been authenticated.

Data destruction by a third party (such as hard drive shredding and CD shredding), which is a sub-scenario of data transmission, must also be analyzed.

After reviewing the entire data lifecycle process from inbound to outbound, vulnerabilities, threats, and risks must be identified for each link in the process and each system involved.

The content to be analyzed is as follows:

(1) What vulnerabilities exist in systems, applications, processes, or people.
(2) What internal, external, environmental and physical threats each vulnerability will cause.
(3) What the likelihood (what is the risk) of each threat triggering a particular vulnerability is.

When considering vulnerabilities, threats, and risks, the following categories should be considered:

(1) Digital, such as setting a weak password on the system.
(2) Physical, such as an attacker plugging in a network cable or using a data cable to gain access to the device.
(3) Internal, such as employees using private email addresses and downloading malware.
(4) External, such as hackers entering the company's internal network by compromising remote access software.
(5) Environmental, such as a fire that destroys a building where data backups are stored.
(6) Accidental, such as an employee accidentally losing a laptop containing critical data.
(7) Deliberate, such as employees abusing their authority to spy on celebrities or spouses/partners.

2.2.3.2 *Draw information system diagrams and data flow diagrams*

The objectives of drawing information system diagrams and data flow diagrams are to perform the following:

(1) Describe the major components of the information system.
(2) Describe how data moves through the system.
(3) Describe user interaction points and authorization boundaries.

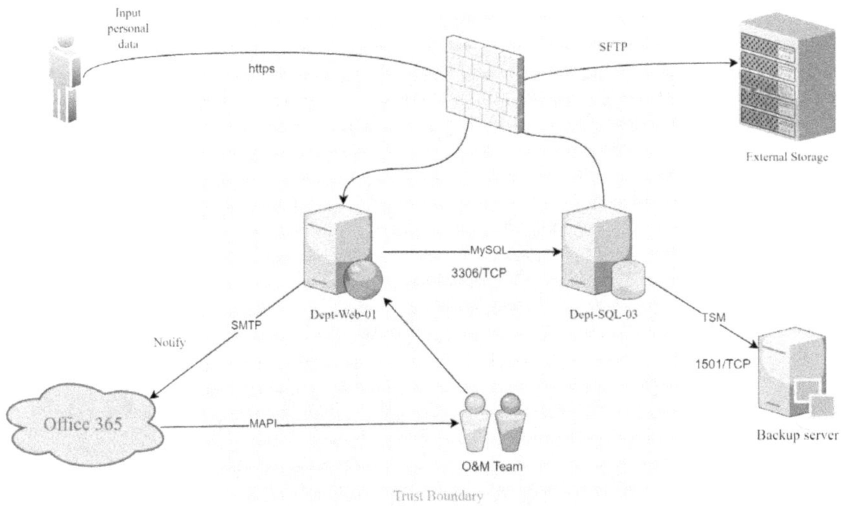

Fig. 2.18. Typical information system diagram and data flow diagram.

Information system diagrams and data flow diagrams should be conceptual diagrams rather than technical diagrams. Multiple systems can be abstracted together, and there is no need to specify the implementation mechanism of each network connection. The authorization boundary describes the scope of the information system that is the key part currently being evaluated. Outside of the authorization boundary, information systems often have dependencies on other information systems that must be independently assessed and risk factored into the current system.

A typical information system diagram and data flow diagram is shown in Fig. 2.18. The diagram shows the components of the system, the system interactions and authorization boundaries, and the data flow.

Take another look at a bad information system diagram and data flow diagram, such as the one shown in Fig. 2.19.

The diagram focuses too much on system components, includes unnecessary information, and does not explain how data moves through the system, what protocols are used, or what system boundaries are being assessed.

Directional arrows indicating data flow and protocols are important in the assessment process because they can highlight which parts of the information system need to be examined during the

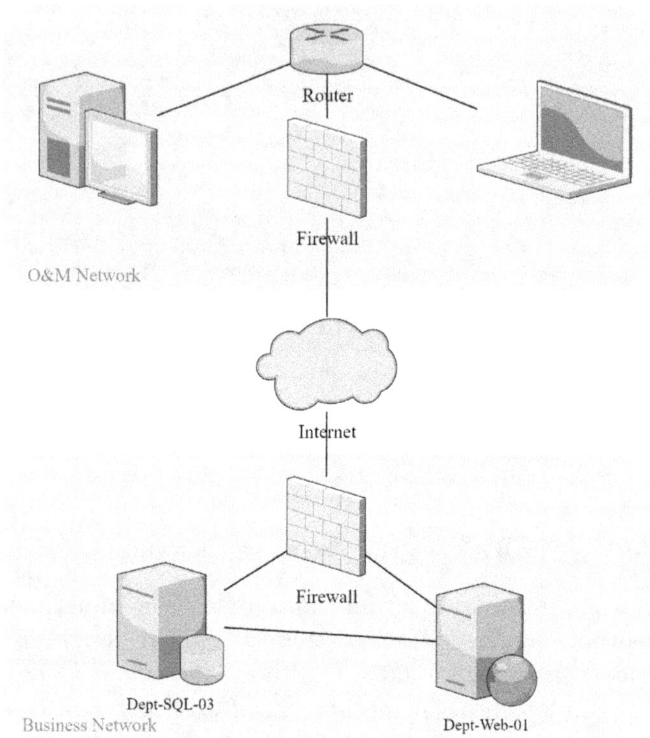

Fig. 2.19. Bad information system diagram and data flow diagram.

assessment. For example, the system description shown in Fig. 2.19 may only show that "data is transferred from the customer service network to the database in the production network via the Internet". In this case, as shown in Fig. 2.19, the Internet protocol is described as "FTP", and the evaluator can ask relevant questions.

Note that information system diagrams and data flow diagrams are not static and will change as businesses and network environments change. Therefore, it is necessary to maintain up-to-date information system diagrams and data flow diagrams to accurately reflect the current state. Implementation-wise, the best way to keep things up to date is to add a step to your change control process and ask after each change:

(1) Do I need to update the information system diagram?
(2) Do I need to update the data flow diagram?

Both diagrams should be reviewed at least every six months and after any major changes to the system.

2.2.4 *Selection, implementation, and evaluation of control measures*

NIST SP 800–37 Risk Management Framework (RMF) (Ross *et al.*, 2010) indicates that risk management should be integrated into the system development lifecycle and should provide a mandatory process for making appropriate risk management decisions.

The risk management framework is shown in Fig. 2.20.

Good preparation is a prerequisite and, in particular, should be done at the organizational and system level, depending on the context and priority of the security and privacy risks.

Step 1 is the classification of the information system, which occurs during the inception phase. The system and the data it processes,

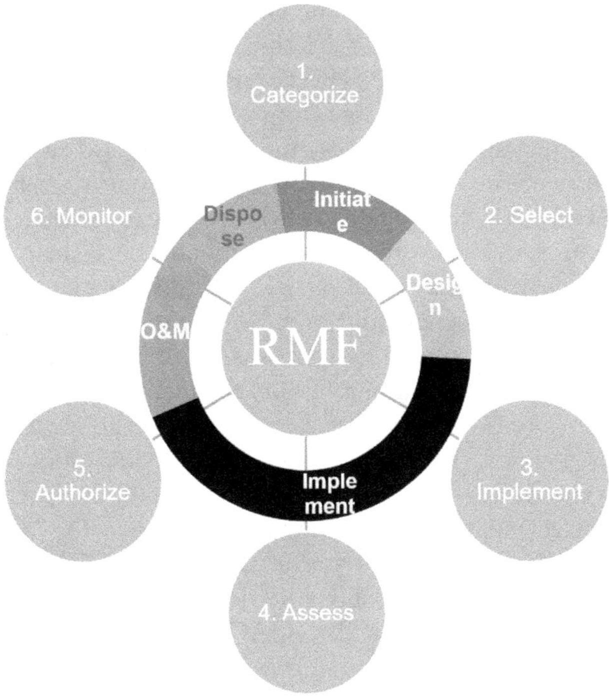

Fig. 2.20. Risk management framework.

stores, and transmits are categorized based on an analysis of the potential impact of loss.

Step 2 is to select a set of security controls for the system and adjust the controls as necessary to reduce risk to an acceptable level based on the risk assessment. This step is typically performed during the design phase.

Steps 3 through 5 are the implementation phase.

Step 3 is to implement the security controls and describe how the security controls will be used in the system and the environment in which they will operate.

Step 4 is to evaluate the security controls to determine whether the controls are implemented correctly, are working as intended, and are producing the desired results in terms of meeting security and privacy requirements.

Step 5 is to approve the implementation of the system with acceptable risk to the organization's operations and assets, individuals, and other organizations.

During the operation and maintenance phase, Step 6 refers to the continuous monitoring of the system and associated controls, including evaluating the effectiveness of controls, documenting changes to the system and operating environment, performing risk assessments and impact analyses, and reporting on the security and privacy status of the system on a periodic or event-driven basis.

The ways in which systems and data are categorized have been mentioned in previous sections and will not be repeated here. It should be noted that the main output of Step 1 is the Security and Privacy Plan. The characterization of the system should be included in the Security and Privacy Plan. The level of detail of the Security and Privacy Plan is determined by the organization and is consistent with the security classification of the system and the security and privacy risk assessment. The security and privacy plan shall be updated as the characteristics of the system change during its lifecycle.

When selecting security control measures, the ISO 27001 standard, the NIST SP 800–53 standard (Force and Initiative, 2013), and the fifth edition of "Control Objectives for Information and Related Technologies" (COBIT) (ISACA, 2021), CIS control sets (Controls) provide relevant guidance and suggestions.

For example, the CIS control set V7.1 (Center for Internet Security, 2022) provides 20 types of control measures and divides these control measures into Basic level, Foundational level, and Organizational level.

The basic level includes the following six items:

(1) **Inventory and control of hardware assets:** Actively manage (inventory, track, and remediate) all hardware devices on the network so that only authorized devices are granted access, and unauthorized and unmanaged devices are detected and prevented from gaining access.
(2) **Inventory and control of software assets:** Actively manage (inventory, track, and remediate) all software on the network so that only authorized software can be installed and run, and all unauthorized and unmanaged software is found and prevented from being installed or run.
(3) **Continuous vulnerability management:** Continuously collect, assess, and act on new information to identify and remediate vulnerabilities and minimize the window of opportunity for attackers.
(4) **Controlled use of administrative privileges:** The processes and tools used to track/control/prevent/correct the use, assignment, and configuration of administrative privileges on computers, networks, and applications.
(5) **Secure configuration for hardware and software on mobile devices, laptops, workstations, and servers:** Establish, implement, and actively manage (track, report, and remediate) the security configuration of mobile devices, laptops, servers, and workstations using a rigorous configuration management and change control process to prevent attackers from exploiting vulnerable services and settings.
(6) **Maintain, monitor, and analyze audit logs:** Collect, manage, and analyze audit logs of events that could help detect, understand, or recover from an attack.

The foundational level includes the following 10 items:

(1) **Email and web Browser Protections:** Minimize the attack surface and the opportunities for attackers to manipulate human behavior through their interaction with web browsers and email systems.

(2) **Malware defenses:** Control the installation, propagation, and execution of malicious code at multiple points in the enterprise, while optimizing the use of automation to enable rapid defense updates, data collection, and remediation.
(3) **Limitation and control of network ports, protocols, and services:** Manage (track/control/correct) the ongoing operational use of ports, protocols, and services on networked devices in order to minimize windows of vulnerability available to attackers.
(4) **Data recovery capabilities:** The processes and tools used to properly back up critical information with a proven methodology for timely recovery of it.
(5) **Secure configuration for network devices, such as firewalls, routers, and switches:** Establish, implement, and actively manage (track, report, and remediate) the security configuration of network infrastructure devices using a rigorous configuration management and change control process in order to prevent attackers from exploiting vulnerable services and settings.
(6) **Boundary defense:** Detect/prevent/correct the flow of information transferring across networks of varying trust levels, with a focus on security-damaging data.
(7) **Data protection:** The processes and tools used to prevent data exfiltration, mitigate the effects of exfiltrated data, and ensure the privacy and integrity of sensitive information.
(8) **Controlled access based on the need to know:** The processes and tools used to track/control/prevent/correct secure access to critical assets (e.g., information, resources, and systems) according to the formal determination of which individuals, computers, and applications have a need and right to access those critical assets based on an approved classification.
(9) **Wireless access control:** The processes and tools used to track/control/prevent/correct the secure use of wireless local area networks (WLANs), access points, and wireless client systems.
(10) **Account monitoring and control:** Actively managing the lifecycle of system and application accounts — their creation, use, inactivity, and deletion — to minimize opportunities for attackers to exploit them.

The Organizational level includes the following four items.

(1) **Implement a security awareness and training program:** For all functional roles in the organization (prioritizing those critical to the business and its security), identify the specific knowledge, skills, and abilities needed to support the defense of the enterprise; develop and execute an integrated plan to assess, identify gaps, and remediate through policy, organizational planning, training, and awareness programs.
(2) **Application software security:** Manage the security lifecycle of all internally developed and acquired software in order to prevent, detect, and remediate security vulnerabilities.
(3) **Incident response and management:** Protect the organization's information and reputation by developing and implementing an incident response infrastructure (e.g., plans, defined roles, training, communication, and management oversight) to quickly detect an attack and then effectively contain the damage, eliminate the attacker's presence, and restore the integrity of the network and systems.
(4) **Penetration tests and red team exercises:** Test the overall strength of an organization's defenses (technology, processes, and people) by simulating the objectives and actions of an attacker.

When selecting security controls, two common methods are the control baseline-based selection method and the organization-customized selection method. If the organization, solution, and product do not have specific security and privacy requirements, you can refer to some established security control baselines (such as NIST SP 800–53) or privacy control baselines and formulate an organizational baseline.

For highly specialized systems (e.g., medical devices) or systems with a specific purpose or scope (e.g., smart meters), organizations cannot refer to a predefined security control baseline and must select and customize security controls and form their own baseline. In this scenario, it may make more sense for the organization to take a bottom-up approach, selecting and applying the appropriate controls and forming a collection, rather than removing them through a customization process based on a broad baseline of controls.

Whether it is a selection method based on a baseline of controls or an organization-defined selection method, organizations can use lifecycle-based system engineering process standards, such as ISO/IEC 15288–2008 (International Organization for Standardization, 2015) and NIST SP 800–160 (Ross *et al.*, 2016). This process generates a set of requirements that can be used to guide the selection of a control baseline. Similarly, organizations can use the NIST CSF to develop a cybersecurity framework profile specific to an organization's security and privacy requirements and then select controls based on the NIST SP 800-53 standard.

During the implementation phase, the organization implements the controls described in the security and privacy plan based on the enterprise architecture and the associated security and privacy architecture. Organizations may use best practices in implementing controls, including methods, concepts, and principles of system security and privacy engineering. Risk assessment guides decisions about the costs, benefits, and risk trade-offs of implementing controls using different techniques or strategies. When an organization does not have direct control over the controls implemented in a system component, such as when shipping a commercial product that is maintained by a customer, it should consider integrating system components that have been evaluated or verified by approved independent third-party testing. Third-party certified components that may be considered include cryptographic algorithm components certified by NIST FIPS or system components certified by Common Criteria (CC). These tests, evaluations, and certifications consider the product in specific configurations and isolation scenarios; the implementation of controls addresses how to integrate the product into a system while maintaining security functionality.

During the control implementation phase, an initial control assessment may also be an option. This assessment can be done in parallel with the development and implementation phases of the Security Development Lifecycle (SDLC) (Lipner, 2004), helping in identifying defects early and initiate corrective actions in a timely manner. Issues found during the assessment can be escalated to the appropriate owner for resolution according to the issue handling process. The results of the initial control assessment can also be used in Step 5 (Information System Implementation Authorization Phase) to avoid delays in the assessment that result in rework or duplication of effort. Subsequently, the assessment results can be reused in other phases of the SDLC to meet the reuse requirements established by the organization.

In Step 4, the Assess Security Controls phase, the goal is to determine whether the controls selected for implementation are properly implemented, functioning as intended, and producing the desired results in meeting the security and privacy requirements of the system and the organization.

In general, the Evaluate Security Controls phase includes the following tasks:

(1) Selecting evaluators.
(2) Formulating an evaluation plan.
(3) Evaluating and performing control measures.
(4) Preparation of the evaluation report.
(5) Analysis of corrective actions.
(6) Developing action plans and milestones.

The following discussion focuses on the evaluation and implementation of controls.

The evaluation and implementation of controls is performed to determine the extent to which the selected controls are implemented correctly, operate as intended, and produce the desired results in terms of meeting the security and privacy requirements of the system and the organization. The system owner, control provider, or organization relies on the skills and expertise of the assessor to evaluate the implemented controls using the assessment process specified in the assessment plan and to make recommendations on how to respond to control deficiencies in order to reduce or eliminate identified vulnerabilities or unacceptable risks. The senior manager responsible for privacy (e.g., the Chief Security Officer) serves as the privacy control evaluator and is responsible for performing an initial evaluation of privacy controls before the system is put into operation and for performing periodic evaluations of controls thereafter on a cycle sufficient to ensure compliance with privacy requirements. At the same time, controls implemented to achieve both security and privacy objectives may require a degree of cooperation between security and privacy control evaluators. The evaluator is responsible for providing factual reports on whether the controls are functioning as intended and whether any control deficiencies were identified during the evaluation process.

An assessment of controls should be performed as early as possible in the security development lifecycle (SDLC), preferably during the development phase. These types of assessments are called development tests and assessments, and they verify that controls are implemented correctly and that they are consistent with established information security and privacy architectures. Development testing and assessment activities include design and code reviews, regression testing, and application scanning. Defects discovered early in the SDLC can be addressed more cost-effectively. Vulnerabilities, like problems, are more expensive to fix the later they are fixed. In addition, assessments may be required prior to vendor selection during the sourcing process, rather than after a contract has been signed, formal contract execution, or development. The results of control assessments performed during the SDLC can also be used in the approval process (in accordance with the organization's established reuse criteria) to avoid unnecessary delays or costly duplication of assessments. Organizations can maximize the use of control assessment automation to increase the speed, effectiveness, and efficiency of assessments and to support continuous monitoring of the security and privacy status of the organization's systems.

In Step 5, the Information Systems Implementation Authorization phase, the purpose is to provide organizational accountability by requiring senior management to determine whether the security and privacy risks to the organization's operations and assets, individuals, and other organizations (including supply chain risks) are acceptable.

In general, the authorization phase includes the following tasks:

(1) Preparation and submission of authorization documentation.
(2) Risk analysis and calculation.
(3) Risk response.
(4) Making approval decisions.
(5) Writing the authorization report.

In Step 6, the Monitoring Security Controls phase, the purpose is to maintain continuous situational awareness of the security and privacy status of information systems and organizations to support risk management decisions.

During this phase, changes to information systems and the operating environment shall be monitored and the effectiveness of controls shall be continuously evaluated in accordance with the continuous monitoring strategy. The results of ongoing monitoring activities should be analyzed and acted upon, and risk management documentation should be updated. In addition, a process should be established for regularly reporting the security and privacy status to the Chief Security Officer or other senior management. The delegation of Step 5 must also be done on an ongoing basis, based on the results of ongoing monitoring activities, to communicate changes in risk assessment and decision making.

2.2.5 *Overview of threat modeling*

As mentioned above, risk management is a core concern of every organization and one that executives should take seriously. Data security is the strategic focus of large organizations or institutions, especially the potential risks of data leakage, infrastructure disruption, regulatory violations, and legal penalties. Risks come in many different forms and can come from inside or outside an organization.

Many laws and regulations focus on risk management and the controls in place to prevent potential threats, such as the European Union's GDPR. Organizations that fail to comply can face significant fines. Threat modeling can play a key role in selecting risk management and control measures to identify and respond to all potential threats.

Threat modeling, combined with risk management, answers the questions of who will attack a system and how or where the attack might come from. Threat modeling provides a systematic view and analysis of the risks an organization faces, and recommends appropriate control measures based on the results of qualitative and quantitative analysis, with the goal of stopping actual threats before they occur.

Threat modeling is a structured process, so it follows a specific set of rules (or methodology). A variety of methodologies have evolved around the concept of threat modeling, of which the following are the most well known and widely used.

1. **STRIDE**

STRIDE is a threat model developed by Microsoft Corporation in 1999 and published in 2005 (Howard and Lipner, 2006). It is categorized based on six key objectives of the attackers. The naming scheme of STRIDE is also based on the English initials of these six key objectives. The key objectives include the following:

(1) (Identity) Spoofing, refers to pretending to have an identity that one does not have.
(2) Tampering, refers to exceeding one's authority to modify certain information.
(3) Repudiation, refers to an operation that can be denied.
(4) Information disclosure, refers to the illegal acquisition of information.
(5) Denial of Service (DoS), refers to preventing the provision of certain services.
(6) Elevation of privilege, refers to obtaining some kind of authority that one does not have.

2. **PASTA**

The full name of PASTA (Ucedavélez and Morana, 2015) is "Process for Attack Simulation and Threat Analysis". PASTA is a risk-oriented methodology that attempts to link business goals and technical requirements. The PASTA methodology is a 7-step process to systematically identify the business or application impact of a threat:

(1) Define business objectives and security requirements.
(2) Define the technical scope.
(3) Decompose the application into use cases and data flow diagrams (DFD).
(4) Threat analysis.
(5) Perform vulnerability analysis.
(6) Analyze possible attack vectors.
(7) Risk and business impact analysis.

3. **VAST**

VAST stands for "Visual, Agile, Simple Threat Modeling" and is the foundation of the first commercial threat modeling tool Threat-Modeler (Automated Threat Modeling Platform). The rationale for VAST is to extend the threat modeling process between infrastructure and security development lifecycles (SDLCs) and enable seamless integration with agile software development methodologies. VAST is designed to provide valuable and actionable insights to all stakeholders including executives, developers, and security professionals.

Regardless of which of the above models is used, there are some common basic concepts and methodologies. The following are explained separately from an attacker's and a defender's perspective.

From an attacker's perspective, there are the following concepts: Vulnerabilities, Exploits, Attack Vectors, and Threats.

The first concept is vulnerability. This concept covers a wide range of meanings. In the threat modeling community, the following definition is commonly used: any violation of assumptions about the trustworthiness of people, processes, or technology that leads to the exploitation of a system.

Typical types of vulnerability are as follows:

(1) **Software implementation vulnerabilities:** Vulnerabilities caused by errors in software design or coding. For example, incorrect input validation assumes the user's input is credible and fails to filter special characters, leading to attacks such as command injection and denial of service.
(2) **Software configuration vulnerabilities:** Improper security configuration or security settings can be exploited by attackers. The most typical vulnerability is the improper configuration of operating system file permissions, which allows attackers to gain unauthorized access or execute commands. Note that security configurations and features may conflict. Absolute "security" may also result in the software not providing the required functionality.
(3) **Software feature abuse vulnerability:** Some software functions are misused, causing damage to the system. For example, in order to display HTML content normally, many mail client software include rendering engines for HTML content. At this point, the attacker can perform various types of attacks on the victim through carefully constructed email content, such as URL click spoofing.

The second concept related to vulnerabilities is exploitation. Exploitation is the use of a vulnerability to violate security objectives, such as confidentiality, integrity, and availability. Source code or other commands used for exploitation are also commonly referred to as exploitation ("EXP") or attack. Note that "exploitation" generally refers to a successful security breach, while "attack" refers to an attempted security breach but does not indicate success or failure.

The third concept is an attack vector. An attack vector is a subset of all possible paths to exploit a vulnerability. A typical attack vector has three components: the source of the malicious content, the potentially vulnerable processor of the malicious content, and the malicious content itself.

Examples of attack vectors are as follows:

(1) Malicious web content (content) downloaded from a website (source) by a vulnerable web browser (processor).
(2) Malicious email attachments (content) downloaded from an email server (source) to a vulnerable email client (processor).
(3) Vulnerable network services (processors) that have externally accessible vulnerabilities (content) and are maliciously used by external clients (sources).
(4) Obtaining usernames and passwords of vulnerable users (processors) from social engineering-based conversations (content) conducted by human attackers (sources) over the telephone.
(5) The attacker (source) enters the stolen user credentials (content) into the web interface of the organization's authentication system (processor).
(6) Obtaining personal information about the user from social media (Content), which the attacker (Source) enters into a password reset website to exploit a weak password reset process (Processor) to reset the password.

The characteristics of the attacks vary greatly. Some attacks use a single attack vector and a single vulnerability, while others use multiple vulnerabilities and multiple attack vectors, or even a single vulnerability and multiple attack vectors. In addition, vulnerabilities and attack vectors may be spread across multiple hosts, resulting in the propagation of vulnerabilities.

The following is an example of an attack using a single vulnerability:

(1) A malicious email attachment (content) is sent from a host (source) to an organization's mail server (processor).

(2) A malicious email attachment (content) is sent from an organization's mail server (source) to an antivirus server (processor).
(3) Malicious email attachments (content) are sent from the organization's antivirus server (source) to the organization's internal mail server (processor).
(4) Malicious email attachments (content) are sent from the organization's internal mail server (source) to the user's mail client software (processor).
(5) Malicious email attachments (content) are viewed (processor) by a vulnerable email client software (source).

Note that although there are five attack vectors in the above attack process, there is actually only one vulnerability, which is the email client software vulnerability involved in the last step. In addition, relevant countermeasures can be applied to each attack vector in the entire attack process to block the attack vector and thereby mitigate the impact of the attack. It is worth noting that the cost and value of choosing which attack vector to apply the appropriate countermeasures to will vary.

Other concepts related to attack vectors include attack models, threat models, and attack surfaces. The attack model includes a scenario that is likely to occur and a single path (one or more attack vectors in chronological order) that can be taken for that scenario. Together, the attack model and the security controls that counter it form a "threat model". Another way to analyze attack vectors is to analyze all attack vectors directly against a particular system, which is the system's "attack surface".

The fourth concept is **threat**. NIST SP 800–30 defines "threat" as "any circumstance or event with the potential to adversely impact organizational operations and assets, individuals, other organizations, or the Nation through an information system via unauthorized access, destruction, disclosure, or modification of information, and/or denial of service" (Stoneburner *et al.*, 2002). Threats can be intentional or unintentional. A "threat source" is the cause of a threat, such as hostile cyber or physical attack, human negligence or error, hardware or software failure within the organization's control, or other failure beyond the organization's control.

Concepts related to threats include threat events. According to NIST Special Publication (SP) 800–30, a threat event refers to "an event or situation that has the potential for causing undesirable consequences or impact".

From the defender's perspective, there are concepts such as risk, security control measures, and security objectives.

"Risk" is generally defined as follows:

A measure of the degree to which an entity is threatened by a potential situation or event, usually with

(1) the adverse consequences that would result if such a situation or event were to occur,
(2) the likelihood of its occurrence.

"Risk management" is defined as "The process of managing risks to organizational operations (including mission, functions, image, or reputation), organizational assets, or individuals resulting from the operation of an information system...". A risk assessment considers possible threats and vulnerabilities and determines what security controls should be used to mitigate the impact of those threats and vulnerabilities so that the risk can be reduced to an acceptable level.

CNSSI 4009 defines security controls as "The management, operational, and technical controls (i.e., safeguards or countermeasures) prescribed for an information system to protect the confidentiality, integrity, and availability of the system and its information". While technical controls can be fully automated and are the primary option for preventing attacks, management and operational controls also play an important role. For example, users must be educated about their security responsibilities so that they are less likely to violate security policies and fall for phishing attacks, as well as avoid other actions that compromise the organization's security. Ultimately, an organization's security depends on a combination of people, processes, and technology.

Security objectives are defined as how an organization protects the confidentiality, integrity, and/or availability of data. Note that security objectives for instances of different data types should not be of equal importance, and in some cases, an organization may want to focus its threat modeling efforts on one objective. For example, in the common threat modeling process, the primary concern is protecting the confidentiality of sensitive data. However, information that has been released to the public may still require integrity and availability protection but not confidentiality protection.

2.2.6 *STRIDE threat model*

STRIDE (Shostack, 2008) is a threat modeling methodology proposed by Microsoft Corporation that is well integrated into the Security Development Life Cycle (SDLC). To build a secure software system, a very important aspect is to consider how attackers exploit design flaws to compromise the system and to build necessary defenses into the system. Threat modeling plays a key role in this regard.

The STRIDE threat model divides threats into six dimensions, and the name of STRIDE comes from the acronym of these six dimensions:

(1) **Spoofing threats:** Spoofing threats refer to the behavior of impersonating the identity of another entity. A common method is to exploit the weakness of the authentication mechanism to bypass authentication. For example, authenticating with stolen authentication information (username and password) from another user.
(2) **Tampering threats:** Data tampering threats refer to the malicious modification of data that destroys the integrity of the system and data. Data tampering can occur in memory, on disk, on the network, or elsewhere. For example, executing scripts to tamper with data in the database or performing man-in-the-middle attacks to tamper with data transmitted over the network.
(3) **Repudiation threats:** Repudiation threats refer to users denying what they have done. In most cases, attackers want to hide their identity and activity records to avoid being identified and blocked. The most common repudiation threat is the manipulation of operational logs.
(4) **Information disclosure threats:** Information disclosure threats refer to information that can be leaked to someone who does not have access to the information, compromising the confidentiality of the information. For example, a database that stores customer information with a default or simple password or is vulnerable to SQL injection attacks can lead to information disclosure.
(5) **Denial of service threats:** A denial of service (DoS) threat is an attacker's attempt to exhaust the resources needed to provide a service in order to affect the availability of the service.

For example, making a web server temporarily unavailable by sending a large number of packets to flood normal requests.

(6) **Elevation of privilege threats:** Elevation of privilege threats refer to allowing users to do unauthorized things. An unprivileged user with privileged access could steal information or compromise the entire system. An attacker can gain privileges by impersonating a user or by manipulating the system to elevate privileges.

The STRIDE threat model can cover almost all current security issues and has detailed processes and methods.

The STRIDE approach to threat modeling is used to ensure that the target system has the appropriate security attributes, as shown in Table 2.1. The correspondence between STRIDE and the security attributes is as follows: spoofing corresponds to authentication, tampering corresponds to integrity, repudiation corresponds to non-repudiation, information leakage corresponds to confidentiality, denial of service corresponds to availability, and elevation of privilege corresponds to authorization.

The steps of STRIDE modeling are as follows:

Step 1: Define the business scenario of the system, including which business functions the target software system provides, which functions are strongly related to security, which require threat modeling, and so on. For example, a hypothetical social networking application can be divided into several functions, such as user registration, user login, message publishing, and message commenting.

Table 2.1. STRIDE threat model.

Threats	Definition	Security Properties
Spoofing threats	Impersonating something or someone else	Authentication
Tampering threats	Modifying data or code	Integrity
Repudiation threats	Claiming to have not performed an action	Non-repudiation
Information disclosure threats	Exposing information to someone not authorized	Confidentiality
Denial of service threats	Deny or degrade service to users	Availability
Elevation of privilege threats	Gain capabilities without proper authorization	Authorization

Step 2: Gather the list of external dependencies. Each application is dependent on the operating system it runs on, the database it uses, etc.; these dependencies need to be defined.
Step 3: Define security assumptions. During the analysis phase, threat decisions are often based on implicit assumptions. Therefore, document all assumptions to gain a complete understanding of the entire system.
Step 4: Create an external security specification. Since every external dependency has an impact on security, the security specification must list the constraints and impacts introduced by all external dependencies. An example of an external security specification would be which network ports the system opens for database access or HTTP communication.
Step 5: Draw a Data Flow Diagram (DFD) of one or more analyzed applications. The analyzed software system is broken down into related (logical or structural) components, and each of these components is analyzed for the corresponding threat. This process can be repeated and progressively refined until an acceptable level of residual threat is achieved.

Data flow diagrams are typically drawn based on specific scenarios. An example of a data flow diagram is shown in Fig. 2.21. This figure is a data flow diagram of an IoT analysis network that

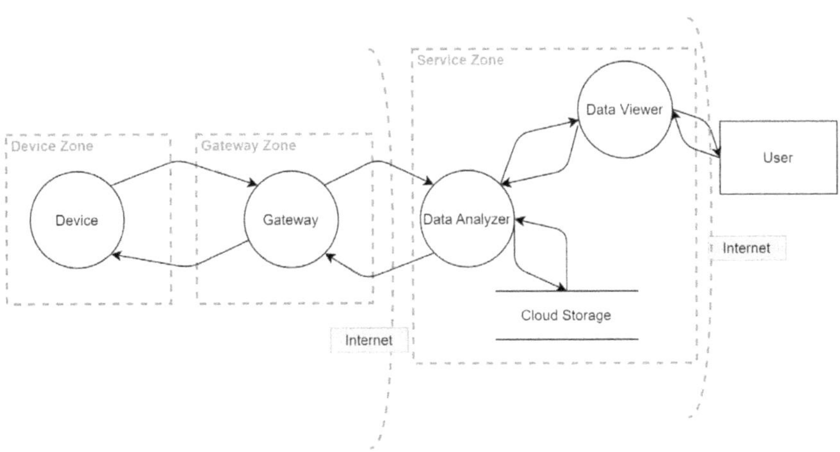

Fig. 2.21. Example of data flow diagram.

includes the device zone, gateway zone, and service zone and indicates trust boundaries (dotted lines in the figure) and data storage (indicated by double horizontal lines in the figure).

Microsoft provides a specialized STRIDE threat modeling tool that includes the ability to draw data flow diagrams. Of course, software such as Visio and Diagrams also provide similar capabilities.

STRIDE threat modeling includes four core components: circle, double horizontal line, line with arrow, and square:

(1) **Circle:** This indicates processes, services, etc., such as application processes, web services, browser plug-ins, virtual machines, etc.
(2) **Double horizontal line:** This indicates data storage, such as database, cache, configuration file, registry, etc.
(3) **Lines with arrows:** These indicate data flow, i.e., the flow of data between components. If the data flow between two components is bi-directional, then the two data flows between these two components should be drawn in a data flow diagram.
(4) **Rectangle:** This indicates external interactors, which are elements that interact with the system but are not under the control of the system, such as users, browsers, and other systems.

In addition to the four core components in the data flow diagram, there are dotted line segments on the diagram that represent the trust boundary, which is the boundary between trusted and untrusted elements. Processes, data stores, and communications (data flows) between two elements are all assumed to be trusted within the same trust boundary.

Step 6: Determine the threat type, which is the STRIDE threat model taxonomy mentioned earlier in this section.
Step 7: Determine the actual threat based on the threat type. After drawing the data flow diagram, analyze the threats that each element in the data flow may face one by one, but it is not necessary to analyze all of the STRIDE threats of each element. Table 2.2 shows the possible threats to each type of element.

Only processes can be exposed to all STRIDE threats and must be analyzed one by one. External interactors only need to analyze the two threats: Deception (S) and Repudiation (R). The data flow only needs to be analyzed for three threats: Tampering (T), Information

Table 2.2. Threats affect each element type.

Element	S	T	R	I	D	E
Process (circular)	√	√	√	√	√	√
Data Store (parallel lines)		√	√*	√	√	
Data Flow (directed arrow)		√		√	√	
External Interactors (rectangle box)	√		√			

leakage (I), and Denial of service (D). Note that the Repudiation (R) threat to data storage may not exist. Analyze the Repudiation (R) threat (see asterisk in Table 2.2) only if the data store being analyzed is used for auditing.

Step 8: Identify risks. Each threat must be assigned an appropriate security risk level that can be used to prioritize remediation. To determine the risk level, you can use the risk assessment method described in Section 2.2.1 "Risk Assessment Method".
Step 9: The final step is the development of mitigation measures or mitigation programs. By implementing appropriate countermeasures and defenses, the risks posed by the identified threats are reduced or eliminated. This is called a "mitigation measures" because each threat identified by the STRIDE threat analysis may not be able to be eliminated or may need to be eliminated for a variety of reasons, and it is the assessment of the threat that is used as the basis for measuring how much it will cost to address the threat.

With the above steps, data assets and their exposure to threat types and threat descriptions can be identified in advance of the development process and mitigation measures can be developed accordingly.

2.2.7 *Data-centric threat modeling*

Threat modeling is a form of risk assessment that models the attack and defense aspects of a specific logical entity, such as data, applications, hosts, systems, or environments.

Comprehensive protection of all logical entities is neither necessary nor possible. Security resources are always limited, so it is necessary to determine how to use those limited resources effectively.

Threat modeling must focus on the comprehensiveness of risk response and the dynamic nature of risks. Attackers can be added, attack methods and motivations can evolve, and new vulnerabilities are discovered all the time. Therefore, threat modeling is not a one-time effort nor can it solve security risks once and for all. In addition, security controls are constantly being improved and enhanced, and new types of security controls are constantly being released, all of which can be used in the analysis and evaluation of threat models. To respond to this constant change, organizations should continually reassess and evolve their defenses. This includes using continuous monitoring practices, security automation techniques, and threat intelligence feeds to detect new vulnerabilities and attack attempts in near real time so that risks can be mitigated quickly.

Based on the organization's business operations and analysis of the product development process, it is clear that following generic "security best practices" is not enough to protect high-value data. Best practices are designed to mitigate common threats and vulnerabilities. By their very nature, such best practices are business-agnostic, especially when applied to generic types of products (Web browsers, web servers, desktop operating systems, etc.), and they do not take into account the uniqueness of each system. Similarly, most best practices are designed to prevent a host or system from being compromised and do not consider the security needs of specific data. Therefore, in some cases, following "best practices" may ignore the security controls necessary to effectively reduce data security risks.

Data-centric threat modeling focuses on protecting specific types of data within a system. This approach to system threat modeling allows organizations to consider security needs on a per scenario basis rather than relying solely on general "best practice" recommendations. Securing data shouldn't be that difficult, but traditionally, security professionals, system administrators, and other security leaders have focused on securing systems rather than securing data.

The system security approach begins with system classification based on FIPS 199 and then selecting appropriate security controls from NIST SP 800–53. Instead, the first step in a data security approach begins with a reference architecture, such as NIST SP 800–60 (Stine *et al.*, 2008) and then classifies data types.

NIST SP 800–154 "Guidelines for Data-Centric System Threat Modeling" (Souppaya and Scarfone, 2016) points out that for data-centric system threat modeling, the security goal of data should be considered first rather than the security of the system. That is, prioritize data confidentiality, integrity, and availability risks caused by attacks. Data security objectives relate to risk management, assessment, mitigation, and security controls. For example, if the risk assessment indicates that the risk to data confidentiality is too high, then additional security controls may be required, or changes to existing security controls may be required to reduce the confidentiality risk to an acceptable level. The same can be done for data integrity and availability risks.

Data-centric system threat modeling creates standardized models for both attack and defense aspects of valuable data, facilitating security analysis, decision-making, and change planning. Data-centric system threat modeling does not attempt to define a new threat modeling methodology but rather defines the implementation framework and steps for data-centric system threat modeling.

Data-centric system threat modeling consists of the following four main steps:

(1) Identify and characterize target systems and data.
(2) Identify and select attack vectors to include in the model.
(3) Characterize security controls to mitigate the attack vectors.
(4) Analyze the threat model.

In Step 1, identify and characterize the target systems and data. Systems and data here refer only to the specific host systems being analyzed and specific datasets identified as high value.

Once the system and data are defined, they must be characterized to understand how the system operates and how the data is used. At a minimum, the data should include the following characteristics:

(1) Authorized locations of data in the system, including some or all of the locations in Table 2.3.
(2) A basic understanding of how data moves between authorized locations within the system. For example, files may be held in memory when created and written to storage when needed by the user. Depending on the complexity of the system, understanding how data moves may require first understanding the functions and processes of the system, user usage, workflow, trust assumptions, and characterization of other aspects of the people,

Table 2.3. Authorized locations for data.

Location	Description
Storage	All places where data may be at rest within the system boundaries
Transmission	All ways in which data may transit between system components and across the system's boundaries
Execution environment	Data held in local memory during runtime, data processed by virtual CPUs, etc.
Input	For example, data typed in using the keyboard
Output	For example, data printed to a physically attached printer, data displayed on the laptop screen, etc.

processes, and technologies associated with the system, such as connectivity protocols, transfer modes, and so on.

(3) Data security objectives (e.g., confidentiality, integrity, and availability). In many cases, some objectives are more important than others; in other cases, organizations may want to focus on a single objective for a particular threat model.

(4) People and processes that are authorized to access data in ways that may impact security objectives. For example, if the organization has selected confidentiality as the only objective for a particular threat model, then authorized people and processes should include all users, administrators, applications, services, etc. that are allowed to read the data.

Consider a simple use scenario. The human resources department maintains a monthly overtime payroll spreadsheet that lists personal information such as name, department, cell phone number, overtime hours, base salary, and overtime pay for a number of employees. Following the methodology above, the first step is to analyze the data authorization locations, the results of which are shown in Table 2.4.

Consider the workflows and relevant roles associated with this form, as well as the privacy objectives. In this oversimplified example, it is clear that the confidentiality of the data is the focus of the consideration, and the authorization model must be considered in light of that confidentiality. For simplicity, define this HR as the only person authorized to access the data.

Table 2.4. Examples of authorized locations.

Location	Description
Storage	A local hard drive, a USB flash drive
Transmission	Sent to a printer via a wireless network
Execution environment	Local laptop memory and processors
Input	Typed in using the laptop keyboard
Output	Displayed to the screen

In Step 2, identify and select attack vectors to include in the model, i.e., possible vectors that could negatively impact the pre-identified security objectives against the data assets. Once the set of attack vectors is identified, a subset of attack vectors can be selected for inclusion in the threat model based on some criteria. One approach would be to include all attack vectors for threat modeling, but this is too resource-intensive and usually not practical. Possible criteria to consider include the likelihood of the attack vector being used and the most likely impact of a successful attack.

Still take the overtime payroll table above as an example. By analyzing the authorized locations of the data, the vast majority of possible attack vectors can be aggregated (see Table 2.5).

Obviously, the different attack vectors mentioned above have different attack difficulties, different costs, and different possibilities, so the difficulty of obtaining data after a successful attack is also different.

After analysis, the following subset of attack vectors can be selected:

(1c) The attacker logs in after stealing the user's password.
(2b) After stealing the user's password, the attacker logs in and accesses the mounted USB flash drive.
(4a) The attacker can take control of the user's device.

In Step 3, characterize security controls to mitigate the attack vectors, i.e., for each attack vector selected in Step 2, identify and document the need for its security controls (additions to existing security controls, reconfigurations of existing security controls, etc.).

Next, for each selected security control, estimate its effectiveness against the particular attack vector; effectiveness can be assessed qualitatively or quantitatively. The negative impact of the security

Table 2.5. Attack vectors of overtime payroll table.

No	Location	Possible Attack Vectors
1	Stored on a local hard drive	(1a) Attacker gains unauthorized physical access to the laptop, uses forensic tools; (1b) attacker gains unauthorized physical access to the laptop, exploits unknown vulnerabilities; (1c) attacker steals and reuses user passwords or credentials; (1d) attacker gains control over user's session; (1e) user forwards the file to an unauthorized recipient (phishing, malicious user, etc.); (1f) attacker accesses unsecured network and gains access to the file.
2	Backup on a USB flash drive	(2a) Attacker gains unauthorized physical access to the flash drive; (2b) attacker steals and reuses user passwords or credentials for laptop; (2c) user forwards the file to an unauthorized recipient.
3	Printed to a nearby printer	(3a) Attacker monitors unencrypted or weakly encrypted wireless network communications and captures the data being sent to the printer; (3b) attacker views a printout of the spreadsheet.
4	Execution environment	(4a) Attacker gains control over user's device.
5	Input locally	(5a) Attacker watches the information being typed in to the laptop; (5b) attacker uses keystroke logger on laptop to monitor keystrokes.
6	Output locally	(6a) Attacker views the information on the laptop screen; (6b) attacker uses malware on laptop to take screen shots.

controls must also be considered, including costs (e.g., acquisition and implementation costs, annual operations and maintenance costs, which can be estimated, but it is of course better to have exact figures) and degradation of functionality, availability, and performance. For security controls that have not yet been implemented, it may be particularly difficult to accurately determine these factors, so it may be necessary to make a very rough estimate using a simple, organization-specific "low/medium/high" multi-dimensional

Table 2.6. Characterize the security controls for mitigating the attack vectors.

No	Security Controls	Effectiveness
1	Uses strong passwords	Countermeasures for vectors 1c and 2b: l Effectiveness: Low l Acquisition and implementation costs: Low l Annual management/maintenance costs: Low l Impact on functionality: Low l Impact on usability: Low l Impact on performance: Low
2	Multi-factor authentication	Countermeasures for vectors 1c and 2b: l Effectiveness: High l Acquisition and implementation costs: Moderate l Annual management/maintenance costs: Moderate l Impact on functionality: Low l Impact on usability: Moderate l Impact on performance: Low
3	Use antivirus software, spam filtering, etc.	Countermeasures for vectors 1c, 1d, 2b, 2c, and 4a: l Effectiveness: High l Acquisition and implementation costs: Moderate l Annual management/maintenance costs: Moderate l Impact on functionality: Low l Impact on usability: Moderate l Impact on performance: Low
4	Security awareness education	Countermeasures for vectors 1e, 2c, and 3b: l Effectiveness: Low l Acquisition and implementation costs: Low l Annual management/maintenance costs: Low l Impact on functionality: Low l Impact on usability: Low l Impact on performance: Low
5	Patch vulnerabilities	Countermeasures for vectors 1c, 1d, 2b, 2c, and 4a: l Effectiveness: Low l Acquisition and implementation costs: Moderate l Annual management/maintenance costs: Moderate l Impact on functionality: Moderate l Impact on usability: Low l Impact on performance: Moderate

security control scorecard. The security control scorecard is shown in Table 2.6.

Step 4, Analyzing the Threat Model, analyzes all the characteristics captured in the previous steps, which together form the threat model to help evaluate the effectiveness and efficiency of each security control against the selected attack vector. In addition to the financial costs of acquisition, implementation, management, and maintenance, security controls can have a negative impact on functionality, availability, and performance. Any evaluation of security controls should consider all important relevant factors.

The most challenging part of threat modeling analysis is determining how to account for all of these characteristics. Comparing individual characteristics (such as annual management/maintenance costs) between mitigating security controls is straightforward. While comparing the entire set of characteristics of one attack vector to the entire set of characteristics of another attack vector is not straightforward, this comparison is critical to determining how to mitigate risk in the most cost-effective manner across all attack vectors, while controlling for negative impact on the organization's operations. Each organization must determine how to compare the characteristics of each attack vector/security control combination as a basis for comparing attack vector characteristics with security control characteristics.

One way to enable these comparisons is to assign scores and weights to each characteristic. For example, narrative descriptions of threat consequences can be converted to numerical values corresponding to high, medium, and low, or even complex characteristics (such as cost) can be mapped to simple scales.

In addition to assigning scores to the possible values or possible ranges of each characteristic, organizations must consider the relative weight of each characteristic. For example, if the characteristic "Effective Defense" is considered more important than the other characteristics, this could be calculated by adding a factor of 2 or 3 to its score. Similarly, a factor can be assigned to each of the other characteristics to adjust the final score. All scores are then summed and a relative score is determined for each attack vector/security control combination.

Another scoring approach that can be used in addition to the above is to set thresholds or rules for certain criteria and eliminate any attack vector/security control combination that does not meet those criteria. A simple example would be to eliminate all portfolio costs of $100,000 or more over three years. A more complex example would be to eliminate all combinations that cost more than $50,000 over three years and have a high impact on availability while having a moderate or low impact on attacks.

Take the overtime pay table above as an example. After weighing the trade-offs, the organization decided to set the following scores for these characteristics:

(1) No security control effectiveness = 0.
(2) Low security control effectiveness = 1.
(3) Moderate security control effectiveness = 2.
(4) High security control effectiveness = 3.
(5) High negative impact = 1.
(6) Moderate negative impact = 2.
(7) Low negative impact = 3.

Calculate the sum of the negative impact scores for each security control. The negative rating table for security controls is shown in Table 2.7.

The total score obtained can then be multiplied by the effectiveness evaluation score to assess the relevance of the security controls to the corresponding attack vectors, see the shaded cells in Table 2.8.

All information is now available for decision-making.

The above example illustrates the complete process of data-centric threat modeling. The system and methodology of data-centric threat modeling is still evolving and developing. The methodologies highlighted above are more qualitative than quantitative and also rely heavily on the experience and judgment of the person performing the threat modeling or risk assessment. Automated tools are not yet available for data-centric threat modeling, and the effectiveness of modeling for large and complex systems has yet to be practically explored and analyzed. In addition, there is no absolute "right or wrong" approach to threat modeling. The most appropriate and effective threat modeling is the one that meets the needs of the organization, identifies the most important threats and risks, and implements appropriate security controls accordingly.

2.2.8 *Anomaly and event detection*

The purpose of anomaly and incident detection is to detect anomalous activity in a timely manner and understand the potential impact of an incident. First of all, an anomaly is anything that deviates from the standard, normal, or expected and therefore has a

Table 2.7. The total of the negative implication scores.

No	Security Controls	Acquisition and Implementation Costs	Annual Management/ Maintenance Costs	Impact on Functionality	Impact on Usability	Impact on Performance	Total Score
1	Uses strong passwords	1	1	1	1	1	5
2	Multi-factor authentication	2	2	1	2	1	8
3	Use antivirus software, spam filtering, etc.	2	2	2	2	2	10
4	Security awareness education	1	1	1	1	1	5
5	Patch vulnerabilities	2	2	2	1	2	9

Table 2.8. Potential impact definitions for security objectives.

		Effectiveness Per Attack Vector				Security Control Effectiveness		
No.	Security Controls	(1c)	(2b)	(4a)	Negative Implication	(1c)	(2b)	(4a)
1	2	3	4	5	6	7	8	9

contextual connection to the two links of "identify" and "protect". Without identifying resources and applying appropriate protection mechanisms to protect them from potential threats, there is no benchmark to determine whether an activity is normal, usual, or anomalous.

For example, suppose a database server in an organization initiates a connection to a foreign IP address. Is this normal? Where is the server physically located, who is using it, and is there endpoint protection on the device itself? As you can see, without identifying, discovering, and implementing security controls for this server, it is impossible to predict whether its activities are abnormal.

Understanding the potential impact of a security incident requires sufficient and accurate information. This is where Asset Intelligence comes in. Where is the device? Is other behavior occurring at the same time (physical access by personnel, IP connections)? Is the current anomaly outside the scope of the security policy? What was happening on the device just prior to the event? Is sensitive data being transmitted over the network or stored locally at risk? Based on asset intelligence, a deeper understanding of devices, data, users, applications, and behaviors is required to better meet the "Detect" requirements of the NIST CSF framework.

This asset intelligence must be continuously operational, resilient, and ideally self-healing. As mentioned earlier, without continuous asset intelligence, it is impossible to determine if an activity is anomalous. Furthermore, without a 360-degree view of hardware and software, there is no way to understand the impact of a security event.

From a data security perspective, the anomaly and event detection phase must focus primarily on the data flow baseline. The key point is to give a reasonable location and flow of data and define the data flow baseline based on the identification and definition of the organization's network architecture, physical environment, device assets, organizational data, and customer data. Activities that violate the baseline can result in a breach of data confidentiality, integrity, and availability. Defining data flow baselines must consider all plausible scenarios and strike a balance between the value of the data and its protection.

Risk analysis and anomaly detection from a threat perspective, such as malicious code and software vulnerabilities, detection of unauthorized devices, personnel, connections, and software, are also important components of protecting an organization's data security. Most of the measures in this section are already covered in traditional organizational security architecture and security response activities.

2.2.9 *Continuous monitoring*

Information systems and assets should be monitored at specified time intervals to identify cybersecurity incidents and verify the effectiveness of protective measures.

The scope of monitoring should include, at a minimum, the following:

(1) Integrity of software, firmware, and information.
(2) Whether unauthorized software has been installed on the device.
(3) The patch status of the device.
(4) Whether security monitoring applications or agents have been deployed.

A typical top-level architecture of continuous monitoring is shown in Fig. 2.22.

The first (lower) layer is the collection of security-related information and events from the organization's assets, including user actions, status or configuration changes, network connections, etc., via a monitored agent or client.

The second layer is a centralized log repository and log analysis engine for log storage and aggregation.

The third layer is the analysis of the behavior of users or other entities by the log analysis engine. That is, based on well-defined organizational policies, correlate and analyze the activities of users to determine their behavior.

The fourth layer is to generate appropriate events and alerts based on the conclusions of the behavioral analysis. The fourth layer

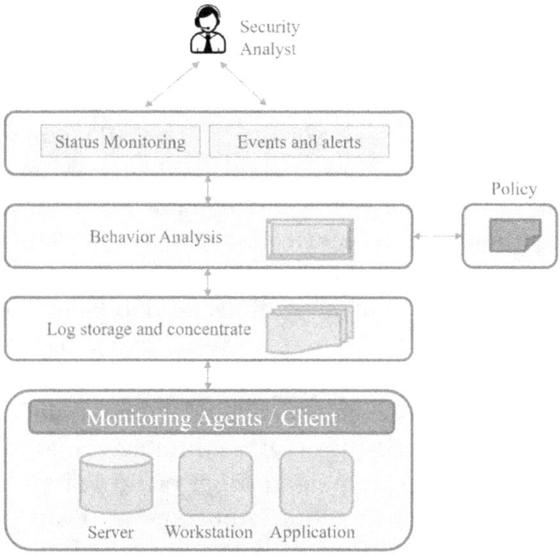

Fig. 2.22. Top-level architecture of continuous monitoring.

also includes the security dashboard, which is the aggregation and presentation of events, logs, and other security-related information.

Finally, security analysts or security operators make decisions and implement appropriate security activities based on the interaction with the security dashboard and the assessment of security events and alarms.

The goal of continuous monitoring is to provide security situational awareness of business data and information systems based on the collection and correlation of security-related data, to evaluate the effectiveness of existing security controls, and to help security leaders make risk-based decisions.

Continuous monitoring is the monitoring of the system and the security controls in place. It does not directly see or touch business data and is independent of the business data flow. Therefore, the term "data-centric continuous monitoring" is generally not advocated. Of course, when performing risk identification and decision-making based on continuous monitoring and event reporting, it is still recommended to first make a data-centric attack target assessment and attack impact assessment and then make a targeted response and recovery plan.

2.2.10 *Response planning and implementation*

Compared to the first three functions in the IPDRR framework (identify, protect, and detect), respond and recover functions are often neglected by organizations. According to the 2018 annual survey on "Cyber Resilient Organizations", sponsored by IBM and conducted by the Ponemon Institute (IBM, 2018), 77% of organizations admit to not having an adequate security incident response plan. The survey also states, "Ensuring the security function is equipped with a proper incident response plan, staffing, and budget will lead to a stronger security posture and better overall cyber resilience".

Indeed, the response function enables stakeholders to take action on detected cybersecurity incidents. This function builds on the organization's efforts to identify, protect, and detect capabilities, including mitigating detected threats and other critical steps.

The response function is more important in terms of outcomes. The processes and activities performed during the response phase can easily affect the outcome of a cybersecurity incident. Timely detection of threats is very useful, but the ability to quickly analyze the problem, contain the damage, and execute a response plan can lead to two distinctly different outcomes: a major breach or an unsuccessful hack.

In addition, in today's threat landscape, targeted attacks against specific victims or industries are becoming more common. Given real-world targeted attack scenarios, rapid and accurate responses are even more critical.

Section 2.1.10, "Major Threats to Data Security", points out that targeted attack scenarios are generally initiated by organized hackers. Their primary goal is to steal data, especially highly sensitive and valuable data assets. The type of data can be an organization's intellectual property or customer data.

The potential damage from a targeted attack can be severe, including the following:

(1) Loss of data or intellectual property.
(2) Disruption of critical business operations.
(3) Loss of customer data.
(4) Reputational and financial impact, including loss of customer confidence and litigation.

Fig. 2.23. Response function.

When an organization may become the target of a targeted attack, security leaders must consider how to guide the planning and implementation of response efforts to prevent the aforementioned breaches.

By following a robust response plan and containing the breach, security leaders and stakeholders can effectively reduce the impact to their organization.

In time series, the response function consists of the following main steps, as shown in Fig. 2.23.

The first step is to develop a response plan (referred to as "the plan" in Fig. 2.23). A response plan can give security teams an understanding of their roles and responsibilities in responding to anomalous cyber activity while protecting an organization's most critical information assets and systems.

The second step is communication. The organization's security leader (typically the chief security officer) and his or her team must coordinate response activities with internal and external stakeholders and seek assistance from law enforcement when necessary. As part of this process, initial threat events and related incidents must be reported and coordinated with stakeholders in accordance with the response plan and role responsibilities.

The third step is analysis. In this step, information from the event is used to analyze the impact of the event, determine if the current response is sufficient, and include forensic work if necessary.

The fourth step is remediation. Based on the conclusions of the analysis steps, appropriate security activities are implemented to contain the incident, prevent its spread, and mitigate potential damage. In addition, any new vulnerabilities not previously identified should be documented as part of the organization's risk management.

The fifth and final step is improvement. Chief security officers and other stakeholders need to examine the lessons learned from responding to threats and incorporate those lessons into future response strategies.

2.2.11 *Recovery plan and crisis management*

Data security incidents, which fall into two categories — massive personal data breaches and breaches of critical corporate information assets, can trigger media, consumer, and legal crises.

Therefore, in addition to a response plan, it is also necessary to develop a recovery plan for data security incidents. A good data recovery plan should include at least the following aspects:

1. **Commitment to eliminating all threats**

The first step in responding to a data breach from various types of vulnerabilities is not to immediately respond to a single attack but rather to take preventative measures to ensure that the root cause of the breach is identified and fixed or mitigated. This stops the damage early rather than being plagued by similar breaches or incidents too often.

An imprecise but vivid analogy for an organizational data breach is having coffee spilled on your white shirt by the kid next to you at a banquet. Most people's strategy for dealing with the situation is to remove the coffee cup to minimize the likelihood of the coffee spilling on their shirt again and to scrub their shirt while ensuring that the occasion is appropriate. These intuitive responses can be vaguely mapped to "vulnerability mitigation" and "environmental isolation" in a data recovery plan.

2. **Emphasize the importance of internal communication**

When developing a disaster recovery plan, organizations must identify all internal stakeholders and ensure appropriate communication.

If there is a major breach of personal data, roles such as the chief security officer, privacy officer, and head of the company's IT department will all receive media attention. But other departments, especially PR, marketing, and customer service, are just as important because they manage public reaction, communicate with the media, and provide customer support.

In fact, miscommunication can lead to inaccurate portrayals of the incident itself and its impact by various key players, which can damage the image of a company already in the spotlight. Companies need to ensure that everyone, from the top down, is informed and up to date to avoid unnecessary speculation and questions.

3. **Restore external trust**

Obviously, a data breach will result in direct losses to the company and even more damage to its reputation and credibility. Many customers will question the company's ability to protect their data, fearing that their information will be compromised again in the future. Or some may be concerned that the company is trying to cover up the actual data breach and minimize the value of describing the damage and stolen data in order to preserve the company's image.

Therefore, when communicating with the public after a problem has occurred, companies should be as transparent and honest as possible to project an image of the company as a proactive problem solver and trustworthy. To some extent, this can lead the public to focus on remediation plans to rebuild customer trust and make customers believe that companies can avoid similar attacks in the future.

4. **Implement and improve recovery plans**

In an ever-changing security environment, it is important to continually improve an organization's data recovery plan because attackers are constantly improving their methods.

Whether an organization has recently been attacked or a similar organization in the industry has experienced a data breach, it is important to analyze what led up to these events in order to optimize the data breach recovery plan. This will ensure that the organization is ready to respond to today's serious threats, not yesterday's common attacks.

Chapter 3

Data Security Architecture and Design

Data is a critical resource for every organization. As businesses move to the cloud and digitize their assets, the importance and value of data will only increase. Lost, compromised, or stolen data can have a devastating impact on business. According to a study conducted jointly by the IBM Corporation Security Team and Ponai Research Institute (IBM Security, 2019), the cost of handling each data breach incident is \$3.92 million, and the average cost of exposing sensitive data is \$150 per record. Verizon's "2012 Data Breach Investigation Report" (Verizon Enterprise Solutions, 2012) summarized 855 security incidents that occurred within one year, involving more than 174 million records. Seven years later, the "2019 Data Leakage Investigation Report" (Verizon Enterprise Solutions, 2019) summarized 41,686 security incidents, of which 2013 was a confirmed data leak, and the total number of records could no longer be aggregated and counted.

Common data breaches are often related to inadequate security controls, granting too many privileges to internal users, and over-reliance on network and perimeter security. Often, organizations still rely on traditional network-based security controls and fail to take a truly security-in-depth approach to protecting their environments. As organizations continue to embrace initiatives such as cloud computing and the need to support various forms of remote access and mobile devices, traditional network-only approaches to securing the IT infrastructure may fail. Today's security threats are diverse and often persistent, and traditional network perimeter security controls cannot effectively contain them.

Organizations must implement a more effective, data-centric, layered security control architecture and take a holistic approach to protecting systems containing sensitive applications and data from external and internal threats.

Establishing and designing a data security architecture requires reference to industry best practices and management methodologies. Establishing a good data security management structure at the organizational level requires not only reference standards but also the interpretation and refinement of standards to meet the needs of different types of organizations.

3.1 Design and Implementation of Data Security Architecture

"Security architecture" is an extremely broad concept. In fact, with respect to "architecture" itself, there are also definitions that describe its various dimensional characteristics.

ISO/IEC 42010–2022 (International Organization for Standardization, 2022) defines "architecture" as "fundamental concepts or properties of an entity in its environment and governing principles for the realization and evolution of this entity and its related life cycle processes". For its purpose, architecture is a practice that enables people to deal effectively with large, complex projects in a consistent, systematic, and structured manner.

The ANSI/IEEE 1471–2000 standard (CEPAL Electronic Engineers, 2000) defines architecture (for software-intensive systems) as "the fundamental organization of a system embodied in its components, their relationships to each other, and to the environment, and the principles that guide its design and evolution".

Referring to the definition in NIST SP 800–37 (Ross *et al.*, 2010), a security architecture is "a set of physical and logical security-relevant representations (i.e., views) of system architecture that conveys information about how the system is partitioned into security domains and makes use of security-relevant elements to enforce security policies within and between security domains based on how data and information must be protected".

The data security architecture is dependent on the overall security architecture. To define and describe the security architecture, it is advisable to adopt the project management methodology and the appropriate development and deployment methodologies according to the different phases of project charter, design, implementation, and deployment. Treat the architecture design as a project to be delivered, and complete it through the four stages of "project charter, design, implement, and deliver". The schematic diagram of the architecture development framework is shown in Fig. 3.1.

The panoramic view of the security architecture framework is shown in Fig. 3.2. In these four phases, different security architecture frameworks can be used as a framework to systematically perform the analysis and design of the security architecture.

The first phase is the project charter phase. What problem will the security architecture solve? What are the objectives of the security architecture? Who are the users and how will they use it? How will it be successful? These questions are critical to the success of the security architecture.

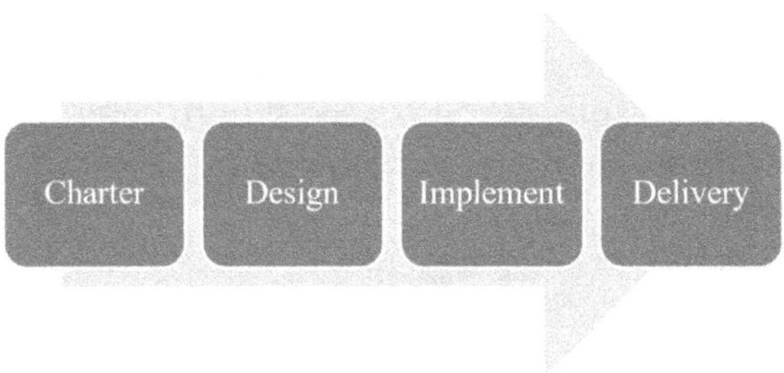

Fig. 3.1. Schematic diagram of architecture development framework.

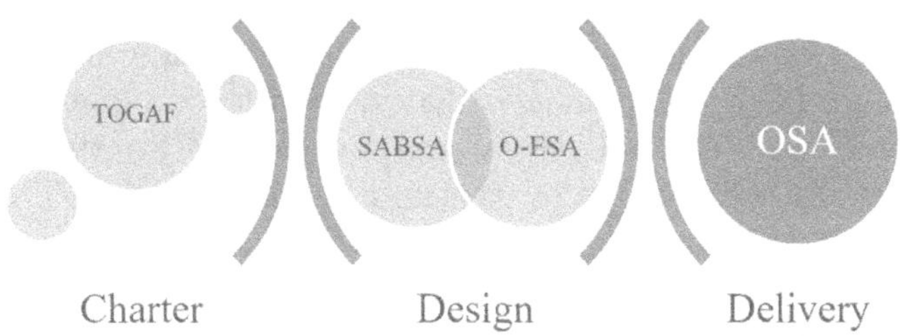

Fig. 3.2. Panoramic view of the security architecture framework.

At the project charter stage, The Open Group Architecture Framework (TOGAF) can help architects answer the above questions. According to the TOGAF-based output, the business objectives can be better aligned with the business managers and the security architecture can be mapped to the business architecture to accurately determine the value of this security architecture project.

The security architecture, with data security as its core, must also refer to the appropriate and applicable business framework, security framework, regulatory framework, and audit framework during the project charter phase. A business framework is the framework that an organization's product or solution must conform to. For example, if it involves cloud computing services, it must cover the Software as a Service (SaaS), Platform as a Service (PaaS), and Infrastructure as a Service (IaaS) cloud computing layer framework. The security framework refers to the security management standards applicable to the organization, such as ISO 27002 "Information Security Control Practice Guidelines" (International Organization for Standardization, 2022). There are many regulatory frameworks, such as PCI DSS and SOX, which are mentioned in Section 5.4. Among the various audit frameworks, the more systematic one is "COBIT 5: Information and Related Technology Control Objectives". The COBIT 5 framework provides the general process of IT audit and governance. After selecting the framework, it is necessary to determine the design and implementation methods of the security architecture based on each reference architecture.

The second phase is the design phase. In this phase, the Sherwood Applied Business Security Architecture (SABSA) model and the Open Enterprise Security Architecture (O-ESA) model can be extensively used to better integrate the security architecture with the business architecture. The SABSA model provides an overall framework for enterprise security architecture. It has a holistic approach that maps from business goals to source code. As an umbrella model, it can be used at an early stage to decide which components are necessary and which are not. Necessary components are drawn in the architectural hierarchy model according to their layering. This architectural hierarchy model in the SABSA model has the advantage of being simple and intuitive. It provides a top-level architectural diagram of the security architecture.

The driving force of architecture should come from the business, and the same is true for the security architecture. The SABSA model also provides support for the actual population of components at each layer of the security architecture, especially at the top layer, which is the business driver layer, where business drivers are linked to logical security services. The power of the SABSA model is that it provides a business attribute model as a model for managing security requirements. "Security is about helping to achieve business goals" is often dismissed as a motto or slogan, but the SABSA model reflects that security is related to business goals and stakeholders, making it easier to get business buy-in for security planning.

Once the security architecture is complete and aligned with the business, the third phase, the implementation phase, begins. The lowest layer in the security architecture includes functional and technical security controls. The term "security" in this context is more in line with the common understanding of "security", such as access control, system hardening, security scanning, and security awareness. In the implementation phase, the truly distinctive model is the O-ESA model. It extends the policy-driven architecture and articulates the vision that operational security will become increasingly digital and operational risk decisions will become more dynamic, requiring security policy automation. Compared to other security control measures, the O-ESA model places great emphasis on access control and security monitoring. Unlike the SABSA model, where a security policy is an action to be defined, the O-ESA model assumes that a security policy already exists.

The fourth phase is the deployment phase. In this phase, the Open Security Architecture (OSA) model allows the content of the security architecture to be presented more intuitively to stakeholders. Once the security architecture is finalized, proper communication is key to ensuring that all roles within the organization can fully leverage the architecture. The OSA model provides a wealth of lessons learned and cases for visualization to help stakeholders fully understand the content of the security architecture.

3.1.1 *Security architecture project establishment and requirements analysis*

The Open Group Architecture Framework (TOGAF) is an enterprise architecture framework that provides a method for designing, planning, implementing, and managing an organization's information technology architecture. TOGAF is a high-level design methodology typically modeled at four levels: business, application, data, and technology. It relies heavily on modularity, standardization, and existing, proven technologies and products.

TOGAF has been developed by The Open Group since 1995 and is based on the U.S. Department of Defense's Technical Architecture Framework for Information Management (TAFIM). At the end of 2016, The Open Group claimed that TOGAF had been adopted by 80% of the Global 50 companies and 60% of the Fortune 500 companies.

TOGAF is a set of tools that can be used to develop a variety of different architectures. It covers the following:

(1) Describes a method for defining an information system in terms of a set of components.
(2) Shows how the components fit together.
(3) Provides a set of tools.
(4) Provides a common vocabulary.
(5) Provides a list of recommended standards.
(6) Provides a list of compatible products that can be used to implement the component.

The TOGAF development cycle is appropriate for an organization that is starting to build its security architecture. The structure of the TOGAF development cycle is shown in Fig. 3.3. TOGAF is similar to other frameworks, with Business View at the top, followed by the Technical and Information Views.

TOGAF can serve as a useful reference when implementing the following types of services:

(1) Defining the structure, goals, and vision.
(2) Performing gap analysis.
(3) Monitoring the whole process.

The organization chooses the appropriate business direction according to its own vision and goals. For example, organizations that provide cloud services can follow the conceptual reference model proposed in NIST SP 500–292 when designing their own products and

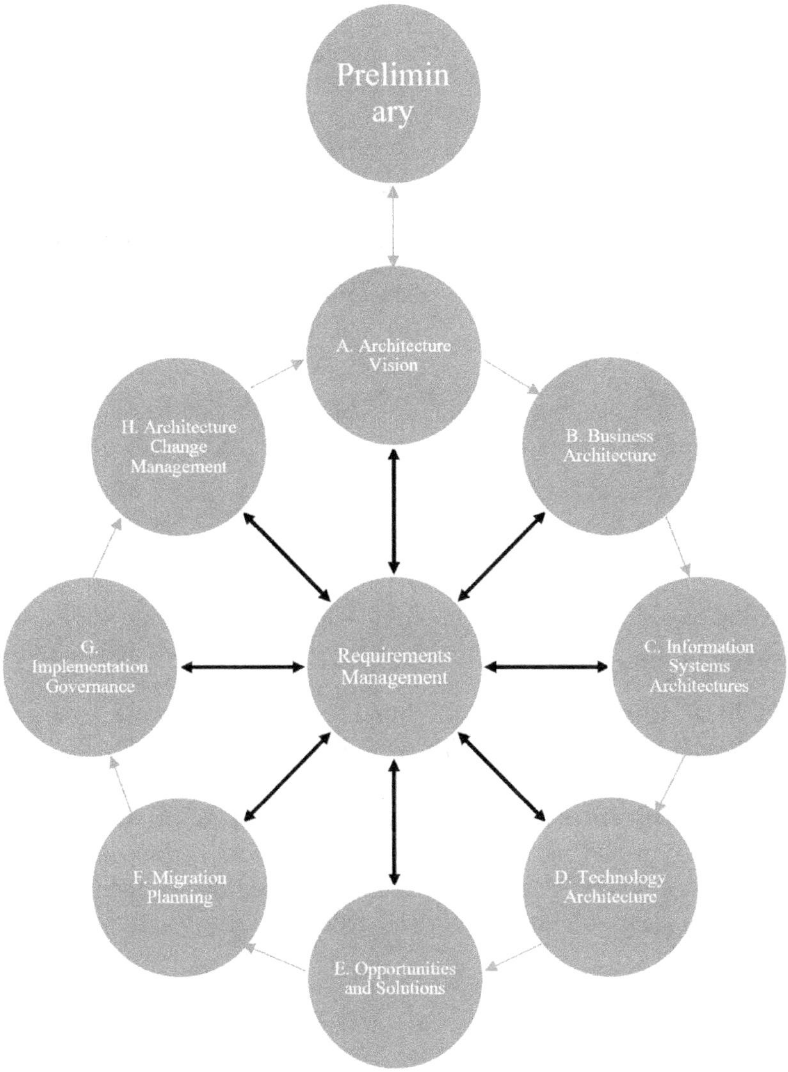

Fig. 3.3. TOGAF development cycle structure.

solutions to determine the organization's business architecture. Business architecture design needs to consider models that have formed international standards, national standards, or industry *de facto* standards so that suppliers, customers, and stakeholders can have the same consensus basis.

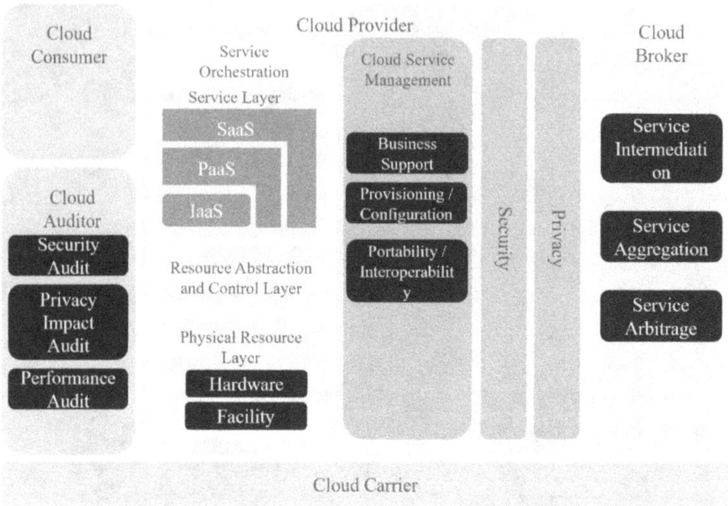

Fig. 3.4. NIST cloud computing conceptual reference model.

As shown in Fig. 3.4, the NIST Cloud Computing Conceptual Reference Model includes the following five important participants:

(1) **Cloud consumer:** An individual or organization that maintains a business relationship with a cloud provider and uses the cloud provider's services.
(2) **Cloud provider:** An individual, organization, or entity responsible for providing cloud services to a Cloud Consumer.
(3) **Cloud carrier:** An intermediary that provides connectivity and transmission of cloud services from cloud providers to cloud consumers.
(4) **Cloud auditor:** A party that can independently assess the performance and security of cloud services, information system operations, and cloud implementations.
(5) **Cloud broker:** An entity that manages the usage, performance, and delivery of cloud services and coordinates the relationship between cloud providers and cloud consumers.

Different participants have different business architectures, which naturally correspond to different information system architectures and technical architectures. This affects the requirements and design of the security architecture.

3.1.2 *Security architecture design and delivery*

Designing and deploying a security architecture in an organization is challenging on many levels. Traditionally, a security architecture consists of a set of preventive, detective, and corrective controls. These controls are implemented to protect an organization's IT infrastructure and applications. Of course, there are also best practices for adding administrative controls, including security policies and security procedures, to the security architecture. Many security professionals with a traditional mindset view a security architecture as a collection of security policies, controls, tools, and monitoring.

The world has changed, security is not what it used to be, and today's risks and threats are not as simple as they once were. Emerging technologies such as the Internet of Things (IoT) and emerging businesses have dramatically changed the way organizations operate, their priorities, and their goals. It is important for all security professionals to understand the business objectives and support those business objectives by implementing appropriate controls. These controls must be linked to business risks and easily understood by stakeholders. Enterprise frameworks such as SABSA, COBIT, and TOGAF can help achieve the goal of aligning security requirements with business needs.

The SABSA model is an enterprise-based, business-driven security framework that enables risk and opportunity-based assessment and identification. The SABSA model is purely a business alignment approach and does not provide specific security controls, relying instead on other security control frameworks, such as ISO 27001 or the COBIT framework. The SABSA model has six layers (five horizontal layers and one vertical layer). Each layer has a different purpose and perspective and covers the entire life-cycle of operational capabilities. From the highest level to the lowest, they are context architecture, conceptual architecture, logical architecture, physical architecture, and component architecture. The sixth layer, the Security Service Management Architecture, is overlaid on the other five layers and further analyzed vertically to create a 5*6 SABSA Service Management Matrix. The layering of the SABSA model is shown in Fig. 3.5.

ISACA's COBIT 5 Framework (ISACA, 2021) is a comprehensive framework that helps organizations achieve their goals for enterprise

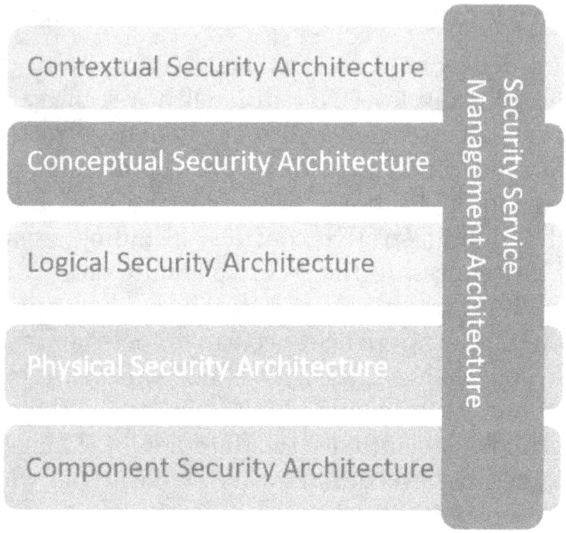

Fig. 3.5. SABSA model layering.

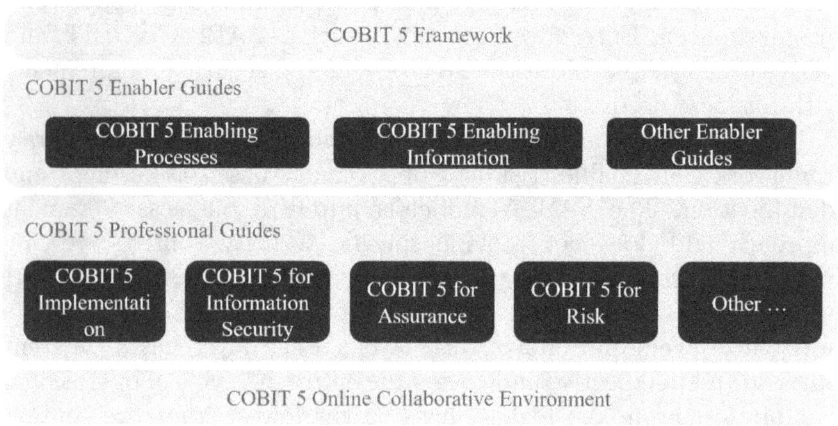

Fig. 3.6. COBIT 5 framework product system.

IT governance and management. The framework includes toolsets and processes that bridge the gap between technical issues, business risks, and process requirements.

The COBIT 5 framework product system is shown in Fig. 3.6. The goal of the COBIT 5 framework is to "enable IT to create the best value by balancing the realization of benefits with the optimization of risk levels and resource usage". The COBIT 5 framework aligns IT with the business while providing governance around it.

The COBIT 5 framework enablers cover factors that independently or collectively affect the effectiveness of an entity.

The COBIT 5 framework is based on the following five principles, as shown in Fig. 3.7:

(1) Satisfy the demands of stakeholders.
(2) Provide end-to-end coverage of organizational objectives.
(3) Apply an integration framework.
(4) Enable a comprehensive approach.
(5) Distinguish between governance and management.

Applying these principles to any architecture ensures consistent business support and optimizes processes.

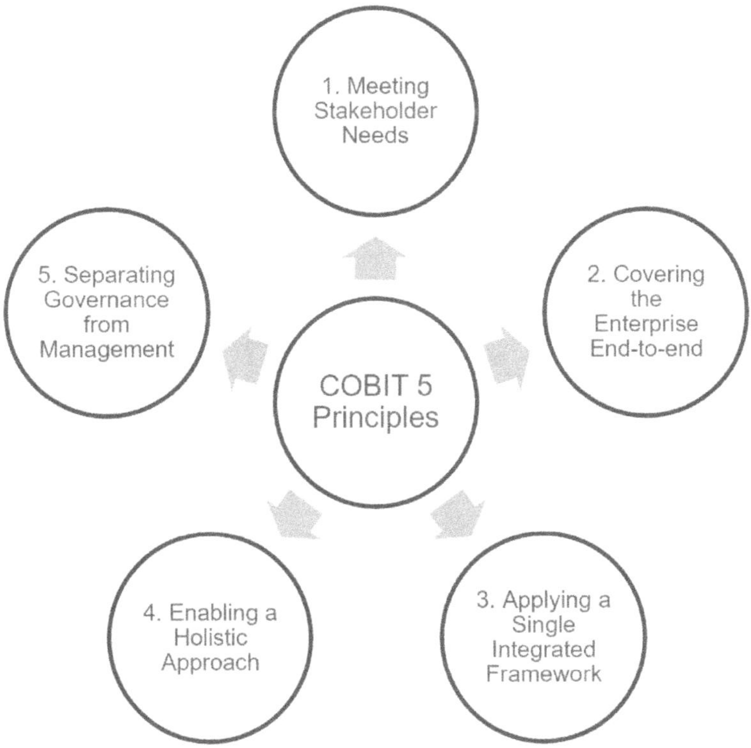

Fig. 3.7. The five principles of the COBIT 5 framework.

By using the SABSA framework in conjunction with COBIT principles, enablers, and processes, a top-down architecture can be defined for each category in the COBIT 5 framework product stack. For example, when developing a computer network architecture, these principles and processes can be used to define a top-down approach from the business context layer down to the component layer.

By using SABSA, COBIT, and TOGAF together, it is possible to define a security architecture that meets business needs and stakeholder requirements. Once the architecture and goals are defined, the TOGAF framework can be used to create projects and steps and monitor the implementation of the security architecture to where it should be.

When designing a security architecture, one of the most common questions is "Where do I start?" Based on an understanding of the above framework, the answer is simple. Security architecture design should be a top-down process. The view of the organization's security architecture is shown in Fig. 3.8.

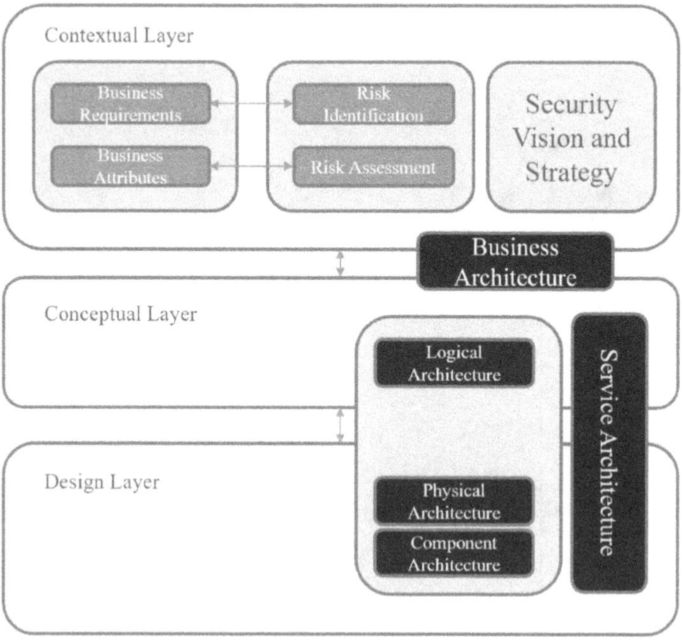

Fig. 3.8. Organizational security architecture view.

Consider the Business Context layer first. From an agile development perspective, the initial steps to initiate an enterprise security architecture development process are as follows:

(1) Establish business vision, goals, and strategies.
(2) Identify the business attributes required to achieve those goals.
(3) Identify any risks associated with the business that could cause it to fail to achieve its goals.
(4) Determine the controls needed to manage the risks.
(5) Establish a process for designing and implementing those controls.

Designing and implementing controls involves defining and mapping the following 4-tier architecture:

(1) Define the conceptual architecture related to business risks.
(2) Define the physical architecture and map it to the conceptual architecture.
(3) Define the component architecture and map the physical architecture.
(4) Define the operational architecture.

Conceptual architecture related to business risks mainly includes governance strategy and domain architecture, operational risk management architecture, information architecture, certificate management architecture, access control architecture, incident response architecture, application security architecture, web service architecture, communication security architecture, and so on.

The physical architecture and mapping conceptual architecture mainly include platform security, hardware security, network security, operating system security, directory and file security, and database security. The conceptual architecture also includes relevant security practices and security processes.

Component architecture and the mapping Physical architecture mainly include the following contents:

(1) Security standards, such as international ISO standards and National Institute of Standards and Technology (NIST) industry standards.

(2) Security products and tools, such as antivirus (AV), virtual private network (VPN), firewall, wireless security, and vulnerability scanner.
(3) Web service security, such as HTTP/HTTPS, Application Programming Interface (API) security and Web Application Firewall (WAF).

The operational architecture mainly includes implementation guidelines, responsible departments, configuration/patch management, monitoring, logging, penetration testing, access management, change management, forensics, etc.

After all risks have been identified and assessed, organizations can begin to design architectural components such as policies, networks, applications, and servers.

The above discussion is rather boring. Here is a specific example to illustrate.

Suppose there is a small bank card processing service provider that wants to serve 2 million users in the next 5 years (business vision and goals). The business attributes involved are as follows:

(1) **Regulatory:** The financial regulatory requirements of the countries and regions served, as well as industry standards, such as PCI DSS.
(2) **Privacy:** User privacy must be protected.
(3) **Accuracy:** User and business information must be accurate to prevent fraud.
(4) **Availability:** The service must be online and available 24/7.

Business risks may include the following:

(1) Failure to meet regulatory requirements (associated with the regulatory business attribute).
(2) Segregation of duties without roles (associated with the privacy business attribute).
(3) Vulnerabilities exist in the application (associated with privacy and accuracy business attributes).
(4) No disaster recovery plan (associated with the availability business attribute).

Possible control measures include the following:

(1) Implement the security control measures required by the PCI DSS.
(2) Establish responsibilities in relevant areas based on the COBIT DSS05 process.
(3) Establish a vulnerability management process and application firewall based on the COBIT DSS05 process.
(4) Establish public key infrastructure (PKI) and encryption control measures based on the COBIT DSS05 process.
(5) Establish a disaster recovery plan based on the COBIT DSS04 process.

By using the above control measures, all security control measures are directly related to business attributes, their meaning is clear, and it is easier to reach agreement with stakeholders and organize their implementation.

Of course, like any other framework, the lifecycle of the enterprise security architecture must be properly managed. It is important to continually update the business attributes and risks, and to define and implement the appropriate controls. As mentioned earlier, the TOGAF framework can be used to manage the security architecture lifecycle. That is, by creating an architectural view and goals, performing a gap analysis, defining the project, and implementing and monitoring the project until it is completed and restarted, the loop is ultimately closed.

3.1.3 *Visual presentation of security architecture*

After completing the design of the security architecture, an intuitive presentation method should be used to explain the security architecture to stakeholders. Among them, a series of architecture presentation methods defined by OSA (Open Security Architecture) can be used as a reference.

In 2007, a large community of security architects began to build and describe the foundations and security patterns of a generalized enterprise security architecture. This is the origin of the OSA organization. The OSA organization provides understandable, reusable parts at different levels of abstraction, the overall landscape, role

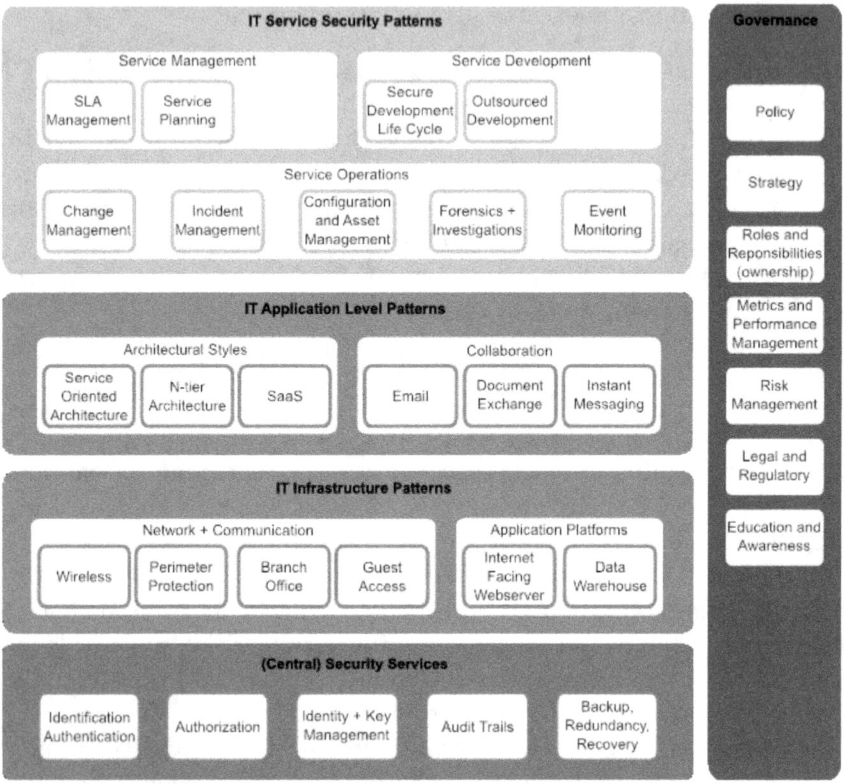

Fig. 3.9. Overall landscape of OSA security architecture.

definitions, terminology definitions, and taxonomies at the highest level, the Security Pattern at the next level, and the threat modeling and security controls (based on NIST) at the lowest level.

The OSA organization considers a security architecture to be a collection of designs applied at different levels. At the level of abstraction, OSA is divided into levels, such as Service Layer Security Patterns, Application Layer Security Patterns, Infrastructure Layer Security Patterns, (Centralized) Security Services, and Governance. The collection of these patterns forms the overall OSA security architecture landscape, as shown in Fig. 3.9.

In terms of defining the roles involved, OSA refers to ITIL (Information Technology Infrastructure Library) and defines various roles of the organization, such as IT Security Manager (or CISO, CSO),

IT Operations and Maintenance Manager, Information Asset Owner, etc. They have different job responsibilities and participate in the security process. They have different responsibilities and participate in the security architecture in different ways.

OSA took a highly visual approach, combining a security design pattern approach with a modular architecture to create a reference library of security architectures. The reference library contains lists of security controls, risks, and patterns. Components such as "Server", "Desktop", "Mobile", and "DMZ" in the reference library are typical components found in almost all enterprise architectures. Reusing these components in security architecture design can increase efficiency and allow security architects to focus more on expressing specific business and architectural parts.

OSA provides many security architecture models based on application scenarios. The more famous ones include the cloud computing scenario model, the public web server model, and so on. Let's take a relatively intuitive mobile device privacy mode as an example, as shown in Fig. 3.10.

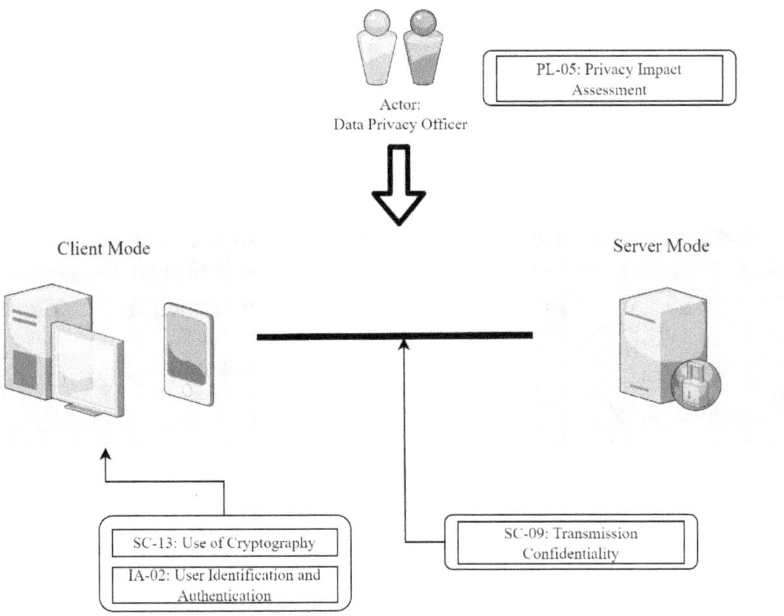

Fig. 3.10. Private mobile device mode.

Privacy laws and regulations in many jurisdictions require organizations to disclose any loss of mobile devices containing sensitive or confidential data. Private data subject to law or regulation includes customer records (such as name and address, financial records, and medical information) or any other personally identifiable information (PII).

Practical security controls to protect this information include the use of encryption on mobile devices (SC-13), combined with strong authentication (IA-02) to ensure that the information cannot be recovered if lost or stolen. Of course, the confidentiality protection required for the transmission of sensitive data must also be considered. The organization's privacy officer must perform a privacy impact analysis and report regularly.

Using the security architecture diagram, architects can intuitively understand the business model, stakeholders, and key security controls.

OSA provides a variety of models, from cloud computing to wireless hotspots to industrial control systems. It is often used as a reference or checklist for similar scenarios. In addition, OSA provides rich visualization support, such as pattern templates and extensive icon libraries. Based on these visualizations, security architects can map an organization's security architecture and place security controls at appropriate levels and locations. The mapped security architecture is visual and intuitive enough to align risks and budgets, align security policies with business stakeholders, and secure the resources to implement these security controls.

OSA's collection of security controls is mapped on a case-by-case basis to existing security standards, such as ISO and COBIT. As a result, the same language can be used when providing evidence to both internal and external auditors.

Unlike other standards such as NIST 800–53 and ISO 27001, which also describe best practices for technical, organizational, and process controls, neither has the ability to present design patterns using visualization. The OSA organization seeks to improve the expressiveness of its best practice standards by combining them with visual (design) patterns. OSA is the equivalent of the Lego building blocks of an enterprise's overall security architecture solution, providing a set of standard components and their interactions that encapsulate common security controls to address common security problems.

3.1.4 *Security architecture from different perspectives*

In modern society, the complexity of various systems is much greater than in the past. In particular, the emergence and rapid evolution of various software systems, web systems, and cloud service systems have brought great challenges in describing complex systems from multiple dimensions and perspectives. Different roles have different focuses on complex systems. For example, developers focus on the software component partitioning and invocation relationships, while operation and maintenance personnel mainly focus on deployment and maintenance methods during runtime.

Since the early 1990s, many efforts have been made to specify methods for describing and analyzing system architectures. "Software Architecture: Perspectives on an Emerging Discipline" published in 1996 (Shaw and Garlan, 1996) proposed the basic elements of software architecture — software components, connectors, and constraints — and attempted to abstract useful system designs, integrate notations and tools for software developers, and study patterns for system organization.

Combining the views of Dewayne Perry, Alex Wolf, and Barry Boehm, the concept of "software architecture" can be expressed in a precise formula:

Software Architecture = Elements, Forms, Rationale/Constraints

In Kruchten's 1995 paper "The 4+1 View Model of Software Architecture" (Kruchten, 1995), it was proposed to use multiple concurrent views to organize the description of the software architecture. A view is used only to describe a specific set of aspects of interest. The paper states: "Software architecture deals with the design and implementation of the high-level structure of the software. It is the result of assembling a certain number of architectural elements in some well-chosen forms to satisfy the major functionality and performance requirements of the system, as well as some other, non-functional requirements such as reliability, scalability, portability, and availability". To accurately describe large and complex architectures, the paper proposes a "4+1" view model.

(1) **Logical view:** Captures the object model of the design (when using object-oriented design methods).
(2) **Process view:** Captures the concurrency and synchronization properties of the design.

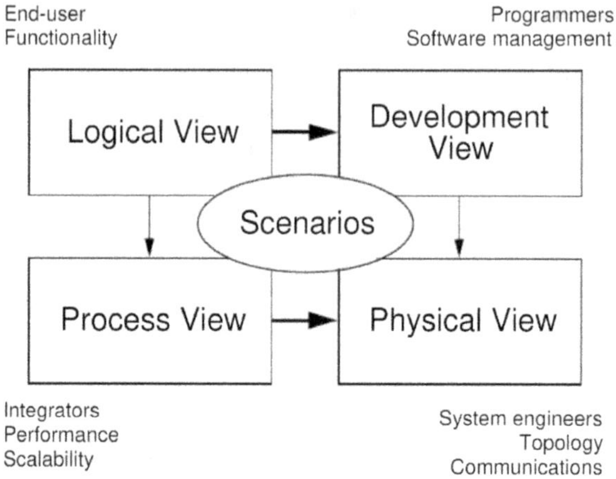

Fig. 3.11. Model of "4+1" view.

(3) **Physical view:** Describes the mapping of software to hardware and reflects the distributed characteristics.
(4) **Development view:** Describes the static organizational structure of the software in the development environment.

The description of the architecture, i.e., the various decisions made, can be organized around these four views, which are then represented by some use cases or scenarios, thus forming the fifth view (scenario view), as shown in Fig. 3.11.

In each view of the model, it is possible to independently apply the above formulas, i.e., to define a collection of elements used (components, containers, connectors), to analyze the working forms and patterns, to capture the relationships and constraints, and to map the architecture to the corresponding requirements, i.e., the five views correspond to the five representations of the different architectures.

The first is logical architecture. Logical architecture focuses on functional requirements, that is, the functions or services that the system should provide to users. From the perspective of the problem domain, the system can first be decomposed into a set of key abstractions. The decomposition is not only for functional analysis but also to identify common mechanisms and design elements throughout all parts of the system; then, through the principles of abstraction, encapsulation, and inheritance, to identify key components and analyze the relationships between components and other design constraints. The so-called "three-tier architecture", which is the abstraction of the presentation layer, business logic layer, and data access layer and their components, is a typical example of a logical architecture.

The second is the process architecture, which is derived from the logical architecture. Its main purpose is to address non-functional requirements, such as performance and usability. In addition, it can address issues of concurrency, distribution, system integrity, fault tolerance, and how the main abstraction of the logical view fits with the processes, i.e., on which control thread the operations of the logical objects are actually performed.

Depending on the problem to be solved, the process architecture can be abstracted and described at different levels. Note that the "process" here does not only refer to the "process" in the operating system but generally refers to the level of policy control process architecture (startup, recovery, reconfiguration, and shutdown). Processes can be reused to improve the ability to handle distributed loads or to improve availability.

The top process architecture can be thought of as running on a group of hardware resources connected by WAN or LAN and realizing some logical collections of specific functions through relatively independent process execution and inter-process communication. This logical collection can be deployed, coexist, or scale across multiple physical resources.

Again, this is the development architecture. A development architecture deals with the organization of the actual modules in a software development environment. Software is packaged into small program blocks (libraries or subsystems) that can be developed by one or a few developers. Subsystems can be organized into layers, with each layer providing a well-defined interface to the layer above it.

The development architecture of the system is represented by module and subsystem diagrams that show the "output" and "input" relationships. A complete development architecture cannot be described until all software elements have been identified. However, it is possible to list the rules that govern the development architecture: chunking, grouping, and visibility.

For the most part, the development architecture takes into account internal requirements related to ease of development, software management, reuse or commonality, and the constraints imposed by the toolset or programming languages. The development architecture is the basis for various development activities, such as requirements allocation, team work distribution, cost estimation and planning, project schedule monitoring, software reusability, portability, and security. These are also fundamental requirements for product development.

Finally, there is physical architecture. Physical architecture deals with the mapping of software to hardware, focusing on the non-functional requirements of the system, such as availability, reliability (fault tolerance), performance (throughput), and scalability. Software runs on computer networks or processing nodes, and the various identified elements (networks, processes, tasks, and objects) must be mapped to different nodes. Architects may want to use different physical configurations, some for development and testing, and others for deployment in different locations and with different customers. Mapping software to nodes therefore requires a high degree of flexibility with minimal impact on the source code.

The physical view of a large system can be very confusing, so it can take many forms, with or without mapping from the process view.

The four architectural views above are more oriented toward static visual descriptions. In practice, the elements of the four architectural views work together to perform an important set of scenario functions. Scenarios are typically represented using object scene diagrams and object interaction diagrams.

The scene view interacts with the other four architectural views and is a complement and explanation of the other views, hence the name "+1". In addition, the scene view can be used as a driving factor to identify and update architectural elements in the architectural design process. Once the architectural design is complete, the Scenario View provides both an additional reference and a starting point for testing.

The different architectural views are not completely orthogonal or independent. Elements in one view are related to elements in other views based on certain design rules and heuristics.

There are arrow connections between the four architectural views in the "4+1" view model shown in Fig. 3.11, which represents the transformation relationship between these views. Important transformation relationships are described as follows:

(1) **From logical view to process view:** It is necessary to identify the autonomy, dependency, distribution, and persistence requirements of each component/object in order to map the components in the logical architecture to the process (or thread) in the process view.
(2) **From the logical view to the development view:** In fact, the Logical View and the Development View are very close to each other, but the focus is not exactly the same. In the development view, the definition of subsystems must take into account many additional constraints, such as the organizational structure of the development team, the code size of the subsystems, reusability, and generality, as well as a strict hierarchical basis (visibility), release strategy, and configuration management.
(3) **From process view to physical view:** Map processes and process groups to available physical hardware in different deployment configurations.

Not all software architectures require a 4+1 view model. For example, if there is only one processor, the physical view can be omitted, and if there is only one process or program, the process view can be omitted. For very small systems, it is even possible that the logical view is so similar to the development view that it does not need to be described separately. The scene view is appropriate in all cases.

A well-designed system architecture can use the "prototype, test, estimate, analyze, optimize" method, similar to agile iteration, to refine and deliver the final architecture and reduce the risks associated with the architecture.

For specific business systems, a scenario-driven approach is recommended. This means that most of the key functions of the system are captured in the form of scenarios (use cases).

The key tasks throughout the lifecycle of scenario-driven architecture design are shown in Table 3.1.

Ideally, after 2–3 iterations, the architecture is basically stable, components, subsystems, and interaction relationships are clear, and the five views are complete. At this point, the software design phase can begin.

Table 3.1. Key tasks in scenario-driven architecture design.

Iterations	Tasks
Iteration 0	Scenarios are chosen for an iteration based on risk and criticality
	Based on the scenarios, identify major abstractions
	Elements discovered are laid out on the four blueprints: logical, process, development, and physical
	Architecture is implemented, tested, measured
	Captured the lessons learned
Prepare for next iteration	Reassessing the risks
	Extending the palette of scenarios to consider
New iteration	Script new scenarios
	Discover additional architectural changes
	Update the four main blueprints
	Revise the existing scenarios based on the changes
	Upgrade the implementation (the architectural prototype)
	Verify all five blueprints
	Upload design guidelines
	Captured the lessons learned

In general, the "4+1" view model allows different roles or stakeholders to view the entire software architecture from their own perspectives. Operations and maintenance personnel start with the physical view and then move to the process view; end users and business analysis experts start with the logical view; project managers and software configuration personnel look at the 4+1 view model through the development view.

3.2 Organizational Data Security Architecture Model

Based on the methods mentioned in Section 3.1.2, "Security Architecture Design and Delivery", data protection in an organization can be abstracted into the organizational data protection model shown in Fig. 3.12.

Data in the enterprise is abstracted into three categories: structured data, unstructured data, and data stored in the cloud. Their protections are not completely consistent.

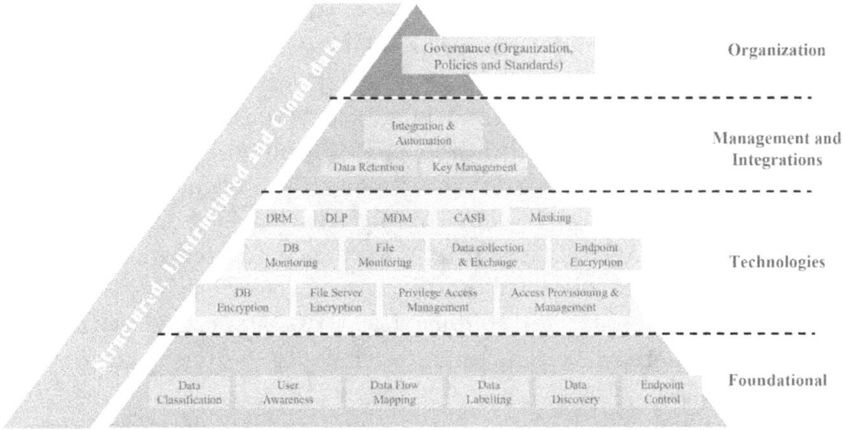

Fig. 3.12. Organizational data protection model.

The organizational data protection model is defined as four layers: governance, management and integration, technology, and foundation, which are described from top to bottom.

The first (top) layer is the organization's unified data governance framework. The data governance framework includes defining the organization, unifying policies, and establishing standards.

The second layer is strategies for managing, integrating, and automating data protection. It focuses on both data retention and key management.

The third layer is the technical layer, which includes key technologies that support data protection management strategies.

The fourth layer is the foundation of data protection, including data classification, data flow mapping, data labeling, data discovery, as well as endpoint control and user right-to-know protection.

3.2.1 *Overview of data security governance framework*

Today's organizations collect and use an increasing amount of data. For an organization's business to be successful, the data must be accurate and its use must be governed by appropriate policies and continuous monitoring.

The first step in implementing the data security governance framework is to define the overall data governance strategy to support the purpose of data security and data compliance; the second step is to build appropriate data governance organizations according to the data governance strategy, such as the data governance management team and working groups to support the achievement of strategies; the third step is to formulate, implement, and maintain data governance standards and data security standards, and ensure that business teams comply with the standards.

An appropriate data security governance framework helps not only achieve data security and compliance objectives but also improve data quality and reduce the costs associated with data corruption, thereby helping organizations to better achieve their business goals.

The first step is to have a clear and unified data governance strategy within the organization. The strategy can be developed by a team of stakeholders and project participants and must include the overall goals, governance model, and scope of data governance.

The second step is to build a data governance team organization that is appropriate for the organization based on the data governance strategy. For large enterprises, refer to industry practices such as open source governance and choose a three-tier architecture model: a group composed of senior managers, usually called the Data Governance Steering Committee (hereinafter referred to as the "Steering Committee"), is responsible for making high-level decisions; a mid-level management group, which could be called the Data Governance Office, provides business and management direction; a group called the Data Governance Working Group, which is responsible for most of the administrative work, also needs to include business and IT experts.

The Steering Committee consists of senior management and stakeholders. The Steering Committee should be chaired by a senior executive who reports to the CEO and has the authority to approve the project budget. Senior management on the steering committee is accountable to their respective business units.

The Steering Committee guides the data governance process across the organization and ensures that policies and procedures established for data (such as data quality policies and procedures) are followed. The Steering Committee is also responsible for approving the data project charter and strategy, approving funding, and nominating projects. In addition, the Steering Committee is responsible for providing guidance on business strategy and resolving conflicts.

The Office of Data Governance is responsible for implementing data governance. Roles in the Data Governance Office include Data Governance Leader, IT Representative and Coordinator, etc. The head of the data governance office is typically one level below the vice president and is a full-time position. He or she works with all business units and IT to unify data governance and strategy across the organization. The data governance office is also responsible for policy enforcement, establishing, monitoring, and reporting on data quality and data governance metrics, and collaborating with business unit leaders and IT resources.

Members of the Data Governance Working Group are managers or senior-level staff from various business departments who are responsible for bridging the gap between business and information technology, driving data management and data quality in specific areas, and requiring expertise in business and IT issues. Roles in this group may include data quality owners, data stewards, and data architects.

The third step is to formulate and drive the implementation of appropriate data governance and data security standards, processes, and best practices in accordance with the organization's data governance strategy once the organizational structure is complete.

It is important to ensure that everyone in the organization adheres to data governance policies and standards, especially when it comes to data quality. Data quality can be compromised if policies and standards are not followed.

Compliance with data governance policies and standards can be facilitated by making them understandable and easily accessible. A policy that is not readily available will rarely be read. Data governance policies and standards should include real-world metrics that demonstrate how compliance has a significant impact on the business.

Data governance policies and standards should be updated regularly and in a timely manner to reflect changes in corporate strategy and objectives, as well as changes in external regulatory requirements.

In summary, data governance is essential to effectively manage data and ensure its quality and security across the enterprise. Defining a unified strategy, selecting the right data governance team members, and ensuring ongoing compliance with data governance policies and standards are essential for organizations to establish an effective and successful data governance framework.

3.2.2 *Data governance strategies and standards*

COBIT (ISACA, 2021) is an internationally recognized IT governance management framework that helps organizations establish a comprehensive framework for information technology security and information technology management and control. IT governance ensures that an organization's IT investments support business goals, manage risk, and meet regulatory requirements. COBIT Best Practices bridges the gap between business risks, control requirements, and technical issues to address many aspects of management, including data governance.

Typically initiated and implemented by an organization's data governance committee, data governance is a set of policies and procedures that govern the business applications and technical management of data across the enterprise. Data governance is a set of continuous improvement management mechanisms that typically include organizational structure, policies and systems, technical tools, data standards, operating procedures, monitoring and evaluation, etc. The role of the Data Governance Council is to facilitate and support these governance activities. It will collect project success criteria and report to data stakeholders, and will continue to "serve" stakeholders in the form of interaction, information access, record keeping, training, etc.

Broadly speaking, data governance builds on the concept of governance in other disciplines (such as management, accounting, and IT) and can be thought of as a set of practices and policies that define data-related roles and responsibilities within an organization and work to improve data quality and enhance data-driven insights within the organization. These policies support the organization's business model through the generation and use of data.

In short, data governance is the management of data as a strategic asset, including the controls in place to ensure its content, structure, use, and security. To provide effective data governance, you need to know what data exists, is it of good quality, is it available, who is accessing it, who is using it, what are they doing with it, and are the use cases secure, compliant, and regulated.

According to the DGI Data Governance Framework (Data Governance Institute, 2024), data governance should achieve the following management objectives:

(1) Meet the needs of data stakeholders and create standardized, repeatable governance processes.
(2) Ensure transparency of the governance process and reduce operational conflicts.
(3) Make better decisions, reduce costs, and improve efficiency through collaboration.
(4) Educate management and staff on common data-related issues.

So far, data governance is very similar to IT governance, except for the asset types. The target audience for data governance includes all individuals required for IT governance, as well as other senior management, including but not limited to the board of directors, finance, operations, marketing, sales, human resources, suppliers, the CIO, and IT management. As the forms and types of data become more diverse, data security must focus not only on data and big data governance but also on protecting the privacy of personal information.

Data governance follows the COBIT IT governance framework, which defines a set of enabling processes for the enterprise. This requires bringing together a common architecture consisting of "business process, information, data, application and technology layers" to effectively implement an enterprise data governance strategy in support of the organization's strategic direction.

Data governance is inextricably linked to IT governance. Only by establishing an IT governance framework at the enterprise level can data governance be effectively managed. Through the organization-level IT governance framework and the establishment of the enterprise data governance framework, business leaders must formulate a value-based data governance plan to ensure that the board and shareholders can use data conveniently, securely, quickly, and reliably for decision support and business operations.

Data governance requires translating policy into concrete practices and guidelines. Where data governance really takes place is in strategy and in the day-to-day between regular operations. Data governance should be the bridge that transforms the strategic vision, recognizes the importance of data to the organization, and codifies it into operationally supportive practices and policies to ensure the delivery of products and services to customers.

By acting as a bridge in the design of the organization, data governance supports the implementation of strategy and encourages innovation, while providing the necessary safeguards to ensure the security and confidentiality of the information owned and/or processed by the organization. Unfortunately, many organizations put data first and data governance second, resulting in inadequate support for strategy and an unnecessary burden on operations.

Here's what organizations need to do to implement their own data governance framework:

(1) **Set data governance goals**

The most important step in creating a data governance framework is to define the goals. After all, it's hard to know what data is valuable without considering business goals. Think separately about goals that can produce long-term and short-term results. It's worth noting that there are scenarios that require trade-offs between goals and risks.

With specific and well-defined long-term and short-term goals, identify and measure key metrics for goal success. Key metrics help measure the team's progress toward the goal.

(2) **Establish an accountable data governance team**

Once the goals are set, the organization needs to achieve them. The most effective approach is to implement data governance through an entire team. This team is typically called the data governance office. Of course, for larger organizations, there may be a data governance committee led by top management. This team should include management, data stewards and liaisons, and any other organizational stakeholders involved in obtaining or securing data. The committee could have a virtual or physical "data governance office" and be responsible for making key management decisions.

(3) **Define the data governance model**

This step involves creating a data governance model for the organization. This model should describe the organizational structure of who can see and distribute different types of data. This ensures that sensitive data remains in the hands of the most trusted employees and is not shared without authorization.

The data governance model should also describe the organization's data collection rules, outlining the standards for securing data and the channels used to access it. This allows for consistency in data collection, resulting in more reliable and defined benefits.

(4) **Create a distribution process**

The final step in creating a data governance framework is to determine how each type of data will be distributed. As mentioned earlier, some data is sensitive and should not be shared throughout the organization. Therefore, organizations need a solid process for classifying data and identifying with whom it can be shared. The process should also define the channels available for distribution, which will allow for smoother and more consistent communication.

The above steps can provide a basic data governance framework for the organization's business. Of course, this is just the beginning; data governance must also consider the following best practices:

(1) **Data governance is not data management**

Importantly, data governance is not the same as data management. Data management is the actions an organization takes to enforce a data governance framework. Data governance takes over the decision-making function of the data management operating authority.

For example, an organization uses a customer relationship management (CRM) system to store customer data, which is a form of data management. Data governance should outline how employees use the information in the CRM. If that information is lost or mishandled, data governance should outline the next corrective actions to be taken.

(2) **Encourage teams to organize themselves**

The data governance office should be staffed with people who know how to best manage user data. These individuals have the ability to determine the optimal process for organizing information. The head of the organization may participate in decision-making, but the input of the data governance office experts must be respected when designing the data governance framework. With sufficient empowerment, data governance professionals can optimize and personalize processes to meet business needs.

(3) **Integrate data governance into every department**

Once a data governance framework is established, it should be integrated into each department of the organization. This will ensure a consistent flow of data collection and provide the organization's marketing, sales, customer service, and product development teams with the tools they need to achieve their goals.

For example, the data governance office should inform the product management team about user behavior and product usage. This information should be easily accessible and influence the way new products are designed. Through data governance, user insights can be streamlined into a series of data reports generated by the tool, increasing productivity across the organization.

(4) **Create risk milestones**

When creating a data governance framework, it is important to consider potential risks. User data is very important, and over time its security can be compromised as it is shared throughout the organization. Risk milestones highlight the risks that can occur when data is shared or moved within an organization. This can help organizations avoid the negative setbacks and high costs that can be associated with user relationships.

(5) **Continuously improve the organization's data governance framework**

A data governance framework should be a consistent process for collecting and distributing data. However, as the business grows and evolves, it is important to adjust the strategy to accommodate changes in the organization. Without adjustments, user data may be overlooked or sensitive information may be inadvertently exposed.

3.2.3 *Security data collection policy*

Collecting user data is certainly good for business and can accelerate business growth, but the act of collecting, storing, and transmitting sensitive and non-sensitive data also carries many risks. On the one hand, a variety of secure data collection strategies should be employed; on the other hand, the collection of user data should be minimized.

For most commercial organizations, collecting and storing users' personal information can provide tremendous ongoing value. Whether it's building an email list or using stored credit card information to speed up checkout, organizations and their users can benefit from data collection.

Data collection is a double-edged sword that can provide business value but also poses serious security risks.

A good mantra to keep in mind when collecting and accumulating user data is as follows: data is toxic. Therefore, collect only what is absolutely necessary — only information that is useful to the business or useful to the user, while respecting the user's right to privacy.

Personal data for which the purpose cannot be determined in advance should not be collected nor should personal data that violates regulations such as the Personal Information Protection Law (China), EU GDPR, PCI DSS industry standards, and CCPA (California Consumer Privacy Act of the United States).

When conducting business, consider maximizing existing data sources to solve business problems rather than adding new sources of data collection. New sources of data collection should only be considered if there are significant near-term benefits and compliance risks are manageable.

Data is a responsibility.

If user data exists on systems managed by the organization, it can be stolen, even with the most comprehensive protection. When data is stolen by attackers such as cybercriminals, breaches of sensitive data can lead to serious problems ranging from fines and lawsuits to complete cessation of business operations.

The costs of data loss can be significant, from fines for violating relevant laws and regulations to damage to brand image, so it's critical to secure the entry point of data — where it enters the organization.

This means that organizations should ensure that gateway portals and web servers are properly configured and choose an appropriate protection scheme based on the type of data being collected.

A number of common data protection schemes can be implemented to more securely protect collected data, including tokenization, encryption, and anonymization (de-identification). It is necessary to be flexible in choosing a privacy solution according to different data types and scenarios.

Sensitive or personal data that has been collected must be properly managed. Once data has been collected, there is a need to always know what it is, where it is stored, and how it is stored. Specifically, this includes the following:

(1) Where the data is stored.
(2) Who has access to it.
(3) How it is protected.

Effective management of collected data not only better protects it but also prepares it to meet the compliance requirements of relevant regulations (e.g., CCPA and PCI DSS).

To control where and how all data is collected, an organization should perform the following:

(1) Establish policies, procedures, and security practices to ensure that access to sensitive data is minimized.
(2) Audit data storage to control access and changes.
(3) Consider vendor and partner access to systems and databases.
(4) Ensure that the data repository is not public.

Organizations must always consider regulatory obligations, i.e., ensuring that the data collected complies with the privacy policy and any related compliance obligations.

Collecting certain types of data, such as cardholder data, is risky and places very high demands on an organization's processes, technology, and people management. If the organization can't do this, it's wise to avoid collecting it.

It is also important not to use data from production systems in test and development environments because design and development environments do not have the same security controls in place as production environments.

In addition, it is necessary to ensure that access to the collected sensitive data has the necessary security configuration and appropriate access control protections, such as role-based access control (RBAC) and multi-factor authentication.

Unused data deserves special attention.

If the organization recognizes that it needs to securely delete certain data in accordance with regulations, this also means that the data subject to the regulations will be collected. At this point, it is also necessary to analyze whether other similar data has been captured that poses a risk.

One of the most important techniques for solving this dilemma is to avoid the idea and practice of "collect it now, it will be useful later". In other words, strictly control the entry of data into the organization.

The entry point of sensitive data into the organization is critical, as is the exit point.

In the secure data disposition phase, data must be securely destroyed when the stored data no longer has business value or is driven by compliance requirements. Data backup and retention strategies are typically further evaluated and updated during this phase.

In summary, when sensitive data is not retained on a commercial organization's systems, the scope of compliance is reduced — making end-to-end compliance faster and easier.

3.2.4 *Data retention strategy*

Data retention defines policies for managing persistent data and records to meet legal and business data archiving requirements.

Data retention policies must balance legal and privacy concerns, economic considerations, and the principle of least privilege to determine retention periods, archiving rules, data formats, and permissible storage, access, and encryption methods.

In the telecommunications industry, data retention typically refers to call detail records (CDRs) and Internet protocol detail records (IPDRs). In the case of government data retention, the data stored is typically phone calls made and received, emails sent and received, and websites visited. Location data is also collected.

The main purposes of government data retention are traffic analysis and mass surveillance. By analyzing the data that remains, governments can determine the geographic location of individuals, their associates, and groups, such as the political opposition. Whether such activities are legal depends on the constitution and laws of each country. In most jurisdictions, government access to such data usually requires judicial authorization.

In the case of commercial data retention, the data retained typically includes records of transactions and website visits. Data retention also includes data collected by other means (e.g., automatic license plate recognition systems) and retained by commercial organizations.

A data retention policy is a recognized agreement within an organization to retain information for operational purposes while ensuring compliance with applicable laws and regulations. The goals of a data retention policy are to preserve important information for future use or reference, to organize information so that it is easily searchable and accessible, and to dispose of information that is no longer needed.

A data retention policy typically addresses what data is archived, how long it is retained, what happens to the data (archived or destroyed) at the end of the retention period, and other factors related to retention.

An effective data retention policy typically includes permanent deletion of retained data; secure deletion of data is often achieved by first encrypting the data while it is stored and then deleting the encryption key after a specified retention period has elapsed. Secure deletion must consider the efficient deletion of all associated data objects and their copies stored in both online and offline locations.

In practice, data retention is a typical conflict scenario between public rights, private property rights, and privacy rights, maintaining a fragile balance between criminal investigations, the commercial interests of organizations, and the privacy rights of individuals. Organizations should pay particular attention to the differing or even conflicting requirements of different jurisdictions when designing their data retention policies. The relevant EU legislation is used in the following as an example.

On March 15, 2006, the European Union adopted the Data Retention Directive (the "Directive") "on the retention of data generated or processed in connection with the provision of publicly available electronic communications services or of public communications networks and amending Directive 2002/58/EC". It requires EU Member States to ensure that communications providers retain the necessary data specified in the Directive for a period of 6 months to 2 years in order to perform the following:

(1) Track and identify the source of communications.
(2) Track and identify communication destinations.

(3) Determine the date, time and duration of the communication.
(4) Determine the type of communication.
(5) Identify communication equipment.
(6) Identify the location of the mobile communication device.

The Directive requires data to be made available to competent national authorities in specific cases "for the purpose of the investigation, detection and prosecution of serious crime as defined by each Member State in its national law".

The Directive covers fixed telephony, mobile telephony, Internet access, email, and VoIP. Member States are required to transpose it into national law within 18 months (no later than September 2007), and the requirements for Internet access, email, and VoIP may be delayed by a further 18 months. Since then, all member states of the European Union have notified the European Commission that they have implemented the Directive. However, in Germany and Belgium, only parts of the Directive are covered by legislation.

In April 2011, the European Commission published a report evaluating the Directive. It concluded that data retention is an invaluable tool for ensuring criminal justice and public protection but that it serves a limited purpose. Service providers have expressed serious concerns about the costs of compliance, and civil society organizations have expressed serious concerns, arguing that mandatory data retention is an unacceptable violation of the fundamental rights to privacy and the protection of personal data. The European Commission has been reviewing the legislation.

In response to the report, on May 31, 2011, the European Union Data Protection Supervisor (EDPS) expressed concerns about the European Data Retention Directive, emphasizing that the Directive does not meet the requirements of the right to privacy and the fundamental right to data protection.

On April 8, 2014, the Court of Justice of the European Union annulled Directive 2006/24/EC for violating fundamental rights (Tracey, 2015). The Council's legal services reportedly stated in closed session that paragraph 59 of the ECJ ruling "suggests that general and blanket data retention is no longer possible".

The situation in the U.S. is more complicated. At the federal level, lawmakers have proposed bills that would require Internet service providers (ISPs) to keep records of users' Internet browsing, but they have never been enacted. At the corporate level, some large commercial companies, such as Amazon, may retain records of users' transactions for long periods of time. Google Inc. also retains users' search histories and other data. These tactics do not violate applicable federal and state privacy laws, but they are causing growing consumer skepticism. In Europe, privacy regulators have also begun to investigate and impose fines.

At the beginning of 2019, after an investigation against Google, the French National Commission on Information and Freedom (Commission Nationale de l'Informatique et des Libertés, CNIL) announced that the Internet giant had violated the GDPR (Netter, 2019), that is, it had not obtained the data subject's consent to use their data for targeted advertising. Consent is not adequate, and the GDPR's transparency requirements regarding the purpose of processing and data retention periods are not met. Google then appealed.

France's highest administrative court (the "Conseil d'État") rejected Google's appeal in June 2020, confirming that Google had provided incomplete information to data subjects, in particular regarding data retention periods and the purposes of the various processing operations. The court also ruled that a fine of 50 million euros was appropriate given the nature of the violations, their impact on users, the persistence and duration of these violations, and Google's financial situation.

In response to the court's decision, Google stated that it would make appropriate changes to its information management mechanisms to comply with the consent mechanism required by the GDPR, which requires affirmative consent (opt-in) rather than implied consent.

3.3 Data Security Governance Framework

"Framework for Improving Critical Infrastructure Cybersecurity", published by the National Institute of Standards and Technology (NIST), provides a systematic framework for organizations of all types to assess and improve their ability to prevent, detect, and respond to cyber incidents. The NIST Cybersecurity Framework (also known as the NIST CSF) was released in version 1.1 in April 2018, and this version has been rapidly adopted across industries.

As a voluntary framework, the CSF provides standards, guidance, and best practices for organizations to manage cybersecurity risks. Many organizations have adopted the framework to help them manage their cybersecurity risks. According to Gartner, approximately 30% of U.S. organizations used the CSF in 2015. The 2019 SANS OT/ICS Cybersecurity Survey shows that the NIST CSF is the *de facto* standard for cybersecurity frameworks in use today. Cybersecurity frameworks are also increasingly being adopted by government agencies and across industries around the world.

The Cybersecurity Framework has three main components: the Framework Core, the Framework Implementation Tiers, and the Framework Profiles:

(1) The Core represents a set of cybersecurity practices, outcomes, and technical, operational, and managerial security controls that support the five risk management functions of Identify, Protect, Detect, Respond, and Recover.
(2) The Framework Implementation Tiers can be used by an organization to clarify to itself and its partners how it views cybersecurity risk and the level of sophistication of its management approach.
(3) A Framework Profile is a list of deliverables intended to communicate the organization's "as is" and "to be" cybersecurity posture.

The NIST Cybersecurity Framework includes five key functions — Identify, Protect, Detect, Respond, and Recover — as shown in Fig. 3.13, according to the acronyms of these five key functions, also known as the IPDRR framework.

The five key functions are the highest level of abstraction included in the framework and cover the key activities of a cybersecurity program. The five key functions approximate each link in the sequential process, are easy to understand, and are key pillars of a comprehensive cybersecurity plan. This top-level perspective allows an organization to accurately represent its cybersecurity risk

Fig. 3.13. Five key functions of the NIST cybersecurity framework.

management model and supports the organization's risk management decisions.

The IPDRR framework takes a "before, during, and after" view of the entire process, shifting the cybersecurity framework from a passive response model that focuses on protection and attempts to "keep the enemy out of the perimeter" to a more proactive model that targets the security of an organization's core business and core data, ultimately building an agile security capability that can adapt to change.

In terms of the organizational information security and data security dimensions, the role of each function of the network security governance framework is shown in Fig. 3.14. In terms of purpose, each function has its own focus:

(1) **Identify:** Understanding the organization's assets, analyzing potential risks, and measuring potential attack surfaces.
(2) **Protect:** Prevent and mitigate damage. Measures include patching, isolation, hardening, operation and maintenance access control, vulnerability mitigation, and remediation.
(3) **Detect:** Discover "bad" factors in the system. Measures include detecting security events, triggering anomalies, looking for intrusions, and analyzing security status.
(4) **Respond:** Accurately identify and remove "bad" factors. Measures include taking targeted actions, eliminating entry points, assessing attack damage, and investigating and collecting evidence.
(5) **Recover:** Restore to a "good" state, including data and service recovery, summary and reflection, etc.

In planning and developing the task list for each function, the objectives of the function must be fully understood. In terms of timing, the first two functions occur prior to the occurrence of an "incident" and focus on identification and protective measures, reflecting a structured and systematic mindset; the last three functions occur after the occurrence of an "incident" and focus on accurate detection, response, and recovery from the incident, reflecting a reactive and situational mindset.

The prerequisite for the effective application of the Cybersecurity Framework is the identification of all the assets of the organization.

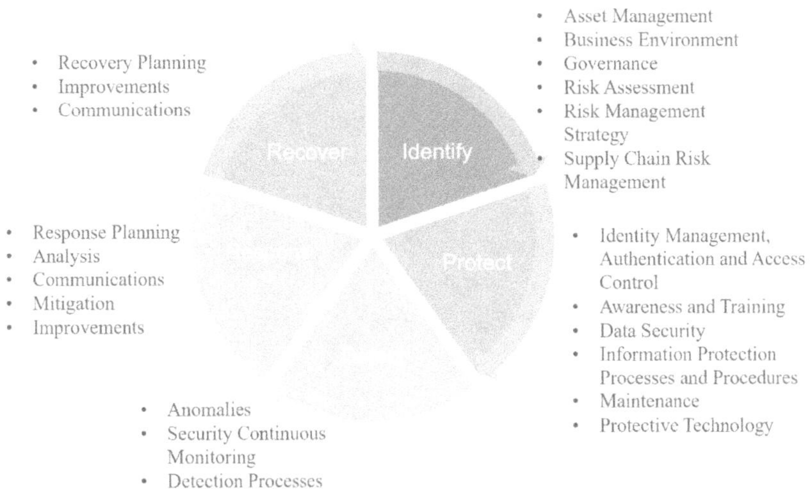

Fig. 3.14. Functions of various functions in the network security governance framework.

The organization's assets can be divided into the following five categories:

(1) **People:** Members of the organization and all subjects who have access to the assets of other organizations.
(2) **Data:** Information that is stored, transmitted, and processed in devices, networks, and applications.
(3) **Application:** Application software that requires attention to the interacting parties of the software and the workflow of the software.
(4) **Network:** Network connections used by devices and applications, and network traffic transmitted.
(5) **Device:** Various workstations, servers, mobile phones, tablet computers, Internet of Things devices, storage devices, network devices, etc.

Based on NIST's five major functions and the above five types, a security defense matrix can be obtained (see Table 3.2). The columns in the table are the five major functions of NIST, the rows are five classifications of organizational assets, and the cells are the protection measures for that dimension.

This section focuses on the application of the NIST Cybersecurity Framework to data security.

Table 3.2. Security defense matrix of CSF.

Assets	Five Functions				
	Identify	Protect	Detect	Respond	Recover
People	Phishing exercise	Awareness	Internal threat analysis, behavior-based anomaly detection	N/A	N/A
Data	Data classification and labeling	Data encryption, access control, DLP	Dark net	DRM	Backup and restore
Application	Configuration management	Application security mechanism (SAST, DAST, etc.)	N/A	N/A	N/A
Network	Traffic classification	Firewall, IPS	DDoS mitigation, IDS	DDoS mitigation	N/A
Equipment	Equipment identification and classification	Endpoint protection, antivirus software, Host IPS	Endpoint visual management platform, endpoint threat detection and response	Endpoint visual management platform, endpoint threat detection and response	N/A

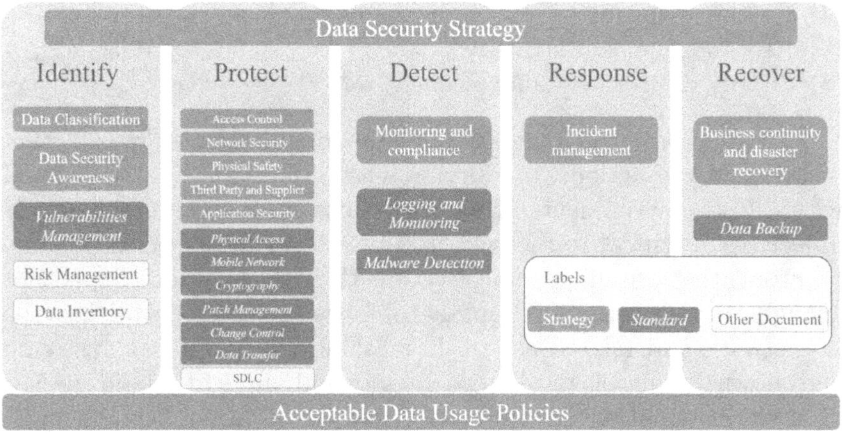

Fig. 3.15. Data security governance framework.

When applied to the field of data security, the data security governance framework based on the NIST Cybersecurity Framework is shown in Fig. 3.15.

Next, each part of the data security governance model framework is introduced separately, and the role of each part in the overall data security solution is presented.

3.3.1 *Identify*

As mentioned earlier, the main goal of the Identify phase is to understand the organization's assets and the risks they may face. In the dimension of sorting out organizational assets, especially digital assets, it is necessary to use the data identification and classification as well as the data security awareness described in Section 1.2 "Principles and Implementation of Data Classification".

After identifying all of the organization's key assets, it is also necessary to analyze the possible risks and measure the possible attack surface in relation to the organization's business and processes.

This step can provide analytical results in the following areas:

(1) Asset management.
(2) Business environment.
(3) Governance.

(4) Risk assessment.
(5) Risk management strategy.
(6) Supply chain risk management.

"Identify" lays the foundation for the organization to take further security-related actions. Identifying what exists, the risks associated with those environments, and how those risks relate to business objectives is critical to the successful use of the framework.

From the overall perspective of the organization, it is not enough in the "Identify" phase to focus only on protecting the IT network and related assets. It must also be extended upstream and downstream of product delivery, and an end-to-end review of asset distribution and potential risks is required. For example, the upstream supply chain, downstream customers, and your own employees all need to be considered in the identification phase. The scope of asset identification is shown in Fig. 3.16.

For example, in the Bring Your Own Device (BYOD) scenario, assets belong to employees but are used to process the organization's business data. Therefore, it is necessary to clarify the asset categories and risks of processing business data, and develop protection mechanisms.

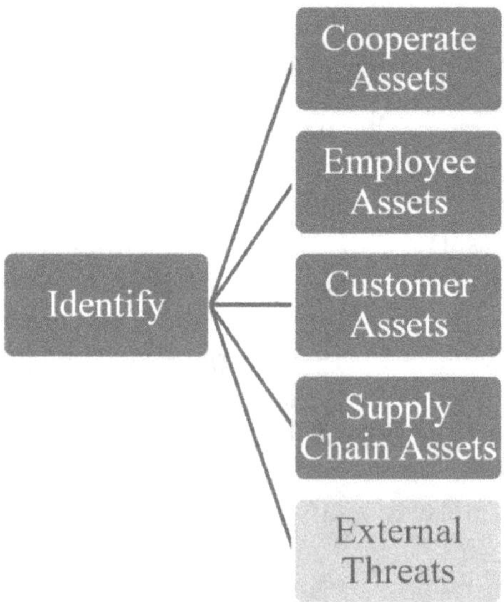

Fig. 3.16. Asset identification category.

Successful implementation of the Identify function provides organizations with a strong understanding of all assets and environments within the enterprise. Defining the current and desired state of controls for securing these assets, and developing a plan to move from the current to the desired state of security, clearly communicates the organization's cybersecurity posture to technical and business stakeholders.

In the risk and vulnerability assessment process, it is worth highlighting vulnerability management and supply chain risk management. Known and unknown vulnerabilities in IT systems, products, and services pose critical risks. In addition, supply chain risk management is often overlooked but is critical to the delivery of holistic solutions, products, and services.

3.3.2 *Protect*

The overall purpose of the NIST Cybersecurity Framework is to help an organization express its management of cybersecurity risk by organizing information, enabling risk management decisions, addressing threats, and improving by learning from past activities.

The "Protect" function is important because its purpose is to "develop and implement the appropriate safeguards to ensure delivery of critical services". A protective function refers to the ability to limit the impact of a potential cybersecurity incident. According to NIST, examples of areas covered by the Protect function include identity management and access control, employee security awareness and training, data security, information protection processes, operational scenarios, and other defensive technologies.

While the Identify function focuses primarily on baselining and monitoring, the Protect function is more proactive. The Protect function covers categories, such as access control, awareness, and training, and is applied in the area of data security, primarily through the use of encryption and access control mechanisms. Examples of access controls include two-factor or multi-factor authentication practices for access to assets and environments, and appropriate access controls. In addition, the Data Loss Prevention (DLP) solution is also an important implementation measure for data protection in the enterprise IT environment and cloud computing scenarios in the "Protect" function.

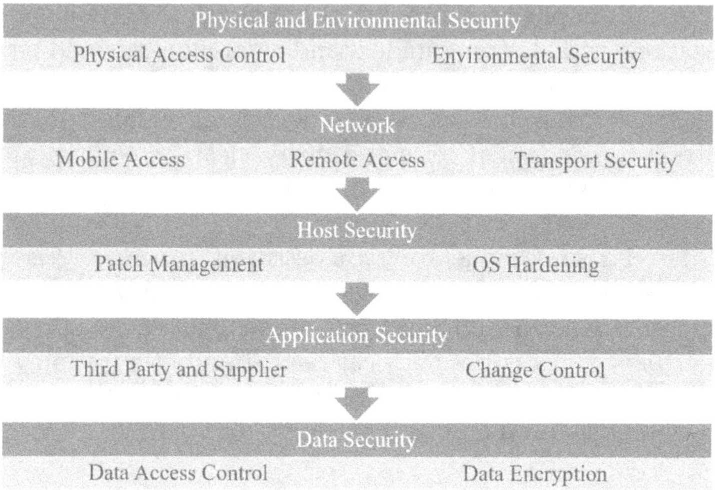

Fig. 3.17. Example of layered protection mechanism.

Adopting a layered defense architecture can help achieve the purpose of data protection. An example of the layered protection mechanism is shown in Fig. 3.17, which describes the protection mechanism of each layer.

In addition, in scenarios involving the product development lifecycle, the Security Development Life Cycle (SDLC) guides the management model of the software development process from a security perspective. The SDLC introduces security into every phase of software development, including requirements, design, coding, testing, and maintenance, to reduce security risks at the source, improve security, and reduce the cost of remediating security vulnerabilities. The SDLC is also an important part of the protection phase.

In summary, as data breach scenarios become more prevalent, it is especially important to adopt appropriate protocols and strategies to mitigate the risk of a breach. The protection features of the CSF framework can be used as a guide to help achieve data protection goals.

3.3.3 *Detect*

Detection refers to the design and implementation of appropriate activities to accurately identify cybersecurity incidents as they occur.

The Detect function can detect network security incidents in a timely manner. Examples of result categories within this function include Incidents and Anomalies, Security and Continuous Monitoring, and Detection Processes:

(1) **Incidents and anomalies:** Anomalous activity is detected as quickly as possible, and the impact of the incident is understood by everyone inside and outside the team.
(2) **Security and continuous monitoring:** Monitor information systems and environments at regular intervals to identify cyber events within the organization.
(3) **Detection process:** Implement the detection process and conduct testing to ensure timely and comprehensive understanding of network events.

The "Detect" function is a critical step in building a strong cyber program — the faster a cyber incident is detected, the faster it can be responded to.

Following is an example of the steps to complete the full detection capability:

(1) **Incidents and anomalies:** Provides teams with the ability to collect and analyze data from multiple endpoints to detect incidents.
(2) **Security and continuous monitoring:** Enables teams to monitor assets 24/7.
(3) **Detection process:** Seeks to understand breaches as quickly as possible and comply with disclosure requirements as necessary; inappropriate access to data should be detected as quickly as possible.

Detecting anomalies or incidents can be a matter of life or death for an organization, so the "Detect" function of the Cybersecurity Framework is critical to security and business success. By following these best practices and implementing these solutions, cybersecurity risks can be mitigated.

3.3.4 *Respond*

The NIST CSF Framework defines the "Respond" function as "developing and implementing the appropriate activities to take action regarding a detected cybersecurity event".

The Respond function refers to containing the impact of a potential cybersecurity incident. This function includes five categories: incident response planning, analysis, communication, mitigation, and continuous improvement to help organizations take the right action immediately in the event of an attack.

Developing an incident response plan is the first step in implementing a response function. An incident response plan is not a technical control measure but a people and process control measure. After an actual security incident occurs, the technical controls of analysis and mitigation are required. The non-technical controls of Communication (RS.CO) and Continuous Improvement (RS.IM) can improve the efficiency and effectiveness of the response.

The objective of the "(RS.AN) Analysis" category is defined as "ensure appropriate response and support recovery activities". Subcategories of this category include investigating alerts from detection systems, understanding the impact of incidents, triaging incidents against response plans, and forensic analysis during incident handling. Access logs and network flow logs in the Detection function play a critical role in event understanding and forensic analysis because they can provide records of data access, transfer, connection, and usage of network systems.

The objectives of the "(RS.MI) Mitigation" category are defined as "activities undertaken to prevent the expansion of the incident, mitigate its impact and eliminate the incident". Subcategories of this category include incident containment and incident mitigation, as well as taking action on newly discovered vulnerabilities and confirming that they are actually mitigated or represent an acceptable risk.

3.3.5 *Recover*

The NIST CSF framework defines "Recover" function as "develop and implement the appropriate activities to maintain plans for resilience and to restore any capabilities or services that were impaired due to a cybersecurity event". It supports the timely restoration of normal operations to reduce the impact of cybersecurity incidents. Lessons learned from incident response (the Respond function) should be incorporated into security operations.

Examples of the "Recover" function include recovery planning, remediation, and communication, as described in the following:

(1) **Recovery planning:** Test, implement, and maintain recovery procedures to mitigate the effects of incidents as early as possible.
(2) **Improvement:** After an incident, identify areas for improvement and enhance recovery plans and processes.
(3) **Communication:** Formulate and implement a comprehensive coordination and communication plan that includes both internal and external organizations.

The ability to recover is important not only to business and security teams, but also to users and the marketplace. The ability to recover quickly puts an organization in a better position internally and externally. If a breach does occur, a recovery plan will help ensure that the organization stays on track to meet security objectives.

3.3.6 *Summary of data security governance framework*

The Data Security Governance Framework, based on the IPDRR model, provides a new perspective that enables a shift from traditional firewall-based perimeter protection to data-centric, full-process protection. The framework recognizes the complexity of the modern enterprise environment (people, applications, data, devices, and networks) and develops complete prevention, control, and recovery measures for the threat lifecycle, avoiding the problem of traditional perimeter protection strategies that lack effective countermeasures after an intrusion.

Implementing a data security governance framework based on the NIST Cybersecurity Framework can be challenging for an organization, whether at the people, process, or technology level. However, it must be worthwhile to accurately assess the current state of the organization and implement key management and control measures based on the data security governance framework. Implementing a data security governance framework not only benefits an organization's cybersecurity and data security but also helps improve ongoing compliance and supports communication between stakeholders, such as technology and business.

3.4 Excellent Cases of Security Architecture

Organizations need a security architecture. Through the standardized guidance of security architecture, the organization's IT systems, information assets, and processes can be built and improved, and risks in business processes can be controlled and reduced through technical means and processes, and organizational efficiency can be improved.

Different organizations have different security architectures to accommodate different organizational goals, business layouts, and even organizational structures and business processes.

The previously proposed defense-in-depth architecture (also called layered defense architecture) was characterized by the construction of multiple lines of defense, similar to those in classical military theory, to protect targets. Existing security architectures and security principles still use this as a reference. Data-centric security architecture focuses on protecting data security rather than network perimeter security and host security and is beginning to have a positive impact. The explanation in this section focuses on specific areas and scenarios, as well as the following security reference architectures: CSA Enterprise Architecture (focusing on the overall security governance framework of the enterprise), NIST Cloud Computing Security Architecture (focusing on cloud computing security), and Zero Trust Security Architecture (focusing on the identification and continuous authentication of resource visitors).

3.4.1 *Defense in depth architecture*

The news of today's society is full of reports of cyberattacks and user data leaks, and there is no shortage of stories about well-known major companies in the IT industry.

In the face of malicious, complex, organized, and new types of cyberattacks, it is unrealistic to use single-point security solutions such as firewalls, antivirus software, and data loss prevention (DLP) to deal with them. Therefore, from an organization or network perspective, it is very necessary to build a unified security architecture.

The principles of "layered security" or "defense in depth" are most commonly used in the design of networks, products, and solutions. These designs use multiple tools and techniques to prevent or mitigate attacks and provide comprehensive risk and security management.

From the perspective of the IT network within the organization, the security architecture is generally divided into 5–6 layers:

(1) **Physical security layer:** Access control/ID cards and other means are used to ensure the security of computer rooms, offices, and physical facilities to prevent attackers from gaining physical access and launching attacks from within the trusted network.
(2) **Perimeter security layer:** With the popularity of mobile office and IoT, the "network perimeter" has become more complex and no longer resembles the traditional internal and external network boundaries based on firewall separation, but most of the network security defense objectives and protection measures are still effective. Intrusion detection systems (IDS), intrusion prevention systems (IPS), honeypot systems, data leakage protection (DLP), and traditional network isolation solutions such as firewalls, quarantine zones, or perimeter networks can still play an important role.
(3) **Network security layer:** This layer includes security measures such as web proxy content filtering, enterprise wireless security, enterprise remote access, VoIP protection, and network access control (NAC). NAC is the process of restricting unauthorized users and devices from accessing a corporate or home network. For example, 802.1X is an IEEE standard for port-based network access control (PNAC), which defines authentication controls for users or devices attempting to access a LAN or WLAN.
(4) **Endpoint security layer:** Every device connected to the network represents a potential entry point for security threats. Laptops, USB drives, tablets, and smartphones all have the ability to store and access sensitive data. Typical countermeasures include timely patching, antivirus, and multi-factor authentication to ensure endpoint security. Corresponding security measures include host IPS/IDS, endpoint firewall, endpoint antivirus, endpoint policy, patch management, etc.
(5) **Application security layer:** Ensure that applications developed and deployed meet the requirements of the OWASP Application Security Specification. The use of third-party services or software also requires a thorough review of their security

before they are integrated with business systems. Such security measures include static and dynamic testing of applications, code review, database scanning and monitoring, and web Application Firewall (WAF).

(6) **Data security layer:** This is the core of the security architecture. This layer includes key technologies such as data classification, data lifecycle protection (storage, use, and transmission), data integrity monitoring, data or disk encryption, data destruction, etc. Broadly speaking, identity and access management (IAM) and public key infrastructure (PKI) also fall under the umbrella of the data security layer.

This layered security architecture is often vividly compared to layered defense, defense in depth, or onion defense. The onion defense diagram is shown in Fig. 3.18.

A layered defense-in-depth security architecture is a static description of the system and network architecture. From a dynamic

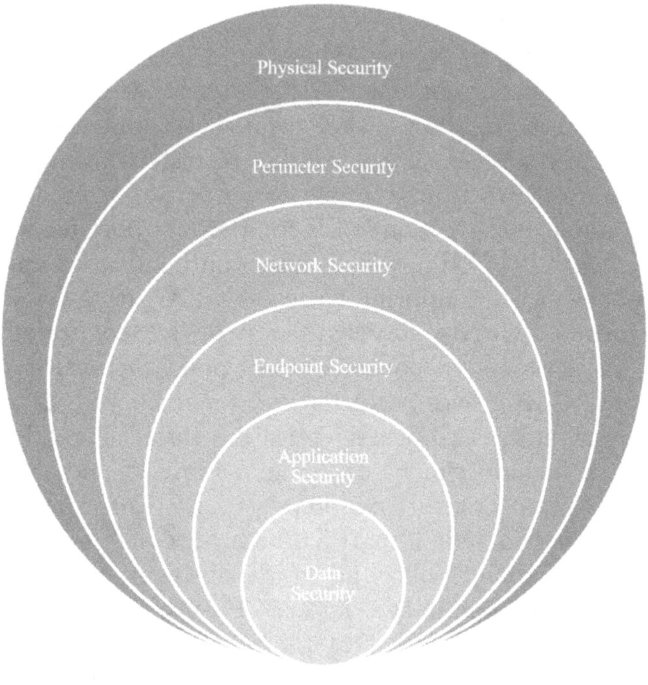

Fig. 3.18. Schematic diagram of onion defense.

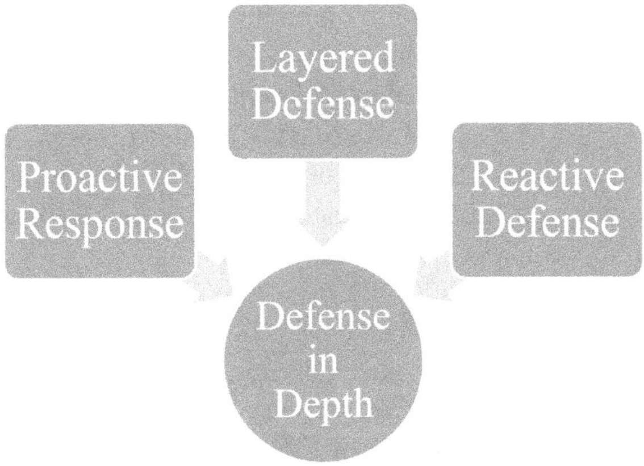

Fig. 3.19. Dynamic defense-in-depth architecture.

perspective, it is also necessary to strengthen the support of the two dimensions of active response (prevention) and passive response (detection and response) in the deployment and operation and maintenance scenarios in IT. The dynamic defense-in-depth architecture is shown in Fig. 3.19.

The active response (prevention) dimension includes IT security governance, security policy and compliance, security architecture and design, network threat intelligence, threat modeling, risk management, security awareness training, penetration testing, etc. The reactive response (detection and response) dimension includes security operations center (SOC), security dashboard, security information and event management (SIEM), continuous behavioral monitoring and asset health awareness, etc.

3.4.2 *Data-centric security architecture*

As mentioned earlier, traditional network security architecture ultimately ensures data security by building layer-by-layer defenses. In a modern complex network environment, people may be unreliable, network boundaries may not be clear, applications may not be trusted, and systems may have loopholes or backdoors, all of which can compromise the ultimate protection goal — data security.

Tyrone Grandison, a security researcher at IBM, introduced the concept of a "Data-Centric Security Model" (DCSM) in his 2007 paper "Elevating the Discussion on Security Management: The Data-Centric Paradigm" (Grandison *et al.*, 2007) and clearly articulated it. The paper suggests that a data-centric security model focuses on deriving an appropriate level of security for data based on a business analysis of that data and based on that level of security, adapting attributes and access control policies to control its use by applications that implement business processes.

In security, the focus of defense has shifted from traditional network-based defense to host-based defense. Going even further, host-based security will naturally extend to the data protected on the host, which is the data-centric security model (DCSM). The data-centric security model is shown in Fig. 3.20. It can be seen that data is at the center of all activities and processes.

After the concept of "Data-centric security model" was proposed, a number of experts and scholars elaborated on it in more detail. Based on the views of several papers, the common processes in the data-centric security model are as follows:

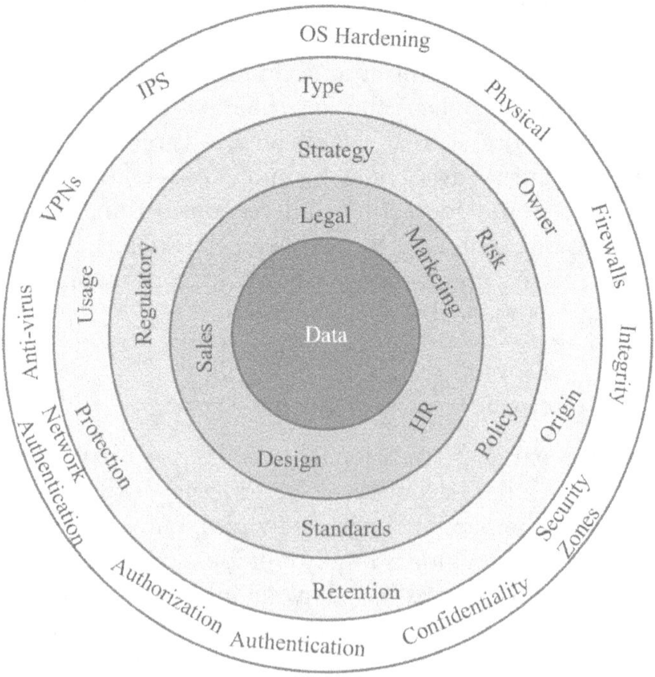

Fig. 3.20. Data-centric security model.

(1) **Discovery:** Know what data contains sensitive information and where it is stored.
(2) **Manage:** The ability to define access policies that determine whether certain data can be accessed or modified by certain users or locations.
(3) **Protect:** Prevent data loss or unauthorized use of data and prevent sensitive data from being sent to unauthorized users or locations.
(4) **Monitor:** Continuously monitor data usage to identify significant deviations from normal behavior that indicate possible malicious intent.

"Discover, Manage, Protect, Monitor" is the business description of the data-centric security model. From a technical perspective, data-centric security relies on the implementation of the following security measures:

(1) Self-describing and protected data.
(2) Consideration of business environment policies and controls.
(3) Data is protected as it moves in and out of applications and storage systems and remains protected as the business environment continues to change.
(4) A unified strategy that can adapt to different data management technologies and implemented layers of protection.

In summary, data-centric security emphasizes the security of the data itself rather than the security of the network, server, or application. Data-centric security models are rapidly evolving as organizations increasingly rely on data to conduct business and as big data becomes the norm. A data-centric security model also allows organizations to overcome the disconnect between IT security techniques and business policy goals by directly associating security services with their implicitly protected data.

In 2014, Gartner Consulting, a globally recognized IT research and advisory consulting firm, stated in "Gartner Says Big Data Needs a Data-Centric Security Focus" (Gartner, 2014) that big data and cloud storage environments are changing the way data is stored, accessed, and processed, Chief Information Security Officers (CISOs) need to develop a data-centric approach to security. At the implementation level, the CISO must work with trusted team members to develop and manage an enterprise data security strategy. The policy defines data retention requirements, stakeholder responsibilities, business needs, risk appetite, data process requirements, and security controls.

"Data-centric security" has gradually become an overall trend in data security and will continue to grow in importance.

3.4.3 *CSA enterprise architecture*

The Cloud Security Alliance (CSA) Security Guidance for Critical Areas of Focus in Cloud Computing proposes a cloud computing security reference model based on the hierarchy and interdependencies of three basic cloud services: software as a service (SaaS), platform as a service (PaaS), and infrastructure as a service (IaaS).

The Trusted Cloud Initiative Reference Architecture (TCI Reference Architecture) is both a methodology and a set of tools, a set of solutions available to security architects, enterprise architects, and risk managers. This suite of solutions addresses a range of needs for risk managers who need to assess the operational status of internal IT security and cloud provider controls. These controls, expressed as security capabilities, are designed to provide a common roadmap for meeting business security requirements. Figure 3.21 shows a model of the TCI reference architecture.

The TCI Reference Architecture must meet business requirements. The requirements referenced in the Trusted Cloud Initiative

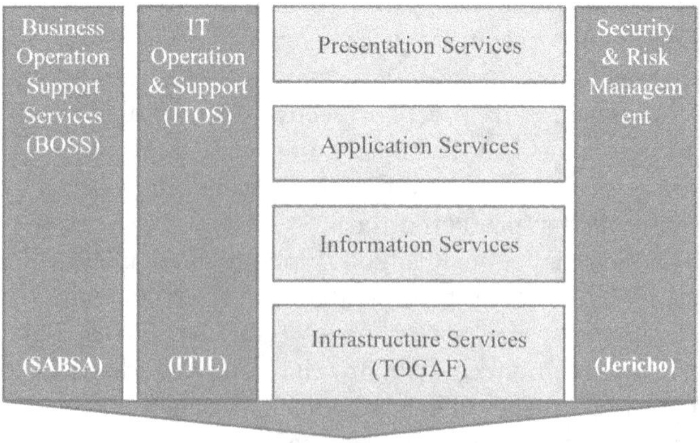

Fig. 3.21. Model of the TCI reference architecture.

come from laws and regulations such as SOX and GLBA, standards frameworks such as ISO 27002 and Payment Card Industry Data Security Standards, and IT audit frameworks such as COBIT. Cloud delivery models such as Software as a Service (SaaS), Platform as a Service (PaaS), and Infrastructure as a Service (IaaS) must first meet compliance requirements.

The Sherwood Application Business Security Architecture (SABSA) defines security capabilities from a business perspective. The Information Technology Infrastructure Library (ITIL) defines the functions required to manage an organization's IT services and therefore the security functions required to manage those services securely. The Jericho Forum defines a set of technical security functions that have evolved from the realities of traditional data center environments to solutions that span multiple data centers and the Internet, some owned by business units and others used purely as outsourced services. The Open Group Architecture Framework (TOGAF) provides an enterprise architecture framework and methodology for planning, designing, and managing information architecture, thereby providing a common framework for integrating the work of security architects with an organization's enterprise architecture. These four parts are detailed in the following.

The first is the Business Operations Support Services (BOSS) domain. This is a business support function that is critical to security programs, such as human resources, compliance, and legal. It is also where the business and its system operations are monitored for signs of abuse or fraud. The design of BOSS is based on best practices and reference frameworks to successfully align the business and transform information security practices across the organization into business enablers. The security architecture not only focuses on technical capabilities but also requires dynamic business collaboration to transform passive practices into active domains, ultimately enabling the business command center to provide relevant information processes about the health of information assets and the business environment. The domain also outlines aspects that must be considered in addition to technical solutions, such as legal policies, compliance and audit activities, human resources, and monitoring capabilities with a focus on fraud prevention.

This is followed by the ITOS (Information Technology Operation & Support) domain. ITOS outlines all of the necessary services an

IT organization must have to support its business needs. When users experience IT problems, the IT organization provides immediate support. This domain provides alignment with industry standards and best practices (CMMI, ISO/IEC 27002, COBIT, and ITIL v3) to enable organizations to support their business needs.

The next domain is security and risk management. It includes encryption and firewalls that protect computer systems and data. It uses simulated hacks and tools to test for vulnerabilities in the system. This is also most people's understanding of network security. The security and risk management domain provides the core components of an organization's information security program to protect assets and identify, assess, and monitor risks associated with operational activities. Its functions include identity and access management, governance, risk and compliance (GRC), policies and standards, threat and vulnerability management, and infrastructure and data protection.

Finally, there is the relatively complex Service Reference Framework. This framework lends itself to being described by the TOGAF framework. In the technology stack of an IT solution, computers and networks are the lowest layer, followed by the data that runs on them, the applications that process the data, and the interactions with users. Based on this layered architecture, four technical solution subdomains are constructed: Presentation Services, Application Services, Information Services, and Infrastructure Services. The CSA Layered Services Reference Framework does not go into all the details of how this architecture works but rather into the details of security issues and required services at each layer in the solution.

The Presentation Services subdomain at the top level (Level 1) of the CSA Reference Architecture refers to services that interact with end users. For example, the page a user sees when accessing online banking or the voice interaction when a user uses an automated hotel reservation system. The security requirements for presentation services are based on different types of users that are very different from the types of services provided by the system. For example, online shopping sites and social media sites may have different security risks. Based on the type of device used by the user, the security requirements will also vary. For example, the data stored in the mobile phone may be at risk of leakage if the mobile phone is lost, and the shared network computer in the library may allow users to access the confidential data of the previous user.

The relationship between this layer and other domains is that the Presentation Services subdomain uses the Security and Risk Management domains to authenticate and authorize end users, to protect data on endpoint devices and in transit to the Application Services domain, and to protect the endpoint devices themselves from tampering, theft, and malware. The ITOS domain provides services to provision and make changes to endpoints, and to manage end user problems and incidents. BOSS domains provide endpoint security monitoring, HR, and compliance policies for IT solutions used by end users.

The second layer of the CSA Reference Architecture is the Application Services subdomain. Application services are the rules and business logic behind the user interface to process data and perform transactions for users. For example, in an online banking transaction, the transaction deducts the payment amount and any processing fees from the user's account and sends the corresponding amount to the payee's account. In addition to application services for IT solutions, the Application Services subdomain also represents the development process that developers go through when creating applications. For example, developers at banks develop application programming interfaces (APIs) for online money transfers. They use a source code analyzer to scan the implementation code of the API, find that the input data may be at risk of injection or out-of-bounds, and make immediate repairs.

The relationship between this layer and other domains is that the Application Services subdomain relies on the Security and Risk Management domain to encrypt messages sent between applications and to authenticate and authorize applications to communicate with each other. The development process of the Application Services subdomain relies on the Threat and Vulnerability Management services of the Security and Risk Management domain to assess the security of the solution being developed. The Application Services subdomain typically receives input from the Presentation Services subdomain and processes data in the Information Services subdomain. The Application Services subdomain also requires servers and network services in the Infrastructure Services subdomain. The ITOS domain is used to manage changes to Application

Services. The BOSS domain provides security monitoring services that allow administrators to monitor application activity for statistically significant anomalies.

The third layer of the CSA Reference Architecture is the Information Services subdomain. This layer provides the transformation of data into meaningful information so that organizations can perform information-based risk assessment, strategy formulation, and day-to-day management. The Information Services subdomain provides data extraction, transformation, cleansing, and loading of common data models for analytical or operational purposes. The Information Services subdomain includes data extraction, transformation, and loading, data mining, balanced scorecard, and other capabilities.

The Information Service subdomain provides the following functions:

(1) **Operational data storage:** This provides a 360-degree view of information assets (application and infrastructure vulnerabilities, incomplete patches, penetration test results, audit results, and controls for each asset).
(2) **Data warehouse:** All historical transactions are used to develop a data warehouse or data mart to measure the results achieved through risk management. The warehouse can also be used to identify behavioral patterns, trends, and systemic gaps across the organization.

For example, when an administrator creates a user account, the user's ID and password are stored in the user directory. When a user logs in to the system, a log recording the logged-in user and time is stored in the security monitoring database.

The relationship of this layer to other domains is that the Information Services subdomain provides contextual support for the Application Services subdomain and the Presentation Services subdomain. The ITOS domain controls the application change and deployment process. The BOSS domain controls the security monitoring of Information Services applications. The BOSS domain also monitors the activities performed by the application for unusual behavior.

The fourth layer of the CSA layered reference framework is the Infrastructure Services subdomain. This layer consists of virtual machines, applications, and databases that support the upper layer. Typically, IaaS services are deployed centrally and run standard machine images with all necessary services pre-configured to support easy integration and reliable connectivity and access.

Infrastructure services are largely invisible to the end user. For example, a user's due diligence may require the physical security of the cloud infrastructure to match the risk profile of their use of cloud services but ignore the operational details of how physical access controls are implemented.

For example, cloud services are not a purely virtual concept; they rely on being physically deployed somewhere. The physical security of these data centers relies on fences, CCTV, security guards, and access controls. Cloud data centers ensure infrastructure availability by connecting to multiple Internet service providers, backup generators, and redundant backup facilities.

The relationship between this layer and other domains is that the Infrastructure Services subdomain provides many core components and functions that support the functions provided by other parts of the reference architecture. The higher-level management provided by the Security and Risk Management domain is largely meaningless without good physical security at the Infrastructure layer. Similarly, the service delivery and support of the ITOS domain depends on the performance and reliability guarantees provided by the Infrastructure Services subdomain.

3.4.4 *NIST cloud computing security architecture*

The NIST Cloud Computing Working Group has developed a cloud computing security reference architecture, the security model of "SP 500–292: NIST Cloud Computing Reference Architecture", and identified the core set of security components needed to build a successful and secure cloud computing ecosystem. The NIST Cloud Computing Security Reference Architecture provides insight into the security interdependencies of the various participants in a cloud service and how to identify and address security requirements to obtain a cloud service that meets the appropriate level of security.

Each component of the NIST Cloud Computing Reference Architecture should implement appropriate security components to enhance security, particularly the security and privacy controls

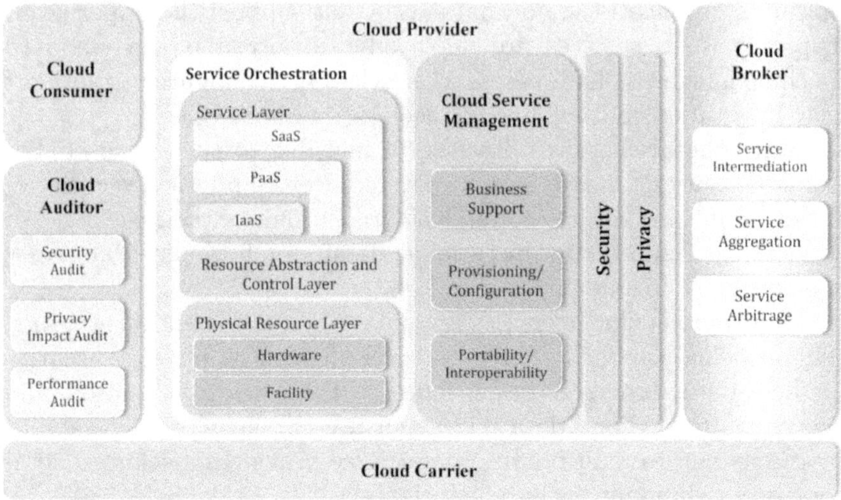

Fig. 3.22. Formal model of NIST cloud computing security reference architecture.

defined in NIST SP 800–53, Security and Privacy Controls for Information Systems and Organizations (Force and Initiative, 2013).

Figure 3.22 shows the formal model of the NIST Cloud Computing Security Reference Architecture, which describes the security architecture components of each cloud computing participant in a layered representation. Components and subcomponents of an architecture may span multiple parties if different parties perform similar or identical functions.

For example, "Secure Provisioning and Configuration" in Fig. 3.22 is an architectural subcomponent that indicates a service that both Providers and Technical Brokers could implement as part of "Secure Cloud Service Management". In addition, the Technical Broker could offer Secure Service Aggregation to its customers. The technical configuration and provisioning aspects associated with aggregation are represented in the diagram by dragging the "Secure Provisioning and Configuration" architectural subcomponent over the graphical representation of the "Secure Service Aggregation" architectural component. It is also important to note that the consumer could implement "Secure Configuration" as part of "Secure Cloud Consumption Management" if needed to complement the services offered by the provider and/or technical broker.

Table 3.3. Components and subcomponents involved in cloud consumers.

Components	Subcomponents	Description
Secure cloud consumption management	Secure configuration	Capabilities, tools, or policies that ensure the secure configuration of cloud resources and compliance with the applicable security standards, specifications
	Secure portability and interoperability	Ensures that data and applications can be moved securely to multiple cloud ecosystems
	Secure business support	Services used to run business operations
	Secure organizational support	Policies, procedures, and processes provided by the organization in support of the overall cloud Secure Consumption Management
Secure cloud ecosystem orchestration		A process that requires various levels of secure participation from all cloud actors
Secure functional layer		The set of Security Components a cloud Consumer implements to secure the cloud Functional Layer depends upon the particular cloud service model used

For Cloud Consumers, the components and subcomponents involved are shown in Table 3.3.

For cloud providers, the components and subcomponents involved are shown in Table 3.4.

For cloud operators, the only issue is secure transport support. Cloud operators generally do not provide services directly to cloud consumers, but the services of cloud providers and cloud service intermediaries require secure transport support from cloud operators.

For cloud auditors, only the secure audit environment is involved. Cloud auditors need a secure audit environment to collect objective evidence from cloud computing participants in a secure and

Table 3.4. Components and subcomponents involved in cloud providers.

Components	Subcomponents	Description
Secure Cloud Service Management	Secure Provisioning and Configuration	All Security Components that ensure the secure configuration and provisioning of cloud resources
	Secure Portability and Interoperability	Ensures that data and applications can be moved securely to multiple cloud ecosystems
	Secure Business Support	Entails the set of business-related services dealing with the Provider's Customers and supporting security processes
Secure Cloud Ecosystem Orchestration	Secure Physical Resource Layer (Hardware & Facility)	Contains the Security Components needed to secure physical computing resources
	Secure Resource Abstraction and Control Layer	Contains the Security Components a cloud Provider would implement to provide and manage secure access to its physical computing resources through software abstraction
	Secure Deployment & Service Layer	The set of Security Components a cloud Provider can implement to secure the Service Layer depends upon the particular type of cloud service offered and the cloud deployment model

trustworthy manner. The cloud auditor's secure audit environment and its components are typically independent of the components of the cloud computing participants.

For Cloud Brokers, the components and subcomponents involved are shown in Table 3.5.

Table 3.5. Components and subcomponents involved in cloud broker.

Components	Subcomponents	Description
Secure Cloud Service Management	Secure Provisioning and Configuration	All Security Components that ensure the secure configuration and provisioning of cloud resources
	Secure Portability and Interoperability	Ensures that data and applications can be moved securely between multiple cloud ecosystems as required by the Cloud Consumer
	Secure Business Support	The set of business-related services that support Consumers' cloud consumption
Secure Cloud Ecosystem Orchestration	Secure Service Layers	The set of Security Components a cloud Broker can implement to secure the Service Layer
Secure Service Aggregation		Addresses the Technical Broker's responsibilities to ensure that all data, including service requests and responses to and from cloud Providers, maintain the required security levels
Secure Service Intermediation	Secure Provisioning and Configuration	Addresses the Broker's responsibilities for ensuring that all capabilities added to existing cloud services offered by Providers maintain the required security levels, and that the Consumer's requirements stated in the contract or SLA are met
Secure Service Arbitrage	Secure Provisioning and Configuration	Similar to the Secure Service Aggregation component, with the exception that the services combined and integrated during arbitrage by the Broker are not fixed

3.4.5 *Zero trust security architecture*

Section 1.1.4 describes the defense-in-depth architecture and "data-centric security". The zero trust security architecture is a complement and enhancement to the above two architectures. Specifically, a zero-trust security architecture shifts network defenses from the broad network perimeter to resources or data.

The concept of zero trust has been around for a long time in the network and information security industry. In 2004, the Jericho Forum proposed the idea of restricting implicit trust based on network location and proposed the concept of "de-perimeterization" (Olovsson and Bodley-Scott, 2005). Later, John Kindervag coined the term "zero trust" while working at Forrester Corporation (Kindervag *et al.*, 2015). This work includes key concepts and a zero trust network architecture model that improves upon the concepts discussed at the Jericho Forum.

Zero Trust Architecture (ZTA) policies are policies that no longer grant implicit trust to a system based on its physical or network location (local area network or Internet). They grant access to data resources only when the resource is needed and authenticate the user or device before establishing a connection.

With the popularity of scenarios such as telecommuting, cross-organizational collaboration, and cloud transformation, the concept of the "corporate intranet" continues to blur. As a result, defenses based on network segmentation and isolation have their limitations, and ZTA focuses on protecting resources from unauthorized access.

Zero Trust Architecture is an end-to-end approach to network and data security, including identity, credentials, access management, operations, endpoints, hosting environments, and interconnection infrastructure. Zero Trust is an architectural approach focused on data protection, with an initial focus on limiting access to resources to those who "need to know". Traditionally, an organization's network has focused on perimeter defense, and authorized users on the internal network have broad access to resources. As a result, unauthorized lateral movement within networks has been one of the biggest challenges organizations have faced.

In essence, a Zero Trust architecture provides granular access decisions for information systems and services. That is, only authorized and approved subjects (users/computers) can access data;

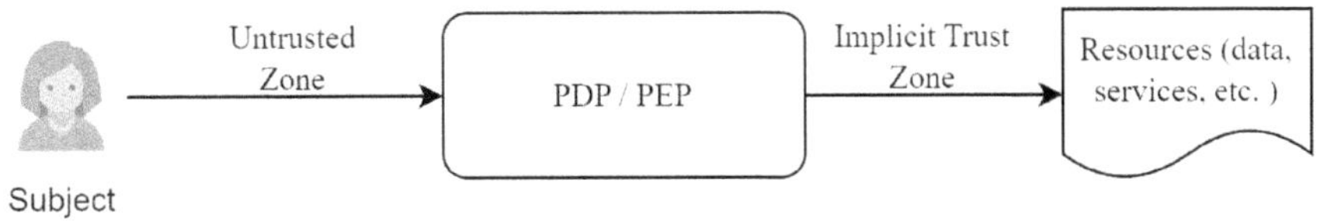

Fig. 3.23. Typical resource access control architecture.

all other subjects, i.e., attackers' access requests, are denied. Furthermore, "data" can be extended to "resources", and the scope of the Zero Trust architecture is extended to include access to printers, computing resources, and IoT actuators.

A typical resource access control architecture is shown in Fig. 3.23. When a subject (user or computer) requests access to a resource, access rights are granted by the Policy Decision Point (PDP) and corresponding Policy Execution Point (PEP) of the access control system. The access control system must complete identity authentication, authorization, and authentication, and the implicit trust zone must be as narrow as possible.

For example, in an airport access control scenario, all passengers present "identity credentials" such as ID cards or passports and a valid airline ticket as "authentication credentials" to enter the waiting area through the airport security checkpoint (PDP/PEP). Passengers are allowed to loiter in the lounge and all passengers share a common trust level. In this model, the implicit trust zone is the lounge. However, in order to board the plane, one must be re-authenticated with a valid ticket for a specific flight. Access to the airport workspace requires an additional means of authentication, such as an access card or airport employee card. Depending on the "resource" to which access is requested, different access control methods are required.

Based on the design concept of access control, access to resources after PDP/PEP is trusted by default. In order for the PDP/PEP to be as granular as possible, the implicit trust zone must be as small as possible. The Zero Trust Architecture provides technologies and capabilities that allow the PDP/PEP to be closer to the resources. The idea is to authenticate and authorize every traffic flow in the network, from participants (or applications) to data.

The design and implementation of a Zero Trust Architecture follows these basic principles:

(1) All data sources and services are considered resources. A network may consist of several different classes of devices. In addition, there are scenarios where personally owned devices access corporate resources. The network may also include IoT devices that send data to aggregators/central storage and systems that send instructions to actuators, etc.
(2) Whether in LAN or WAN, all communications are secure. Location within an organization's LAN does not imply trust. Access requests from systems inside the corporate network infrastructure (e.g., inside the perimeter) must meet the same security requirements as access requests from other networks. In other words, there should be no automatic assumption of trust for devices located within the corporate network infrastructure, and all communications should be conducted in a secure manner (encrypted and authenticated).
(3) Access to individual corporate resources is granted on a per-connection basis. Trust in the requester is assessed before access is granted. Successful identity authentication indicates only that access to the resource is permitted, not that all subsequent visits are permitted.
(4) Access to resources is determined by a policy that measures user identity and the observable state of the requesting system and may include other behavioral attributes. A user identity includes the network account used and any associated attributes assigned to that account by the organization. The state of the requested system includes device characteristics, such as installed software versions, network location, previously observed behavior, installed credentials, and more. Behavioral attributes include automated user profiling, device profiling, and measured deviations from observed usage patterns. Resource access policies should be based on the principle of least privilege and may vary based on resource/data sensitivity.
(5) Organizations should ensure that all systems are in the most secure state possible. A zero-trust architecture relies on an organization's continuous diagnostic and mitigation systems to monitor overall system health and apply patches as needed. Systems that are discovered to be compromised, vulnerable, and/or not owned by the enterprise must be treated differently than systems that are owned or associated with the enterprise and

are considered to be in a secure state. Actions in severe cases may include denying all connections from insecure systems to corporate resources.

(6) User authentication is dynamic and strictly enforced before access is granted. It is a cycle of continuous access, threat scanning and assessment, tuning, and continuous authentication. Organizations implementing a Zero Trust architecture must use multi-factor authentication (MFA) to authenticate users for access to some (or all) enterprise resources. Effective implementation of Zero Trust requires continuous monitoring and authentication of identities during user interactions based on resource access policies (e.g., time-based, new resource requests, and resource changes) in an effort to balance security, availability, usability, and cost efficiency.

The idealized model of the conceptual framework shown in Fig. 3.24 illustrates the basic relationship between the components of a Zero Trust architecture and their interactions. The Policy Enforcement Point (PEP) component is responsible for enabling, monitoring, and terminating connections between principals and enterprise resources. Depending on the deployment scenario, the PEP can be divided into two distinct components: the client side (such as the agent on the user's laptop) and the resource side (such as the gateway component that controls access in front of the resource).

The Policy Decision Point (PDP) is divided into two logical components — Policy Engine and Policy Manager — each of which implements corresponding functions:

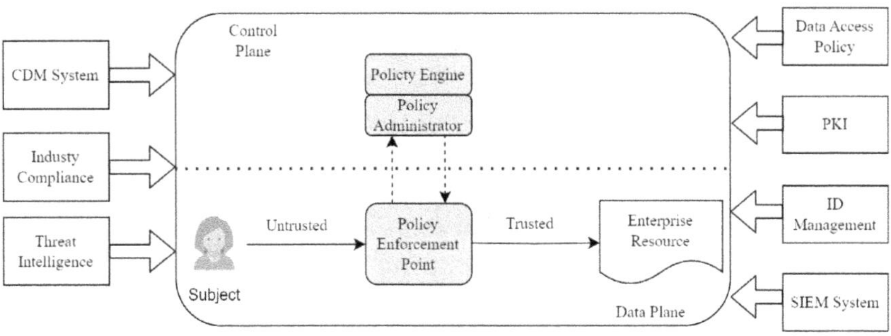

Fig. 3.24. Zero trust architecture.

(1) **Policy Engine (PE):** This component is responsible for ultimately deciding whether the specified access subject should be granted access to resources (objects). The Policy Engine uses corporate security policies and input from external sources such as threat intelligence services as input to a "trust algorithm" to decide whether to allow or deny access to the resource. The Policy Engine component is paired with the Policy Manager component. The Policy Engine makes (and records) the decisions, and the Policy Manager enforces (approves or denies) the decisions.
(2) **Policy Administrator (PA):** This component is responsible for establishing the connection between the client and the resource (it is a logical responsibility, not a physical connection). It generates all authentication tokens or credentials that clients use to access corporate resources. It is tightly coupled with the Policy Engine and relies on the Policy Engine to ultimately decide whether to allow or deny the connection. The Policy Engine and the Policy Manager can be implemented as a single service. The PA communicates with the PEP when a connection is established. This communication takes place through the control plane.

In addition to the core components that implement ZTA policy in an enterprise, there are several data sources that provide input and policy rules that the Policy Engine uses to make access decisions. These data sources include local data sources and external (not controlled or created by the enterprise) data sources:

(1) **Continuous Diagnostics and Mitigation (CDM) system:** This system collects information about the current state of enterprise systems and applies existing updates to configuration and software components. The CDM system provides the policy engine with information about the system making the access request, such as whether the system is running an appropriately patched operating system and applications or whether the system has any known vulnerabilities.
(2) **Industry compliance system:** This system ensures that organizations comply with applicable regulations (such as FISMA, HIPAA, and PCI DSS). This includes all the policy rules that an organization has put in place to ensure compliance.

(3) **Threat intelligence system:** This system provides external information to help the policy engine make access decisions. Threat intelligence is a collection of services that takes data from multiple external sources and provides information about newly discovered attacks or vulnerabilities. Vulnerabilities include DNS deny lists, detected malware, and covert command and control (C&C) systems.
(4) **Data access policy:** A set of data access attributes, rules, and policies created by the organization. This set of rules can be encoded in the policy engine or dynamically generated by the policy engine. These policies are the starting point for granting access to resources and provide the basic access rights for actors and applications in the enterprise. These access rights should be based on user roles and organizational mission requirements.
(5) **Enterprise public key infrastructure (PKI) system:** This system is responsible for generating certificates for subjects, applications, and resources. All processes are documented. The PKI system may interface or integrate with the CA ecosystem.
(6) **Identity management system:** This system is responsible for creating, storing, and managing enterprise user accounts and identity records. The system contains necessary user information (e.g., name, email address, and credentials) and other enterprise characteristics (e.g., roles, access attributes, or assignments). The system typically uses other systems (e.g., PKI) for the artifacts associated with user accounts.
(7) **Security information and event management (SIEM) system:** This system provides enterprise systems that aggregate system logs, network traffic, resource authorizations, and other events that provide feedback on the security posture of enterprise information systems, which in turn can be used to optimize policies and warn of possible active attacks on enterprise systems.

3.5 Security Design Principles and Cases

In 1974, JH Saltzer and MD Schroeder proposed mechanisms to protect computer-stored information from unauthorized use or modification in their paper "The Protection of Information in Computer Systems" (Saltzer and Schroeder, 1975).

The paper pointed out that with the development of computers, there are more and more applications, and their use is more and more extensive, and it involves the storage of a large amount of information. The computer usage model involves the reuse of a large number of resources. In order to ensure the security of information, it is necessary to define and implement the resource authority control model of the computer system and its related software architecture.

In order to systematically eliminate the defects in software design and implementation and prevent users from bypassing access restrictions to obtain unauthorized information, this paper summarizes eight security design principles of information protection mechanism, including the principle of Economy of mechanism, the principle of Fail-safe defaults, the principle of Complete mediation, the principle of Open design, the principle of Separation of privilege, the principle of least privilege, the principle of Least common mechanism, and the principle of Psychological acceptability, as shown in Fig. 3.25.

3.5.1 *Economy of mechanism*

The principle of Economy of Mechanism means that a system should be designed to meet requirements, keeping the design as simple and compact as possible without introducing additional complexity. This principle applies not only to software design but also to security.

One factor in evaluating system security is the complexity of the principle. If security mechanisms are designed and implemented

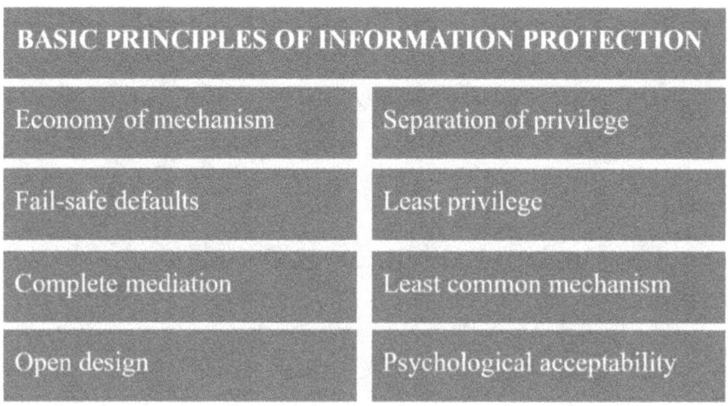

Fig. 3.25. Eight principles of security design.

simply, the likelihood of errors is reduced. At this point, the process of inspection and testing becomes less complex because there are fewer components and use cases to test. If the security mechanism designed or implemented is very complex, the likelihood of security holes in it increases. Since complex mechanisms often make assumptions about the system and environment in which they operate, and if these assumptions are incorrect, the operating behavior of the system can produce unexpected results, leading to security problems.

In addition, subtle problems in complex systems are often difficult to detect, creating potential hidden dangers. For example, if there are different entries and implementation codes for the login function of the system, it will be a challenge to find the processing logic for login failure. To this end, it is necessary to simplify the system with the principle of Economy of mechanism as a goal, such as abstracting and reusing software components, and reducing the amount of code in the system through code sharing. Another strategy for simplifying the system is to remove unneeded code and functionality to avoid introducing additional attack surfaces.

Although the principle of Economy of mechanism is simple and easy to understand, there are still many negative cases caused by non-compliance with this principle:

(1) In February 2017, Cloudflare, a well-known network service provider, exposed the "Cloud Bleeding" vulnerability, which caused user information to be leaked on the Internet for several months. According to the analysis, the vulnerability is caused by an HTML programming error. Specifically, a programmer misspelled the token, resulting in a memory leak. As a result, the personal information of users of the Cloudflare website was widely compromised, including the user privacy information of many companies, including Uber, password management software 1Password, and sports bracelet company Fitbit. The problem arises because the complexity of the system increases the likelihood of a security incident.

(2) In October 2017, the increasingly popular Wi-Fi revealed a logical flaw in the standard protocol, resulting in almost all wireless devices that support WPA/WPA2 (Wi-Fi Protected Access) encryption facing intrusion threats, which attracted widespread

attention. The discovered WPA2 protocol vulnerabilities mainly target Wi-Fi clients (such as mobile phones, tablets, and laptops), causing the aforementioned clients to reinstall keys through key reinstallation attacks in order to complete mutual authentication and crack the WPA2 encrypted wireless network.

(3) In late November 2017, Intel Corporation admitted that it had found multiple serious security vulnerabilities in Intel processors sold over the past two years, including the latest 8th-generation core processor series, and that these security vulnerabilities were primarily focused on the "management engine" function of Intel chips. This chip-level vulnerability would allow hackers to load and run unauthorized programs, compromise systems, or impersonate systems for security audits.

3.5.2 *Fail-safe defaults*

The principle of Fail-safe defaults states that for protected objects, the default access decision should be set to Deny rather than Allow. This principle indicates that all access states of system information should be set to deny access by default and allow the correct users to access the information that should be obtained by setting specific permission conditions.

When designing a system, if we focus on the circumstances under which user access should be denied, then the security of such a design for a system will be inadequate and will often leave many potential logical loopholes. Therefore, a conservative strategy should be adopted — focusing on the conditions under which users can access rather than the conditions under which users cannot access.

The Fail-safe defaults principle requires that if a subject fails to perform any of the tasks it has set, it should undo the changes it has made and restore the system to a stable and consistent state. In this way, the system is secure even if it fails.

This design principle not only enhances security, but its effectiveness is also valuable in practical applications. If the allowed conditions of the design are unreasonable when normal users are denied access, the problem is quickly and accurately fed back so that improvements can be made. If the conditions for denying access are designed to be unreasonable, it is often difficult to detect logical problems or security vulnerabilities in a timely manner because they are not triggered by normal users.

There are many examples of implementations of the Fail-safe defaults principle in consumer electronics or networking products:

(1) In today's electronic products (such as mobile phones, portable computers, and tablets), the design of login passwords is a manifestation of this principle. Users are allowed normal access only after entering the correct password, and by default, user access is denied. An example of the Fail-safe defaults principle (password login) is shown in Fig. 3.26.
(2) The Allow/Deny list is another typical application of the failed default security principle. Allow list and deny list are a set of related concepts with a very wide range of applications. Allow list and deny list rules are applied to many IT systems, such

Fig. 3.26. Fail-safe defaults principle: device password login.

as firewalls, antivirus software, email systems, application software, and so on. Specifically, when a deny list is enabled, requests from blocked users (or IP addresses, IP packets, emails, viruses, etc.) are not allowed through. When an allow list is set up, users (or IP addresses, IP packets, emails, etc.) on the allow list are not affected by access control rules or specific restriction rules. Allow/Deny lists, depending on how they are implemented, do not fully comply with the Fail Safe Defaults principle. If a subject outside the deny list is "passed by default", the requirement of "denied by default" is violated. The allow list mechanism is more reasonable, but it cannot grant additional privileges to the subjects in the allow list. When used properly, the allow list mechanism can defend against attacks by unknown malware. This defense can be achieved by running only processes or software on the allow list on the endpoint.

3.5.3 *Complete mediation*

The complete mediation principle is shown in Fig. 3.27. The principle of complete mediation is that permissions must be checked each time a protected object is accessed. This principle is considered to be the basis of the system's security protection, which forces the system to perform security checks every time information is accessed. In addition to regular access operations, data access permissions must be confirmed again, especially after system initialization, shutdown, and restart. This principle requires that when a caching mechanism is used to speed up access, the system must detect changes in visitor sources or visitor permissions in a timely manner and make appropriate adjustments.

When a request source attempts to read an object, the operating system should first analyze the operation and determine whether the request source can read the object. If the request source has the correct permissions, the system will allocate resources for the read. If the request source tries to read the object again, the system should check again to see if it can still read the object. However, most systems do not perform a second check. They tend to cache the results of the first check and allow the second access based on the results of the previous cache, which often leaves some potential security holes.

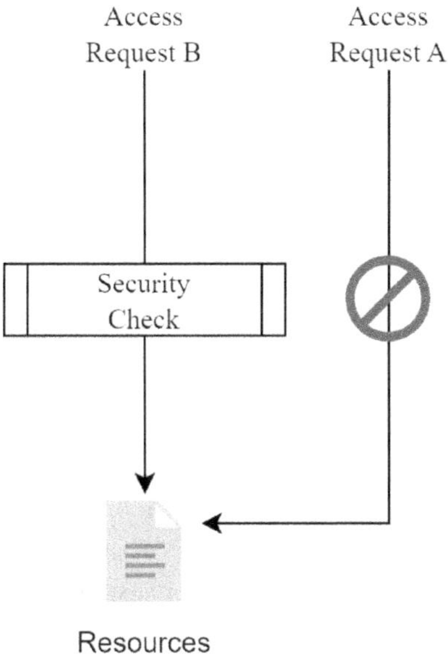

Fig. 3.27. Complete mediation principle.

The complete mediation principle has some failure cases:

(1) When a UNIX process tries to read a file, the operating system first determines whether the process is allowed to read the file. If the permission check is correct, the process receives a file descriptor indicating that access is allowed. Whenever the process wants to read the file, it passes the file descriptor to the kernel, which then grants it access. If the owner of the file tries to stop the process from reading the target file after it has been given the "allow access" file descriptor, the kernel only checks the file descriptor, so the claim is not satisfied and the process can still read the target file. This design violated the principle of complete mediation by not performing a security check on the second access and using only the cached value, making it impossible to restrict subsequent accesses by the process.

(2) In January 2018, security researcher Dhiraj Mishra identified a high-risk vulnerability in the Samsung mobile browser (CVE-2017–17692). Samsung acknowledged the vulnerability, which allows attackers to steal critical information such as passwords or cookie sessions after a victim visits a malicious website. The role of the cookie session is the same as the Unix file descriptor

that indicates "allow access". Both are an access credentials sent to the user to simplify the authentication process. However, if the access credential is stolen due to a breach without the complete mediation principle, the security of information cannot be guaranteed.

3.5.4 *Open design*

The philosophy of open design is that the security of a mechanism should not depend on the secrecy of its design or implementation. The security mechanism itself should not be a secret, and the content of its specific design should be open. As Shannon said, the design itself should not be secret (Shannon, 1949). The specific content of the open design principle is as follows:

(1) The security of the system should not depend on the ignorance of potential attackers but on the possession of specific, protected keys, or passwords.
(2) The security design of the system should allow auditors to perform inspections without worrying that the inspection itself might compromise the security mechanism.
(3) Public disclosure of the security mechanism may convince skeptical users.
(4) Most systems are intended to be widely used, and for these systems, it is impractical to maintain the secrecy of their design.

Many cases in cryptography exemplify the open design principle. The positive case concerns a lattice-based digital signature scheme and its cryptanalysis. In 1997, Goldreich, Goldwasser, and Halevi, three top cryptography researchers, published a paper "Public-key Cryptosystems from Lattice Reduction Problems" (Goldreich *et al.*, 1997) at the cryptography conference CRYPTO, proposing a lattice-based digital signature scheme known as the GGH scheme (named after the initials of the three authors). In 2001, cryptography researchers Hoffstein, Howgrave-Graham, Pipher, Silverman, and Whyte proposed an instance of a GGH scheme: the NTRUSign signature scheme (Hoffstein *et al.*, 2003). In 2002, the IEEE P1363.1 version 8 standard officially adopted the NTRUSign signature scheme as a standard for digital signature algorithms. Oded Regev presented his paper "Learning a Parallelepiped: Cryptanalysis of GGH and

NTRU Signatures" (Nguyen and Regev, 2006) at EUROCRYPT, the premier conference in cryptography, in 2006. In this paper, Oded Regev proposes an algorithm to attack the GGH scheme, which can crack the GGH scheme and the NTRU signature scheme in a few minutes under the setting of high security constants when 400 message/signature pairs are known, and the result of the cracking is to directly obtain the private key of the signature. Cryptography researchers make their cryptographic algorithms public based on the open design principle, which states that the security mechanism itself should not be kept secret. It is this concept that allows the security problems of the GGH scheme to be discovered earlier, giving academia and industry sufficient response time to avoid more serious losses.

A well-known negative example of open design principles is the Content Scrambling System (CSS) used to protect DVD (Digital Video Disc) playback. CSS is an encryption scheme that protects the contents of a DVD from unauthorized copying. The CSS design includes DVD player keys and three DVD keys. The DVD contains an authentication key, a disk key, and a title key. The title key is encrypted with a disk key. A block on a DVD contains multiple copies of the disk key, each copy encrypted by a different player key. When a DVD is inserted into a DVD player, the authentication key is read first. Then the DVD player key is used to decrypt the disk key. If a decryption key with the correct hash is found, it is used to decrypt the title key, and the title key is used to decrypt the contents of the DVD. The authentication key and disk key are stored on the DVD, not in the content file, so no matter how a malicious user copies or distributes the file, the DVD is still required to play its contents. This design can effectively prevent piracy. However, it can be seen from the algorithm description of the CSS system that the system relies on the confidentiality of the algorithm and the confidentiality of the player key. In 1999, a Norwegian research group obtained a software-based DVD player program with an unencrypted DVD player key. The team also derived an algorithm from the software that is fully compatible with the CSS algorithm, which allowed them to decrypt files on any DVD. It can be seen that the security design of the CSS system has limitations based on the confidentiality of the design scheme and the hiding of key information on the device. Neither of these can be fully guaranteed for large-scale distribution systems.

3.5.5 *Separation of privilege*

The principle of separation of privileges is a restrictive principle for the privileges of subjects. When privileges are separated, important and sensitive privileges cannot be granted to a single user based on a single condition. It is necessary to consider that a single user cannot independently complete certain specific tasks or independently complete the whole process of application and approval.

In real life, there are many intuitive expressions of the principle of separation of privileges, such as the cashier and the accountant of a company cannot be the same person, then payments can only be completed with the approval of two people at the same time. Another example is that the bank safe deposit box for storing high-value items must be opened with two keys at the same time.

The advantages of the separation of privileges principle are mainly summarized in the following three points:

(1) Access credentials (such as keys) can be physically or logically separated and held by different programs, organizations, or individuals.
(2) A single accident, deception, or breach of trust is not sufficient to compromise protected information.
(3) In a computer system, when access to resources is subject to multiple conditions, key separation can be used to achieve separation of privileges.

Increasingly, organizations are focusing on separation of privileges and dual controls to pass audits and demonstrate that internal controls are in place to protect against breaches by malicious or unauthorized employees. Many different regulations require companies to implement separation of privileges.

As an example, on Berkeley-based versions of the UNIX operating system, a user may not make his or her account the root account unless two conditions are met. The first condition is that the user knows the password for the root account; the second condition is that the user is in the privileged users group with GID 0.

Another more common and intuitive case is that in the online banking transfer operation, small transfers can be completed using the withdrawal password. However, large transfers must be completed using a withdrawal password plus additional identity authentication methods, such as facial recognition.

3.5.6 *Least privilege*

The principle of least privilege states that a subject (user, account, process, system, IoT device, etc.) should be granted only the privileges it needs to accomplish its task. Privileges here refer to permissions or authorizations to bypass certain security restrictions.

If a subject does not need a privilege, then the subject should not have it. Privileges should be assigned based on a subject's function rather than its identity. The figurative metaphor is "the CEO does not need the privileges of an asset manager". If a specific task requires an increase in the subject's privileges, then these additional privileges should be relinquished or revoked immediately after the task is completed. The principle of least privilege can be compared to the "least-know" principle in the military field.

The principle of least privilege relies on fine-grained privilege definitions and access controls. For example, if a subject needs to query an object but does not need to modify the information already contained in the object, then the subject should be given query privileges rather than modification privileges. In practice, most systems do not have the granularity of permissions required to apply this principle precisely, and in such systems, the consequences of security problems tend to be more severe.

The principle of least privilege and the principle of separation of privileges are related but also distinct. The principle of separation of privileges is concerned with the fact that the implementation of a privilege requires the participation of multiple subjects.

A well-designed system that adheres to the principle of least privilege can help in several ways:

(1) **Reduce the attack surface:** Limiting the privileges of people, processes, and applications means reducing the points of entry and paths of attack.
(2) **Reduce malware infection and spread:** If malware is installed or executed with privilege escalation, it can be blocked due to lack of appropriate privileges.
(3) **Improve system performance:** Least privilege increases control over process behavior and reduces compatibility issues between applications or systems, helping to reduce the risk of downtime.

(4) **Facilitate compliance:** By limiting the potential activities that can be performed, least privilege helps create an audit-friendly environment. Numerous regulations, including HIPAA, PCI DSS, FISMA, and SOX, require organizations to apply the principle of least privilege to ensure effective data management and system security.

An example of least privilege is role-based access control. A role-based security mechanism should identify and describe the different roles of users or processes and assign each role the minimum privileges necessary to perform its tasks. The access control mechanism allows each role to have only its authorized privileges. The minimum set of privileges assigned to each role describes the resources that each role can access. This prevents unauthorized roles from accessing protected resources. For example, users accessing a database only have permission to retrieve data, not to modify it.

In a POSIX (Portable Operating System Interface) system, such as Unix or Linux, the root user has almost all privileges, its operations are allowed by default, and the authorization mechanism is unclear. This is a typical failure of the principle of least privilege. Modern POSIX system implementations use capabilities, sudo, and other mechanisms to mitigate the risk of potential applications running with root privileges. For example, a web server must bind to a network port number less than 1024, such as port 80 of the HTTP protocol. Under the traditional implementation mechanism, the Web server must be started with root user privileges, which introduces a large attack surface. Instead, by using the least privilege mechanism, a normal user can be used to start the web service process.

3.5.7 *Least common mechanism*

In simple terms, the principle of Least Common Mechanism states that the mechanisms used to access resources should not be shared. It can be understood from three aspects:

(1) Minimize the number of mechanisms shared by multiple users.
(2) Any sharing mechanism (especially those involving shared variables) represents a potential information transfer path between users and must be designed very carefully to ensure that these sharing mechanisms do not compromise security.

(3) Any mechanism that provides services to all users must be authenticated.

The principle of least common mechanism is a basic criterion for implementing security design and protecting data security. It is widely used in various operating systems and software platforms, among which Docker is an excellent case. The Docker system architecture is shown in Fig. 3.28. Docker is an open source platform for building, deploying, and managing containerized applications. Developers can package applications into a portable, standardized unit called a container. These containers contain all the functionality needed to run software, including libraries, system tools, code, and runtime environments. With Docker, you can quickly deploy and scale applications across environments such as clouds, servers, and more. The implementation mechanism of Docker is to achieve the purpose of the least common mechanism by isolating the access of different applications to system resources, and each application can only access limited system resources.

3.5.8 *Psychological acceptability*

The principle of Psychological acceptability is one of the basic principles of human–computer interaction design, and it is not

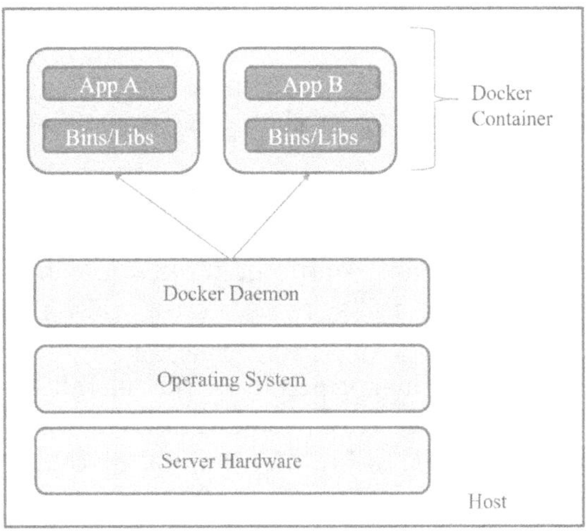

Fig. 3.28. Docker system architecture.

particularly intuitive as a principle of security design. In fact, the principle of psychological acceptability has a very important place in the principles of security design. Since security design measures are inseparable from human participation and interaction, even for automated security processing, the rules are formulated by humans.

The Principle of Psychological Acceptability recognizes the importance of the "human factor" in system and information security. The principle states that security mechanisms should not make access to resources more difficult than it would be in the absence of security mechanisms. When a balance must be struck between usability and security, security must be provided while minimizing the impact on usability.

The principle of psychological acceptability dictates that the configuration and execution of an application should be as simple and intuitive as possible. If the configuration of security-related software is too complex, system administrators may inadvertently disable security-related settings. Similarly, security-related application programs must be easy to use and provide clear, useful, and understandable information. If the login password is incorrect, it should state why the login was denied rather than giving an internal error code. If the configuration parameters of the security-related software are incorrect, the error message should provide instructions on how to configure the correct parameters.

In short, the psychological acceptability principle mainly includes the following contents:

(1) When designing security mechanisms, human factors must be taken into account.
(2) If the user is asked to make judgments based on content he does not understand, he is likely to make mistakes.
(3) If the results of human–computer interaction match the user's expectations, the possibility of user error will be reduced.

A negative case of this principle is the installation certificate warning dialog box of the early version of Windows, as shown in Fig. 3.29. This dialog box is full of terminology that is difficult for users to understand, leading to user confusion.

As can be seen from Fig. 3.29, the Windows operating system tries to allow users to make judgments based on content they do

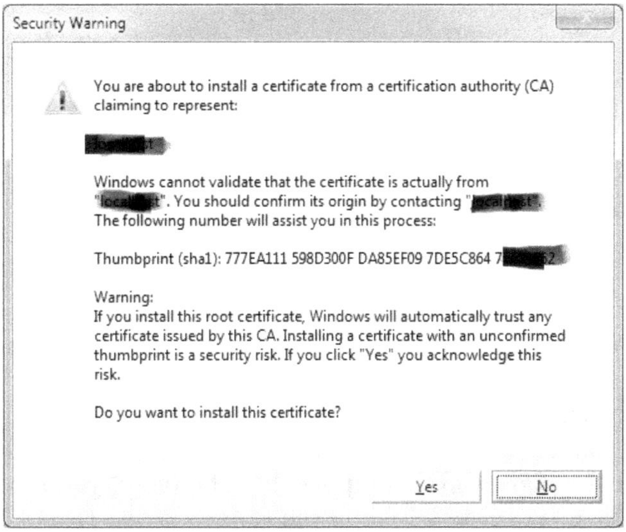

Fig. 3.29. Windows installation certificate warning dialog box (1).

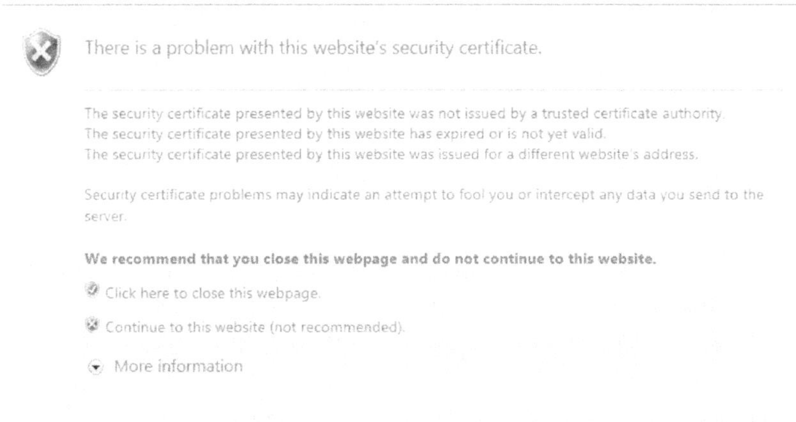

Fig. 3.30. Windows installation certificate warning dialog box (2).

not understand and are unfamiliar with, and its warnings and risk prompts also put more pressure on users.

In fact, the new version of the web browser included in the Windows OS provides more user-friendly and friendly prompts for similar certificate errors, as shown in Fig. 3.30.

By using scenario-based rather than technical descriptions, and by providing suggestions to the user, product vendors can help the user make judgments and adhere to the principle of psychological acceptability.

3.6 Trust Basis for Data Security

Modern computing devices contain a variety of hardware, firmware, and software components that can be abstractly divided into many layers. Currently, many security and protection mechanisms exist in software, and there is unconditional trust that the underlying architecture is secure and reliable. Any vulnerability of these components can affect the credibility of the security mechanisms that depend on them and the higher-level components. Therefore, you should build the security mechanism based on the Root of Trust (RoT) and build trust layer by layer to achieve stronger security guarantees.

The root of trust is a highly reliable piece of hardware, firmware, or software that performs certain critical security functions. The root of trust is the basis for trust and dependency of other components to derive security functions. Security must be provided by design. Most roots of trust are implemented at the hardware level and cannot be tampered with at the software level, so they can provide some defense against malware. The root of trust provides a solid foundation for establishing system security and trust relationships.

Good cryptography and key management engineering practices require hardware as the foundation of trust and building a security architecture layer by layer to ultimately ensure data security.

There are two common types of hardware security devices, each suitable for different application scenarios:

(1) **Hardware Security Module (HSM):** It is an independent hardware device, usually deployed in the network, providing cryptography algorithms and key management services for cloud services, network services and database storage, PKI (Public Key Infrastructure) system, etc.
(2) **Trusted Platform Module (TPM):** It is a stand-alone chip that can be embedded in personal computers or servers, network devices, and mobile terminals and provides cryptography and key management services.

In cloud computing scenarios, most cloud service providers support key management and other security services related to encryption and decryption based on Hardware Security Modules (HSMs).

In all types of modern smart devices (such as laptops, desktop computers, smartphones, and Internet of Things devices), hardware security chips such as Trusted Platform Module (TPM) are often embedded or CPU-based logical security partitions such as Trusted Execution Environment (TEE) are used as the basis for hardware trust.

3.6.1 *The principle and application scenarios of HSM*

The Hardware Security Module (HSM) is a self-contained hardware encryption device. It is typically deployed in data centers or computer rooms. There are also some HSMs in the form of plug-in cards that connect to servers through SCSI, USB ports, etc.

HSMs use physical security measures, logical security controls, and strong encryption to protect sensitive data in transit, in use, and at rest.

As a dedicated encryption device, the HSM is used to manage and protect keys throughout their lifecycle, providing encryption and decryption capabilities to enable services, such as identity verification, digital signatures, and protection of sensitive data. To support key generation and protection, some HSMs also provide functions, such as true random number generators and trusted time sources.

An HSM provides a trusted environment for performing various cryptographic operations, including key exchange, cryptographic key management, and encryption. "Trusted" in this context means free of malware and viruses, and protected from attack or unauthorized access.

As a device with extremely high security requirements, the HSM was designed with many related factors in mind:

(1) Built on certified, well-tested, dedicated hardware.
(2) Runs a security-focused operating system.
(3) The entire design can actively protect and hide password information.
(4) Restrict access to the network through an arbitration interface that is strictly controlled by internal rules.

Without the use of hardware security modules, normal operations and encryption operations are performed in the same computing environment, and attackers can access common business logic data and sensitive information, such as keys and certificates. In extreme cases, attackers can install arbitrary certificates, gain unauthorized access, modify code, decrypt sensitive data, and more.

Encryption is a core function of HSMs and is the process of making sensitive data inaccessible without authorization. Secure decryption and message authentication are also part of the HSM functionality. Random numbers are used to generate encryption keys and are critical to the encryption process. Decrypting sensitive information is easy as long as the operator has the encryption key. Therefore, it is critical to store encryption keys in a secure environment.

Hardware Security Modules generate and store encryption keys that are used between different devices. It uses a hardware True Random Number Generator (TRNG) to create entropy and generate high-quality random keys. Larger organizations can run multiple HSMs simultaneously, not just one. Whether deploying one or more, any centralized key management system based on external regulatory requirements and internal security policies can improve security and compliance.

As a critical function for securing IT infrastructure and applications, HSMs typically require certification to internationally recognized standards, such as FIPS 140 or Common Criteria (CC). Standard certification can also ensure the rationality of the design and implementation of user products and encryption algorithms. The highest level of certification that FIPS 140 security can achieve is Security Level 4. Users in the financial industry typically authenticate to the HSM as defined in the Payment Card Industry Security Standards Council's HSM Security for Financial Payment Applications.

HSMs can be tamper-resistant. For example, HSMs can display visible signs of logging and alerting, or they can become inoperable if records are tampered with. Some HSMs can delete keys if tampering is detected. An HSM is typically protected by a tamper-resistant or tamper-responsive package and contains one or more cryptographic processor chips or combinations of chips to prevent bus probing and bus tampering.

HSMs can often be clustered for high availability because they are often part of mission-critical infrastructure, such as online banking applications or public key infrastructure. Some HSMs enable business continuity and meet the high availability requirements of data center environments. For example, they may have field-replaceable components or dual power supplies to ensure availability in the event of a disaster.

In terms of application scenarios, any application that uses keys can use a hardware security module. The value of using an HSM is maximized only when the compromise of the key would have serious negative consequences.

The main application scenarios of HSMs are as follows:

(1) For Certificate Authorities (CAs), HSMs provide encryption key generation and secure key storage, especially for the master key or the most sensitive root key.
(2) In the PKI environment, Registration Authority (RA) and Certificate Authority (CA) can use HSM to generate, manage, and store asymmetric key pairs.
(3) Verify the integrity of sensitive data stored in relatively less secure locations (e.g., databases) and support encrypted storage of sensitive data.
(4) Manage the keys of storage devices such as tapes or disks and the transparent data encryption keys of databases.
(5) Provide physical and logical protection for sensitive information (including encryption keys) to prevent unauthorized use, disclosure, and potential attacks.
(6) Some HSM systems provide hardware encryption acceleration for SSL connections, significantly reducing CPU load. Most HSMs today support Elliptic Curve Cryptography (ECC). Although ECC has a shorter key length, it provides stronger encryption.
(7) For performance-critical applications that need to use HTTPS (based on the SSL/TLS protocol), SSL/TLS operations (encryption and decryption) can be accelerated by an HSM (using the HSM instead of the host CPU).
(8) A bank hardware security module or card payment system hardware security module is a dedicated HSM used in the payment card industry. These HSMs support typical hardware

security module functions as well as specialized functions for compliance processing and industry standard transactions.
(9) HSM can be used to encrypt digital currency wallets.

3.6.2 *The principle and application scenarios of TPM*

Trusted Platform Module (TPM) can refer to the international TPM standard (also known as ISO/IEC 11889) or a security chip that conforms to the TPM standard, and its meaning must be determined in context. ISO/IEC 11889 defines a secure cryptographic processor (a dedicated microcontroller) designed to protect hardware and software with embedded cryptographic keys.

The TPM standard is created and maintained by the Trusted Computing Group (TCG) and was standardized as ISO/IEC 11889 by the International Organization for Standardization (ISO) and the International Electrotechnical Commission (IEC) in 2009. The latest revision of the TPM, version 1.2, was released on March 3, 2011. TPM version 2.0, which includes its main components, was released on September 29, 2016, and revised on January 8, 2018.

TPM security chip refers to a security chip that conforms to the TPM standard and is commonly embedded in mainstream portable computers, tablet PCs, and other various devices.

The Trusted Platform Module security chip provides the following functions:

(1) **Dedicated hardware encryption accelerator and true random number generator (TRNG):** These two components are necessary for the performance power consumption of the encryption function and the entropy level required to generate the encryption function.
(2) Secure key generation facilities.
(3) **Remote attestation:** Create a nearly tamper-proof hash digest of hardware and software configurations, allowing third parties to verify that the software has not been tampered with.
(4) **Binding:** Use the TPM's binding key (the unique RSA key derived from the storage key) to encrypt data.
(5) **Sealing:** Similar to binding but can specify the TPM state of the data to be decrypted (unsealed).
(6) Other Trusted Computing functions.

Computer programs can use TPMs to authenticate hardware devices because a unique and confidential Endorsement Key (EK) is burned into each TPM chip at the time of manufacture. Embedding security at the hardware level provides greater system protection than a software-only key solution.

TPM application scenarios include platform integrity protection, disk encryption, and password protection.

For the platform integrity protection scenario, "integrity" means consistent with expectations", and "platform" can be any computing device, regardless of its operating system. This is to ensure that the boot process begins with a trusted combination of hardware and software and continues until the operating system is fully booted and applications are running.

The responsibility for providing the above integrity through the TPM rests with the firmware and operating system. For example, the Unified Extensible Firmware Interface (UEFI) can use a TPM to create a root of trust: the TPM contains multiple Platform Configuration Registers (PCRs) for secure storage and reporting of security-related metrics. These metrics can be used to detect changes to previous configurations. For example, the Linux Unified Key Setup (LUKS) and Windows BitLocker disk encryption tools both support the TPM as a root of trust.

Another example of using a TPM for platform integrity is Trusted Execution Technology (TXT), which creates a chain of trust that remotely proves that a computer is using specific hardware and software.

For disk encryption scenarios, full disk encryption applications such as dm-crypt and BitLocker can use the TPM to protect keys for encrypted storage devices and provide integrity verification for trusted boot paths, including firmware and boot sectors.

For password protection scenarios, operating systems typically require authentication (password or otherwise) to protect keys, data, or the system. Password validation is vulnerable to dictionary attacks if the authentication mechanism is implemented only in software. The TPM is implemented in a dedicated hardware module with a built-in dictionary attack prevention mechanism that effectively prevents guessing or automated dictionary attacks while still allowing a sufficient and reasonable number of user attempts.

3.6.3 *Principles and application scenarios of TEE*

With the rapid development of the Internet and the rapid popularization of mobile devices, the use of mobile devices has gradually expanded online and offline, affecting many aspects of work and life. The openness of the mobile ecosystem has led to a rapidly increasing risk of data theft and identity theft. As we all know, the number of mobile applications in the mobile ecosystem is growing exponentially, but the quality of mobile applications varies, and even some malicious applications can send sensitive data to untrusted third parties without authorization. The rich connectivity of today's mobile devices, such as Wi-Fi, Bluetooth, and NFC, not only provides users with rich experiences and convenient functions but also opens the door to network attacks and user data leakage.

In addition, mobile devices play an increasingly important role in people's production and lives. They can store not only general personal data, such as contacts and text messages, but also personal financial and payment data, social media data, etc. Mobile devices are also often used as online authentication tools or as other authentication factors to access highly sensitive domains and resources. Malware can infect mobile devices through unintentional or malicious actions by users. In addition, operations such as jailbreaking, flashing, and loading untrusted applications can also damage the device or compromise data security. Users often turn off security updates for mobile devices, making them easy targets.

The variety of mobile operating systems (OS) and the different mechanisms of OS platform security features make it difficult for application vendors to manage application security. Traditional software-based protection technologies are difficult to defend against today's vulnerabilities, Trojans, viruses, and malware. Chip-based solutions, such as Secure Element (SE), can also provide excellent protection for sensitive code and data, but in addition to added cost, security has scenario limitations. For example, digital rights management (DRM) and enterprise applications for consumer devices may require larger, faster secure storage or more secure access to peripherals. Trusted Execution Environments (TEEs) can be an alternative for these scenarios.

The prototype and concept of a TEE was proposed in 2004 by companies such as Texas Instruments and was then called the "universal trust environment". In 2006, ARM developed Trust Zone technology.

Subsequently, TEE gradually moved toward industry standardization. In 2006, the Open Mobile Platform Organization (OMTP) published a set of standards for trusted execution environments, which were revised in 2008.

Since 2010, the international standards organization Global Platform has been responsible for driving TEE standardization on behalf of the industry. Global Platform has published many TEE-related specifications and offers TEE functional and security certification programs to provide certification to application and software developers and hardware manufacturers to ensure that TEE products comply with Global Platform specifications.

A previous business case for TEE was published in 2011. Netflix manages and protects premium high-definition content on smartphones and tablets through secure digital rights management. Content owners, such as movie studios, need hardware security support to enable service providers to display HD content on Android mobile devices. Only a TEE can meet all the requirements of this business case, in particular the following:

(1) Extremely high processing power (download, decrypt, and display streaming content in real time).
(2) Content decryption and processing independent of hardware.
(3) Display of hardware-independent content (via secure access to peripheral outputs).
(4) Secure storage of hardware-protected sensitive data, such as decryption keys and license files.
(5) Data and application isolation (data cannot be copied or intercepted by other applications).
(6) Standardized API (application portability).

As a result, TEE has been widely used in various scenarios, such as finance, payment, multimedia, BYOD devices, and even Internet of Things (IoT) applications.

The initial goal of Global Platform TEE is to create a trusted environment on mobile devices such as mobile phones to support applications such as digital rights management and secure payment. Subsequently, other consumer electronic devices began to require similar capabilities, such as smart TVs, smart speakers, and even drones.

A fundamental principle of TEE security is hardware isolation between the TEE and the mobile device operating environment. The relationship between the rich OS application environment (also known as the Rich Execution Environment, or REE) and the TEE is shown in the three-tier architecture of the TEE, as illustrated in Fig. 3.31. The Global Platform specification requires that the TEE be separated from the REE through a hardware-based system but does not require that the TEE utilize a hardware-independent chip, so hardware-based security does not affect the cost and benefits of hardware.

TEE can run multiple applications; these applications are called Trusted Applications (TA). Applications in the REE send commands and requests to the TA through the TEE client API, which is connected to the TEE communication agent through the hardware system (see the horizontal arrow in Fig. 3.31). The TEE communication agent forwards these commands and requests to the TA through the TEE internal API.

The trusted operating system in the TEE can connect to hardware secure resources such as touch screen, keyboard, camera, secure storage, and other peripherals through trusted drivers (see vertical

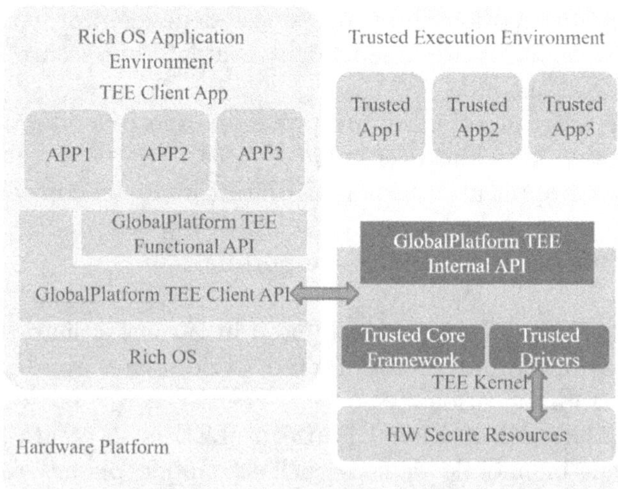

Fig. 3.31. TEE three-layer architecture.

arrows in Fig. 3.31). Peripheral services will be available to TA. Two types of peripheral devices can be used:

(1) Peripherals (such as secure storage and biometric sensors) that only the TEE can access.
(2) Peripherals (such as screen and keyboard) that are shared with the rich execution environment.

The connection between the shared peripheral and the TA does not need to go through the rich execution environment, and all communications to and from the shared peripheral are secure and confidential to the TEE.

The Global Platform specification requires hardware platform protection to separate the TEE implementation from the REE. TEE providers can use the same processor and memory on REE and TEE systems (such as Trust Zone) and run TEE instances on the device's main hardware platform or main processor. Of course, TEE providers can also use separate processors and separate resources.

When application developers need to store, transmit, and process sensitive or personally identifiable information, they must fully consider a risk-based assessment to determine security requirements. For example, a weather forecast application may not involve personal data, but a mobile payment application may involve sensitive information such as payment credentials and authentication information. One way to meet security requirements is to use the data encryption, access control, and file isolation capabilities provided by the operating system. But in many scenarios, this security guarantee is not enough. This is where the TEE becomes a hardware-based secure trusted environment solution. A Secure Element (SE) is another security option that securely hosts the application and its associated sensitive data (such as keys). An example of a secure element is the chip in a payment card, where EMV applications and data are stored. Using secure elements can make some security assessments and certifications easier. However, the disadvantage is that the memory space on the secure element chip is limited and the computing speed is slow, which is not suitable for applications that require graphical display and fast user interaction.

There are also several shortcomings in TEE application scenarios. Of course, like the chip on a payment card, the TEE is embedded in the device, which means that the TEE must be integrated into the device before it leaves the factory. In addition, unlike payment cards, which are typically embedded and loaded with applications and data before being shipped to customers, embedded TEE devices are typically consumer-owned devices on which applications and data may need to be installed remotely. As a result, the process of loading data has complex security requirements. For example, it must be possible to remotely prove that the device is authentic and enable secure data transmission and remote management of the device. To achieve this goal, the upstream and downstream TEE industry ecosystem must work together. For example, TEE vendors must work directly with chipset and device manufacturers.

The main chips used in most Android smartphones and tablets support the TEE environment. However, the integration of the secure TEE operating system is done by the device manufacturers, and different companies have their own implementation methods, such as Huawei HiSilicon's iTrustee, Qualcomm's QSEE, and Trustonic on the ARM platform. Although the GlobalPlatform specification defines the expected functionality of a TEE implementation, not all manufacturers fully adopt it. On many non-Android devices, these features are not open to third parties, such as the iPhone, which provides a similar Secure Enclave. This practice is not conducive to the standardization and popularization of the TEE security execution environment.

3.6.4 *Summary of trust basis for data security*

This section describes the hardware and software implementations used for security purposes, namely the Hardware Security Module (HSM), Trusted Platform Module (TPM), and Trusted Execution Environment (TEE). Their implementation and applicable scenarios vary.

The TPM is a security chip that provides a hardware root of trust through embedded keys. These keys can even withstand some physical attacks, such as side-channel attacks to try to discover the key, prying open the TPM chip package to read the key directly, or removing the TPM from the computer motherboard to which it is soldered to break the cryptographic protection. The TPM is not designed to provide regular secure computing capabilities. It provides some slow, basic cryptographic functions, such as generating random keys and encrypting small amounts of data using its embedded or generated keys. The TPM can also be used to measure the integrity of system components and support secure boot.

Many of the functions of a TPM can be implemented in a TEE, but creating a "full" TPM implementation in a TEE does not make sense: one of the key use cases for a TPM is secure boot, while a TEE is more about providing a runtime processing environment. Unlike a TPM, a TEE does not have a hardware root of trust. The functionality of the TPM must also meet the requirements of the Trusted Computing Group (TCG; the standards body responsible for the TPM), which are more stringent than those of the TEE.

An HSM is an external physical device specifically designed to perform encryption operations, typically by receiving plaintext, encrypting it with the key it holds, and then returning ciphertext (encrypted text) so that the operating system does not need to touch the encryption key. Like TPMs, HSMs are designed with features to detect and deter physical tampering, making the storage of secret data more secure. HSMs typically provide a higher level of protection than TEEs, but they are separate hardware from the main CPU and motherboard, typically accessed through a PCI bus or even a network connection.

All TEE instances and some HSMs (depending on the model) can provide processing power for general computing tasks or can be programmed for specific purposes (such as PKCS#11 modules). Programming an HSM is often difficult and requires a great deal of experience. The cost of an HSM is high (usually at least tens of thousands of dollars) compared to a TEE, which is in the normal price range of a chipset.

HSM, TPM, and TEE can be summarized as follows:

(1) TEEs provide the general processing environment; they are built into the chipset.
(2) TPMs provide a physical root of trust. It can support measurement of other components and secure boot function. TPMs always have limited processing power; they are inexpensive chips built into many computers.
(3) HSMs provide a secure environment for storing confidential information and data; they are expensive external devices that often require specialized knowledge to use properly.

Table 3.6. Comparison of hardware trusted devices.

Device	Performance	Complexity	Cost
HSM	Medium	Very high	High
TPM	Low	Medium	Low
TEE	Medium	High	No

Table 3.6 shows a comparison of hardware trusted devices (HSM, TPM, and TEE).

There is a visual analogy that can make it easier to understand the difference between the three. If you cannot hold an object, or if you accidentally bump your foot while holding it, it is an HSM. If you can see a chip on the motherboard, it is probably a TPM. If you cannot see it with the naked eye, it is probably a TEE.

3.7 Key Technologies of Encryption and Access Control

Data security technology is comprehensive. From the perspective of technical classification, data security technology can be divided into encryption technology, access control technology, monitoring technology, and so on; from the perspective of data type stratification, data security technology can be divided into disk level security technology and file level security technology, field level security technology, and so on.

Take encryption technology as an example. Based on different levels and different environments, the panorama of different encryption technologies is shown in Fig. 3.32.

The key technologies of encryption and access control are selected and introduced in detail in the following.

3.7.1 *Endpoint encryption technologies*

Endpoints face a variety of threats, and different types of endpoints face different types and priorities of threats.

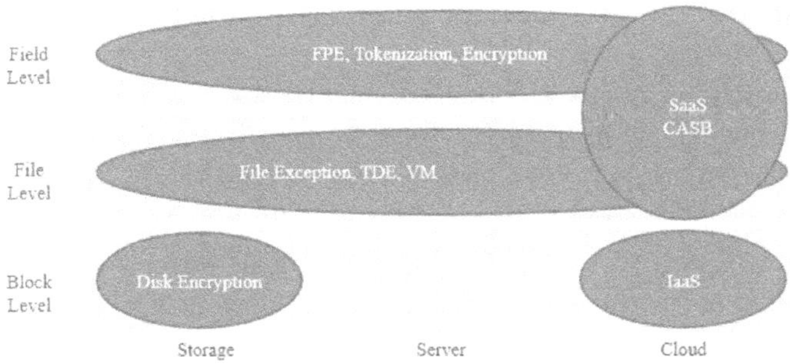

Fig. 3.32. Panorama of various encryption technologies.

Table 3.7. Differences of endpoint device threats.

Threat	Attack	User Informed	Impact
Malware	Logical	Generally unaware	Data confidentiality and integrity
Physical theft	Physical	Aware	Data confidentiality and availability
Evil maid	Physical	Unaware	Data confidentiality, system integrity

For common endpoint devices such as laptops, tablets, and smart terminals, threats can be divided into three types:

(1) Malware threats.
(2) Physical theft threats.
(3) "Evil Maid" threats.

The differences between the three types of threats are shown in Table 3.7.

Malware generally does not require the attacker to physically touch the target device and is therefore a logical attack. Traditional malware is more likely to affect the confidentiality of data. Examples include computer viruses, spyware, or Trojans that attempt to steal a user's login password or other sensitive data. Newer types of malware, such as ransomware, essentially make data unavailable by encrypting it with the goal of compromising its integrity.

Physical theft scenarios are physical attacks where the attacker physically touches the target device and gains physical access to it. Afterwards, the user is usually notified of physical theft. However, in most scenarios, the stolen device is not returned to the user. Without proper encryption and backup, data confidentiality and availability are compromised.

The evil maid scenario is a physical attack where the attacker physically touches the target device and temporarily gains physical access to it. An analogy is a user who forgets his laptop in a hotel room. A malicious hotel attendant (literally, an "evil maid") enters the room, logs on to the laptop, steals data or installs malware, etc., and then restores the scene. The user may not be aware of the "evil maid" attack. In most cases, the attacked and exploited device is still in the user's hands. Without proper access control and encryption, data confidentiality and system integrity are compromised.

The primary purpose of the endpoint encryption solution is to protect against "evil maid" attacks, and it is also suitable for theft scenarios. The attack could be to install a keylogger, manipulate boot files, or steal data stored on a laptop, files on servers, tablets, and other endpoints. From a purpose perspective, endpoint encryption solutions primarily consider encrypting sensitive data at rest and encrypting sensitive data in transit to another endpoint.

An organization's employees may store large amounts of valuable data, such as financial information, customer lists, and business plans, on external storage drives, cloud storage services, network drives, and other storage media, and even share this data via email and file servers. This data is vulnerable to security breaches.

As a result, organizations need to encrypt their data for a variety of reasons. For example, companies in high-tech industries such as pharmaceuticals need to protect their research from theft by competitors, organizations in regulated industries such as healthcare and financial services need to encrypt patient and consumer data to comply with laws and regulations, and the Payment Card Industry Data Security Standard (PCI DSS) requires retailers to encrypt consumer credit card data to prevent unauthorized use.

Endpoint encryption is an important part of a layered data security strategy. Organizations often use endpoint encryption in combination with multiple layers of protection, including firewalls, intrusion prevention, anti-malware, and data loss prevention. Often, encryption is the last layer of protection for data before it falls into the wrong hands.

Basically, encryption is the process of encoding or scrambling data so that it cannot be read or used unless the user has the correct decryption key.

Two commonly used cryptosystems are the RSA cryptosystem and the Advanced Encryption Standard (AES) cryptosystem. The RSA cryptosystem was jointly proposed by Ron Rivest, Adi Shamir, and Leonard Adleman in 1977. RSA is made up of the initials of their three last names. The mathematical basis of RSA's security is the difficulty of factoring large integers into prime factors. RSA can be used for both encryption and digital signing and is resistant to most cryptographic attacks.

In endpoint encryption scenarios, RSA is often used to encrypt data as it is transmitted from one endpoint to another. It uses asymmetric encryption, which means it uses one key to encrypt the data and another key at the recipient's endpoint to decrypt the data.

AES-256 is a symmetric encryption standard that is often used to encrypt data on storage devices such as hard drives or USB flash drives. Government agencies and regulated industry organizations that require advanced levels of encryption often use AES-256. This standard replaces the older Data Encryption Standard (DES), which was vulnerable to brute force attacks.

Endpoint encryption has the following two basic encryption methods:

(1) **Full disk encryption:** This encryption method renders a laptop, server, or other device unusable and accessible only to those with the correct PIN. BitLocker in Microsoft Windows is a typical implementation of this type of encryption.
(2) **File, folder, and removable media encryption:** This type of encryption locks only specific files or folders.

Full disk encryption protects the operating system and data on laptops and desktops by encrypting the entire drive except for the master boot record. The master boot record is not encrypted, so the computer can boot and find encrypted drivers to unlock the system. If a computer with an encrypted drive is lost, it is unlikely that anyone will be able to access the data on it. Full disk encryption is automatic, so everything stored on the drive is automatically encrypted.

Before a user can use the data on an encrypted drive, the data must be decrypted. There are two different ways to do this. The first is to boot the operating system first and decrypt the data on the encrypted drive after the user successfully logs in. The second is to require the user to enter credentials, such as a PIN or password, before the operating system boots. The second solution significantly reduces the attack surface and has higher security, which can handle the attack scenario of brute-force cracking of the login password by the background process. Therefore, most modern disk encryption schemes fall into the second category.

Specific file encryption encrypts selected content on local drives, network shares, or removable media devices. Encryption software deploys agents that encrypt files according to organizational policies. File-based encryption supports both structured and unstructured data, so it can be applied to databases, documents, and images.

File-based encryption keeps data encrypted until an authorized user opens it. This is different from full disk encryption, which decrypts all data after the user is authenticated and the system is booted. Therefore, file-based endpoint encryption continues to protect data even after the files leave the organization. For example, if an encrypted file is sent as an email attachment, the recipient must be authenticated in order to decrypt the file. Recipients who do not have appropriate encryption/decryption software can use a link sent with the email that authenticates the recipient and decrypts the file.

Modern endpoint encryption solutions, especially native endpoint encryption solutions introduced by device manufacturers, have evolved from pure software encryption solutions to software and hardware encryption solutions. Such solutions typically support hardware chips (such as TPM technology).

A comprehensive approach to endpoint encryption allows IT to centrally manage all encrypted endpoints, even those from multiple vendors. For example, IT can monitor and audit encrypted endpoints and manage encryption policies and keys from a single console. IT should unify the management of native endpoint encryption solutions on devices, such as Apple FileVault on OS X and Microsoft BitLocker on Windows, and ideally support third-party endpoint encryption solutions as well. Unified management of endpoint encryption solutions from multiple vendors helps reduce management overhead and costs, and improves management efficiency.

In addition, a unified approach to endpoint encryption provides greater visibility into the health of all endpoints and the ability to audit encryption usage on each endpoint. If a laptop or USB drive is lost or stolen, organizations can demonstrate compliance with a unified endpoint encryption policy.

Endpoint encryption policies can include several management capabilities:

(1) Centralized dashboard with status reporting capability.
(2) Support for mixed encryption environments.
(3) Key management capabilities, including key creation, distribution, destruction, and storage.
(4) Centralized encryption policy creation and management.
(5) Automatically deploy software agents to endpoints to enforce encryption policies.
(6) Identify devices that are missing encryption software.
(7) Ability to lock down endpoints that cannot be automatically checked in.

When selecting endpoint encryption solutions, organizations can take a comprehensive look at the management capabilities listed above and evaluate their integration capabilities with existing security products and solutions.

Encryption is an important part of an organization's security infrastructure. Security measures such as firewalls, intrusion prevention, and role-based access controls all help protect data within an organization. Encrypting data at the endpoint allows data to be protected even after it leaves the organization and is a key defense against data theft and breaches.

When choosing an endpoint encryption solution, you can also consider whether the solution has passed an internationally supported, authoritative certification. The primary standards for encryption software are the National Institute of Standards and Technology (NIST) Federal Information Processing Standard (FIPS) 140–2 and the Common Criteria for Information Technology Security Evaluation (CC).

3.7.2 *File encryption technologies*

Endpoint encryption techniques are described in Section 3.7.1. Endpoint encryption is primarily intended to prevent the loss or theft of devices from affecting stored data. Endpoint encryption technology does not solve the problem of actively sending files via email or file sharing nor does it solve the problem of malicious data theft after the user temporarily leaves the logged-on device. In addition, endpoint encryption focuses on consumer or user-owned devices and is not suitable for server scenarios, especially file server types.

Therefore, there is also a need for more fine-grained file encryption technology that can be used in combination or separately in different scenarios to better protect confidential data.

File encryption schemes are presented in different ways. Some appear as an encrypted drive, and users simply drag and drop the files they want to encrypt into the drive. Some are encryption schemes for common file types, such as compression tool software that supports setting passwords and encrypting files during compression. Others support encryption of specific documents. For example, the Microsoft Office suite supports setting a password to encrypt a single document that is currently open. Email encryption schemes based on Pretty Good Privacy (PGP) and True Crypt encryption software also implement various file encryption techniques.

Individual file/folder encryption depends on user awareness and best practices to ensure that all appropriate information is encrypted.

Assuming an appropriate file encryption scheme is selected based on usage scenarios, file encryption technology can provide protection when information is transmitted over the network to prevent data leakage. For example, a single file can be encrypted and sent as an email attachment, assuming the recipient can decrypt it.

In a multi-user scenario, the file encryption scheme becomes very complicated. This is because you must either allow each user to use a different key or use a shared key. At the same time, it is also necessary to deal with the file locking problem caused by multi-user access.

Encryption mechanisms at the application layer vary. For example, the Microsoft Office series of office software supports the setting of encryption passwords for individual documents. This encryption mechanism protects both the storage and transmission of documents.

There are different ways to encrypt emails. One way is to encrypt only the attachments of the message, and the other way is to encrypt the entire message. The implementation method of the former is simple, and any solution for encrypting a single file can be implemented. For example, compress a file using 7-Zip compression software, set a password to extract it, and send it by email. The latter is slightly more complicated to implement. Obviously, one of the most basic requirements is that the recipient of the message needs to know how to open the message. In fact, there are only two mainstream solutions for email body encryption: S/MIME and PGP.

3.7.2.1 *S/MIME*

S/MIME, or Secure Multipurpose Internet Mail Extensions, is a standard protocol for public key encryption and MIME data signatures. S/MIME was first proposed in 1995 and became a unified standard for the Internet through the IETF in 1998. The second version of the S/MIME standard includes two related Request For Comments (RFC, published by the IETF) standard documents: RFC 2311 defines the message; RFC 2312 defines the certificate handling specifications. RFC standards ensure interoperability of message encryption solutions between different vendors.

An extended standard for S/MIME was proposed in 1999: RFC 2632 is an extension of RFC 2311; RFC 2633 is an extension of RFC 2312; RFC 2634 adds additional services to S/MIME, such as signed receipts, triple wrapping, and security labels, extending the overall functionality.

S/MIME uses asymmetric encryption to protect the confidentiality of email. In terms of integrity, the S/MIME solution also supports digital signatures on emails to verify the legitimate sender of the message. This provides effective protection against phishing attacks, especially those initiated via email. S/MIME encryption schemes are built into many email programs, but it is clear that this scheme requires the recipient to have a trusted certificate. Certificate distribution is a more complex issue.

3.7.2.2 *PGP*

PGP email encryption solutions typically require additional software installation and support for many free open source or commercial solutions.

Both S/MIME and PGP provide options for signing only, without encryption. A signature assures the recipient that the message came from a particular sender and that the content has not been tampered with. However, the signature itself does not provide confidentiality protection. If someone can obtain the message, they can read the contents of the message.

The shortcomings of encryption technology are most obvious with file encryption. Although encryption can ideally prevent attackers from accessing the target data, in certain scenarios, it can also prevent legitimate visitors from accessing the target data. However, file encryption solutions rely heavily on the security and availability of key management. File encryption keys must be stored somewhere on the system, and hackers are very good at finding that location. In addition, the key management system's support for backup and recovery, as well as its own protection mechanism, high-availability mechanism, etc., are all issues that system administrators need to consider.

3.7.3 *Database encryption technologies*

Database encryption technology is currently an industry hot spot. According to a forecast by MarketWatch, the global database encryption market space will grow at a compound growth rate of approximately 29.8%, from US$572 million in 2018 to US$4.62 billion in 2026 (Market Search, 2021).

In a sense, database encryption is a classification of storage state encryption. Storage state encryption is the easiest scenario to solve, but it is also a scenario that covers a wide range and complex technical solutions. Storage state includes the storage of data on hard disks, flash memory, databases, removable media, and even backup devices. It is an attractive target for data theft due to its static nature and the value of the data stored.

When designing storage state encryption schemes, it should be assumed that attackers have access to data storage at rest.

The two main types of static data storage are structured data storage and unstructured data storage. This section primarily discusses structured storage, which refers to database encryption.

Database encryption should provide confidentiality, authentication, data integrity, and anti-repudiation protection for data and ideally should not affect usability.

According to the database operation mode, database encryption is divided into storage-time encryption and transmission-time encryption.

According to the database encryption type, database encryption can be divided into different types, such as key management, transparent encryption, file system encryption, column-level encryption, and application-level encryption.

Depending on the target industry or application scenario, database encryption includes healthcare, government, IT and telecommunications, banking and financial services, retail, aerospace, and other industries.

The different types of database encryption are described in the following.

In a typical modern database, applications access the database through the database engine to manipulate the data. The database engine itself stores one or more files on storage hardware, such as hard disks, through the file system mechanism of the operating system.

Database encryption can be divided into four different levels:

(1) Application layer encryption, as shown in Fig. 3.33, means that the data is encrypted by the application that generates or modifies the data and then written to the database. The database engine and database storage have no knowledge of whether the data is encrypted or not. The advantage of application-layer encryption is that encryption methods can be defined for each user based on user roles and privileges.
(2) Database layer encryption, as shown in Fig. 3.34, means that the entire database (or part of the database, depending on the implementation mechanism) is encrypted. The database engine

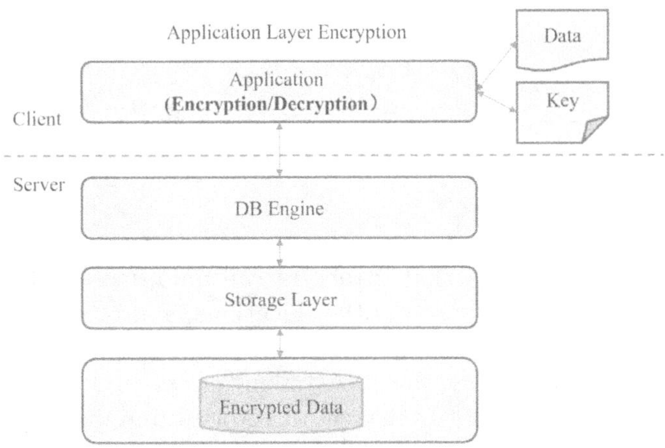

Fig. 3.33. Application layer encryption.

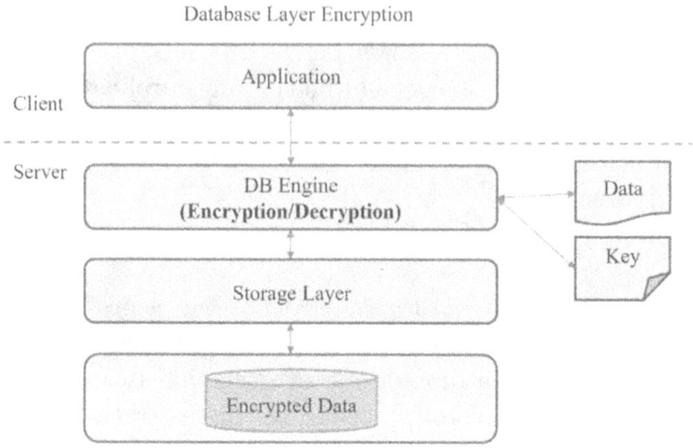

Fig. 3.34. Database layer encryption.

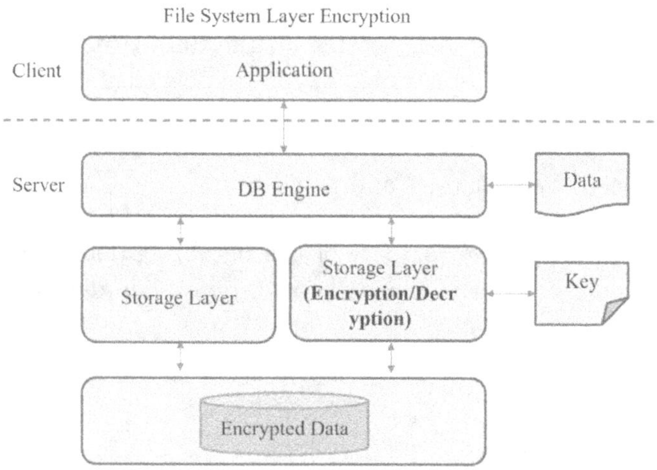

Fig. 3.35. File system layer encryption.

is responsible for managing the keys used to encrypt the data. The main threat is that attackers can gain access to database files, copy the database to other locations, and read its contents.

(3) File system layer encryption, as shown in Fig. 3.35, refers to file encryption using the file system access proxy mechanism, which allows users to encrypt directories and individual files. The agent interrupts read/write calls to the disk and uses policies to determine whether data needs to be decrypted or encrypted.

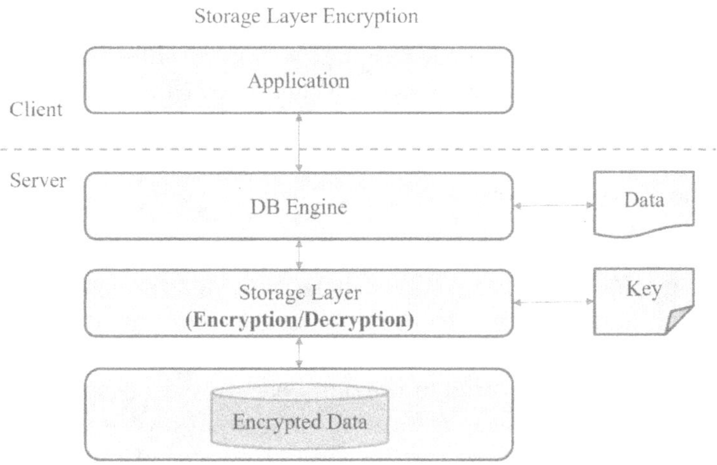

Fig. 3.36. Storage layer encryption.

Like full disk encryption, file system layer encryption encrypts databases and any other data stored in folders but with a more granular level of control.

(4) Storage layer encryption, also called full disk encryption (FDE), is as shown in Fig. 3.36. Storage layer encryption automatically converts data on disk into a form that cannot be decrypted without the key. Databases stored on disks are encrypted along with all other data.

Each of the above four levels of encryption has its own application scenarios and are not mutually exclusive, and even in some demanding situations, these encryption methods can be used together.

Other typical database encryption techniques include the following:

(1) **Column level encryption:** This encrypts specific columns of data in the database. Each column has its own unique encryption key, which can improve data flexibility and security.
(2) **Field-level encryption:** This encrypts data in specific data fields. The database creator can designate sensitive fields so that data entered by users in those fields is encrypted. These fields can include Social Security Numbers (SSNs), credit card numbers, and bank account numbers.

(3) **Hash:** The conversion of a string into a shorter, fixed-length key or a value similar to the original string. Hashes are commonly used in cryptographic systems. When a user first creates a password, it is stored as a hash. When the user logs in again, the password they used is compared to the unique hash to determine if the password is correct.

Transparent Data Encryption (TDE) is a term commonly used in technology and marketing. "Transparent" generally means that the encryption is transparent to the applications that use the database, but encryption can be used to effectively defend against information theft by stealing files stored in the database. In most scenarios, transparent data encryption is a form of file system layer encryption. It effectively protects data at rest by encrypting the entire database. At this point, the backup of the database is also encrypted to prevent data loss if the backup media is stolen or destroyed.

A typical transparent data encryption architecture is shown in Fig. 3.37. Each database instance has its own database encryption key (DEK). The DEK is usually stored with the database instance as encrypted ciphertext, or even stored in the database instance, and the encryption key (DMK) of the DEK can be shared between different database instances. If it is a two-tier key architecture, the DMK can be protected directly by the device hardware encryption and decryption engine (such as Trust Zone) or an external HSM. If it is a three-layer key architecture, there is also an SMK that can be used to derive the DMK or participate in protecting the DMK. At this point, the SMK is hardware-protected.

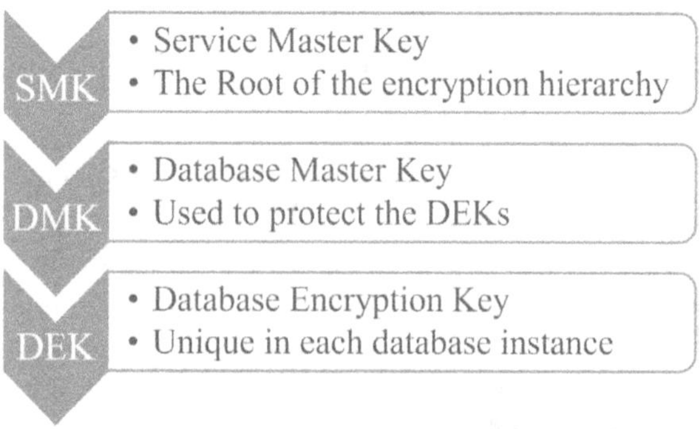

Fig. 3.37. Typical transparent data encryption architecture.

Database encryption has some inherent drawbacks:

(1) Database encryption can cause performance degradation, especially when column-level encryption is used. Therefore, organizations may be reluctant to use database encryption or apply it to all data at rest.

Many relational database management systems provide built-in encryption and key management capabilities. Therefore, database encryption is easier to implement if the data center uses only one vendor's database. Key management can be problematic when managing databases from multiple vendors, and key management failures can lead to security breaches.

(2) Database encryption can cause data loss. If data is encrypted using a strong cryptographic algorithm and the key is lost, the data cannot be recovered. Accidental loss or mismanagement of keys can be catastrophic.

3.7.4 *Format preserving encryption technology*

Encryption is an important data protection tool. However, in real-world applications, when personal data or sensitive data such as credit card numbers and ID numbers are encrypted, the length and type of these data are usually changed.

For example, the credit card number 1234 5678 1234 5670, after being encrypted in AES-128-ECB mode, becomes an array 0xd7b9e6744633c9a0689defbb7513e998 with a length of 16 bytes, which can also be expressed as the string "17nmdEYzyaBone+7dRPpmA==" in BASE64 encoding. Either way, the original code that handles credit card numbers may not be compatible with this, requiring extensive modification. Also, if the credit card number is stored in a database, the format of this field in the database may need to be adjusted. These adjustments add cost and complicate compatibility.

To solve such problems, Format Preserving Encryption (FPE) technology was developed, which requires that the ciphertext and plaintext have the same "format". The meaning of format varies depending on the specific scenario but generally includes character set (numbers and letters) and length requirements.

The characteristics of format preserving encryption are as follows:

(1) The data format is the same before and after encryption.
(2) The length of the data remains unchanged. For example, the length of the data before encryption is N, and the length after encryption is still N.
(3) The data type is unchanged. For example, it is a number type before encryption, and it is still a number type after encryption.
(4) Encryption is reversible. The encrypted data can be decrypted and the original data can be recovered.

Figure 3.38 shows an illustration of format-preserving encryption.

FPE was originally proposed to solve the problem of encrypting sensitive information in database systems or applications. The database encryption techniques described in Section 3.7.3 still have shortcomings in some scenarios. For example, storage layer encryption and file system layer encryption are inefficient and take a long time to execute. FPE can provide a database-independent application-layer encryption mechanism that does not increase the processing load on the server and can be compatible with the existing database system structure with little system modification cost, and it is one of the suitable database encryption methods.

FPE is suitable for fixed-format data encryption and can be used in the following scenarios:

(1) **Enhancing the security of database systems in production environments:** Many of the data stored in large systems such as financial, social security, and e-government

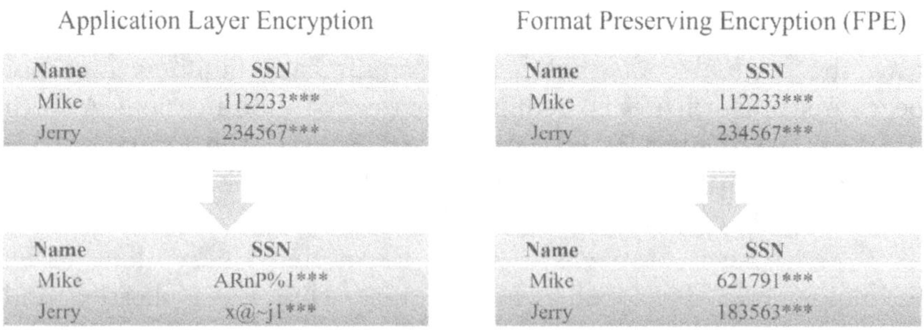

Fig. 3.38. Schematic representation of format-preserving encryption.

systems are database systems. If a large amount of sensitive user information (such as bank card number, social security card number, and user name) stored in the database is stolen, it will cause fatal damage. The introduction of FPE technology can improve the security of the database.

(2) **Data masking in test environment:** Data security problems may occur when data is imported from the production environment to the test environment (or development environment) because the test environment and development environment may not have strict access control and encryption policies. Data masking can encrypt real user data through FPE and output data with the same format and association as the original data, which can be used for functional testing, performance testing, and simulation testing. Sensitive data in the test database (such as ID numbers, phone numbers, and credit card numbers) appears to be real data but is actually masked data, eliminating the potential for sensitive data leakage.

FPE keeps ciphertext and plaintext in the same format, making it particularly suitable for data desensitization scenarios. Data desensitization based on FPE can be used as an extension and complement to traditional mask-based data masking schemes. When using FPE for data desensitization, if decryption is not considered, a random key that does not need to be stored can be generated for FPE to achieve better security.

(3) **Format-compliant encryption field:** In addition to encrypting sensitive personal information, FPE is also very suitable for encrypting data that follows existing protocol formats. For example, the 2010 paper "Efficient Format-Compliant Encryption of Regular Languages" (Stütz and Uhl, 2010) points out that a block-based encryption method similar to FPE can be applied to encrypt data in JPEG 2000 image format.

FPE encryption requires the input and output formats to be the same, resulting in a limited range of input and output domains. As a result, there are inherent weaknesses in encryption strength that must be analyzed in the context of specific use cases. In 2016, NIST published SP 800-38G "Recommendations for Block Cipher Modes of Operation: Methods for Format-Preserving Encryption" (Dworkin *et al.*, 2016). NIST originally considered three modes of operation: FF1, FF2, and FF3. All three modes use the AES block cipher algorithm. However, FF2 was never officially approved. FF3 has also been subject to crypt-analysis attacks, leading NIST to update the FF3 algorithm to FF3–1 in 2019 to address vulnerabilities caused by the limited range of the input and output domains. Therefore, the adopted FPE algorithm must consider the agility of updates in the corresponding application scenarios.

3.7.5 *Data-centric encryption*

In today's era, with the widespread popularization of cloud computing, big data, mobile Internet, and the Internet of Things, the amount of data in various organizations has expanded rapidly, and the speed of data flow has also increased. Traditional data protection measures, especially those based on network isolation and network access control, are finding it difficult to meet the growing demands of data protection. As early as 2004, a group called the Jericho Forum predicted the demise of the network perimeter. People, things, and data can easily cross organizational boundaries, so network perimeter defense will gradually become less relevant.

In recent years, the popularity of scenarios such as enterprise Wi-Fi, guest access mode, and the bring-your-own-device office has further validated this view. In particular, since the beginning of 2020, almost all organizations have started to work from home. Employees use home computers, tablet computers, or mobile devices to access the corporate system via the Internet at home. As a result, IT network security architectures that tie data security to corporate-owned devices or networks are challenged like never before.

In addition, the nature of data itself is rapidly evolving. Organizations are processing far more semi-structured and unstructured data than ever before, and data is more distributed than ever. Data is not only scattered across disparate systems, but it is also stored, transmitted, and consumed in the cloud.

To address these challenges, there is an urgent need for a data-centric security model. A data-centric security model is based on the idea that data, not applications, is an organization's critical asset. It advocates that data should be self-describing and self-protecting and that data security and access control should be the responsibility of the data layer.

The data-centric security model includes key measures, such as data discovery, management, protection, and monitoring:

(1) **Discovery:** Identify what data exists, which data is sensitive information, and where it is stored.
(2) **Management:** Define data access policies to determine whether certain data can be accessed and edited based on certain conditions (such as user account or even user location).
(3) **Protection:** The ability to protect against data loss, destruction, or unauthorized use of data.
(4) **Monitoring:** Continuous monitoring of data usage to identify possible anomalous behavior or malicious attempts.

Among the four key measures mentioned above, data protection is particularly important. In data protection measures, data encryption is a common means. Data-centric encryption focuses on realizing the following aspects:

(1) Self-describing and self-defending data to reduce the potential attack surface.
(2) Fine-grained security logic and protection mechanisms combined with business environment access control policies.
(3) Data can still be protected as it flows between different business systems and applications.
(4) Implementation strategies that are compatible with different types of data management systems.

Data-centric encryption, which shifts the perspective from data security and access control to the security and access control of keys (used to encrypt data), relies on fine-grained access control policies and requires that the access control decision point of the key be as close as possible to the endpoint where the data is used to fully maintain the confidentiality of the data during transmission.

Data-centric encryption is a well-established system. Fine-grained access control and end-to-end encryption are core principles of data-centric data protection. Sometimes, data-centric encryption is specifically referred to as end-to-end encryption (E2EE) technology. End-to-end encryption means encryption on the sender's system or device and decryption on the receiver's system or device. As a result, neither the Internet service provider, the application service provider, nor the hacker can read or tamper with the data during transmission.

A number of new data-centric encryption techniques are emerging. For example, time-stamping data and ensuring the integrity of metadata make data traceable and retrievable in real time, eliminating the possibility of human copying and tampering and reducing the attack surface. It is even possible to put critical data on the blockchain and use security mechanisms such as the blockchain ledger's integrity protection and consensus algorithms to ensure that data use and changes are traceable and auditable. This is also a scenario where the data itself proves to the user that it has not been tampered with or falsified, in line with the definition of data-centric encryption.

3.7.6 *Access control technology*

Access control can be divided into Discretionary Access Control (DAC), Mandatory Access Control (MAC), Role-Based Access Control (RBAC), and Attribute-Based Access Control (ABAC).

Discretionary Access Control (DAC) is an identity-based access control policy that grants or restricts access to an object through an access policy determined by the object's owner group and/or subject. It is called "discretionary" because subjects (owners) can transfer authenticated access to objects or information to other subjects. In other words, the owner determines the object's access rights. In practice, discretionary access control is usually implemented through Access Control Matrix (ACM) and Access Control List (ACL).

The discretionary access control mechanism is relatively flexible and widely used, such as Windows and UNIX. It is currently the most widely implemented access control mechanism in computer systems.

Mandatory Access Control (MAC) is an access control strategy that is defined by a centralized authoritative entity and uniformly enforced on subjects and objects in the system. It controls the authority and operation of the subject on the object by the system according to the established rules, such as the security attributes of the subject and the object. The subject has no right to change the access control rules, cannot transfer its authority to other subjects, and has no right to change the security attributes associated with the subject or object.

In a system, the subject is usually a process or thread. Objects can be files, directories, TCP/UDP ports, shared memory segments, I/O devices, and so on. Both subjects and objects have a number of security attributes. When a subject attempts to access an object, authorization rules enforced by the system kernel check these security attributes and decide whether access is allowed. Database management systems can also use mandatory access controls. In this case, the objects are tables, views, stored procedures, etc. in the database system.

Multi-Level Security (MLS) is a type of mandatory access control. It is based on military concepts and divides the security level of documents into four levels: Top Secret, Secret, Confidential, and Unclassified. All subjects and objects in the system are assigned security labels to identify their security level. Commonly used security models based on MLS include the Bell-LaPadula security model and the Biba security model, which focus on ensuring the confidentiality and integrity of data, respectively.

Traditional mandatory access control mechanisms, especially the hierarchy-based MLS and classification-based Multi-Category Security (MCS) methods, lack flexibility in their configurations and are commonly used in military or specialized systems, but for general-purpose systems, there are difficulties in using them, such as inefficiency and control difficulties.

Role-Based Access Control (RBAC) is the control through access to roles. A role is a collection of actions that a user or group of users can perform within an organization. Each role is associated with a set of users and related actions that the users belonging to the role may be authorized to perform. Since permissions are associated with roles, users can only get permissions if they are members of the appropriate role. Roles are defined by system administrators, and permissions are also enforced by system administrators and are imposed on users. Permissions cannot be transferred independently, so role-based access control is not a type of discretionary access control.

Role-based access control has the advantages of facilitating authorization management, reducing management overhead, facilitating classification and separation of duties according to work needs, facilitating the granting of least privilege, facilitating object classification and file hierarchy management, and so on. It is mainly used in large organizations and enterprises and can improve the flexibility of enterprise security policies. The structural difference between role-based access control and traditional access control is shown in Fig. 3.39.

The above three access controls are not mutually exclusive and can be applied comprehensively. When conflicts arise, authoritative entities must coordinate. The relationship between access control is shown in Fig. 3.40.

The concept of Attribute-Based Access Control (ABAC) has been around for many years. Traditionally, access control is based on the user's identity (role or group) to determine whether the user's request to access a resource should be granted. Such access control is

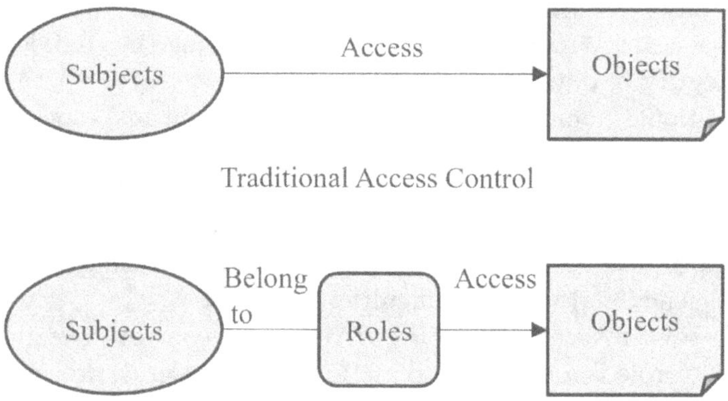

Fig. 3.39. Role-based access control and traditional access structural differences in control.

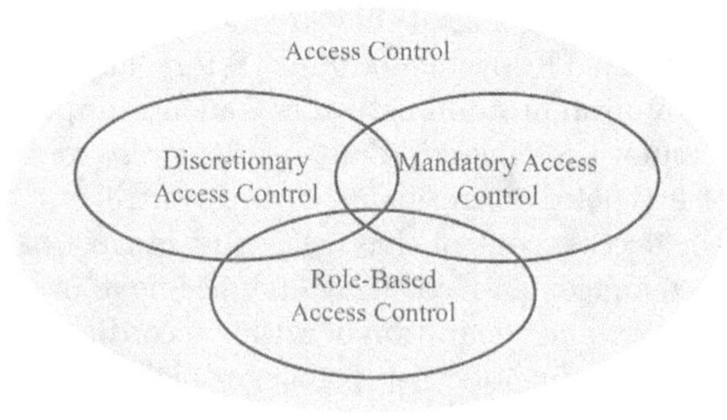

Fig. 3.40. Relationship of access control.

often difficult to manage because of the need to associate capabilities directly with users or their roles or groups. In addition, the requesting identity, group, and role qualifiers are often insufficient to express the actual access control policy. An alternative is to grant or deny user requests based on arbitrary attributes of the user and arbitrary attributes of the object, as well as environmental conditions that can be globally recognized and are more relevant to the current policy. This approach is often called attribute-based access control.

ABAC is a logical access control model that controls access to objects by evaluating rules against attributes of the entity (subject and object), operation, and environment associated with the request. Attribute-based access control is shown in Fig. 3.41. When an access request occurs, attributes and access control rules are evaluated by the Attribute-Based Access Control Policy module to provide access control decisions. In the basic form of ABAC, the Attribute-Based Access Control Policy module contains Policy Decision Points (PDPs) and Policy Enforcement Points (PEPs).

In general, ABAC is capable of implementing both Discretionary Access Control (DAC) and Mandatory Access Control (MAC)

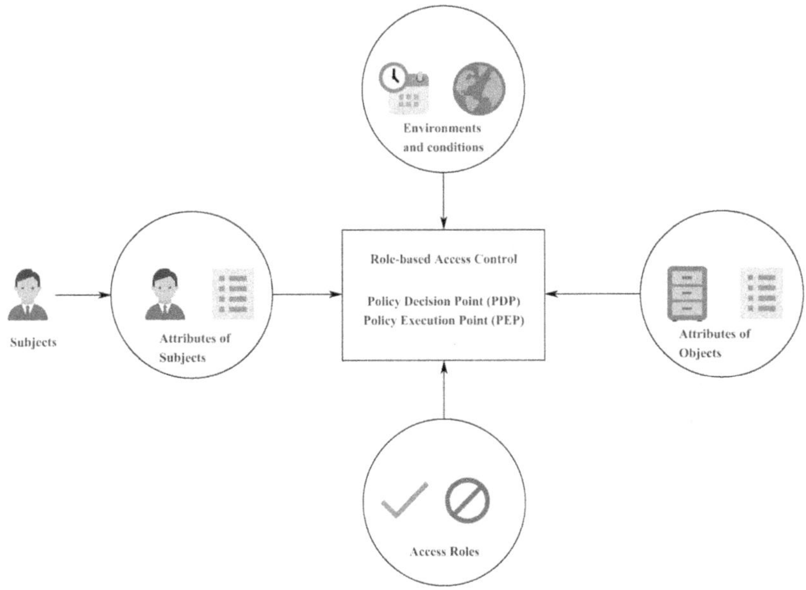

Fig. 3.41. Attribute-based access control.

concepts. In addition, ABAC allows for more granular access control, where a greater number of discrete inputs can be added to the access control, thus providing a greater number of possible combinations of these variables, reflecting a greater and more deterministic likelihood of representing the policy rule set.

The access control that can be implemented in ABAC is limited only by the richness of the language and the attributes available. The flexibility of attributes allows the widest range of subjects to access the widest range of objects, without having to specify a separate relationship between each subject and each object. For example, an employee is assigned a set of subject attributes when hired (e.g., Nancy Smith is a cardiology nurse). When an object is created, it is assigned object properties (e.g., a folder with the object property "Cardiology Patient Records").

Objects may receive their properties directly from their creator, or they may be the result of automated scanning tools. The administrator or owner of the object uses the subject and object attributes to create access control rules to control the set of permitted capabilities (e.g., all nurses in the cardiology department can view the cardiology patient's medical records). In the ABAC access control model, access control can be changed between requests by simply changing attribute values without changing the subject/object relationships that define the underlying rule set. This provides more dynamic access control management capabilities and limits the long-term maintenance requirements for object protection.

In addition, ABAC access control allows object owners or administrators to apply access controls without prior knowledge of a particular subject, which can be infinite. Rules and objects do not need to be changed when new subjects join the organization. As long as the object is assigned the attributes necessary to access the desired object (for example, all nurses in the cardiology department), there is no need to modify existing rules or object attributes. This benefit is often referred to as accommodating external (unexpected) users and is one of the main benefits of using ABAC access control.

When ABAC is deployed in an enterprise, its effectiveness and security depend on the accuracy of subject and object attributes, the automation of access controls, and effective integration with existing IT systems.

In addition to basic policy, attribute, and access control mechanism requirements, organizations deploying an ABAC access control system must support administrative functions for enterprise policy development and distribution, enterprise identity and subject attributes, subject attribute sharing, enterprise object attributes, authentication, and access control mechanism deployment and distribution. The development and deployment of these functions requires careful consideration of a number of factors that can affect the design, security, and interoperability of enterprise ABAC solutions. These factors can be summarized in the following set of activities:

(1) Establish a business case for implementing ABAC.
(2) Understand operational requirements and the overall enterprise architecture.
(3) Establish or improve business processes to support ABAC.
(4) Develop and procure a set of interoperable functionalities.
(5) Operate efficiently.

NIST SP 800-162 "Guide to Attribute Based Access Control (ABAC) Definition and Considerations" (Hu *et al.*, 2013) can help planners, architects, managers, and implementers select and deploy appropriate ABAC systems.

3.7.7 *File monitoring technologies*

In today's cyberspace, the ever-increasing variety of cyberspace threats and the emergence of innovative attack methods pose a real and growing threat to every organization. As anyone can see from the news, cybersecurity attacks often cause significant damage to the reputations of many prominent organizations.

The use of encryption and access control alone is not enough to prevent this damage and requires the identification and detection of potential threats and security incidents, and a targeted response in security operations efforts.

One of the most important ways to handle suspected or actual cybersecurity incidents is to log cybersecurity-related events, monitor them continuously, and thoroughly investigate suspected cybersecurity breaches to remediate any problems.

There are many types of events that can be classified as cybersecurity incidents, ranging from serious cybersecurity attacks

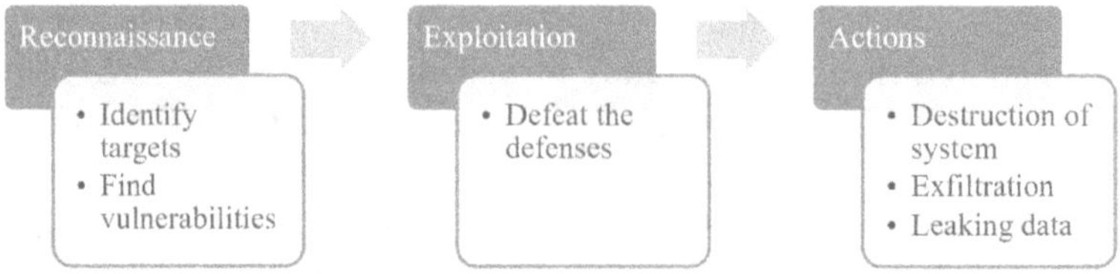

Fig. 3.42. Network attack steps.

to organized crime by cybercriminals, hacking and malware attacks, internal misuse of systems, and software failures.

Closely related to cybersecurity is monitoring and logging, which refers to recording and monitoring cybersecurity events to quickly and effectively locate potential threats and exploit activity to prevent cybersecurity incidents.

Taking a typical network attack as an example, a network attack is generally divided into three steps, as shown in Fig. 3.42:

(1) **Reconnaissance:** First, the target is identified, and then the target is analyzed for vulnerabilities.
(2) **Attack the target:** Exploit relevant vulnerabilities and bypass or destroy other defenses.
(3) **To accomplish the goal:** It may be to destroy the system, manipulate data, or leak sensitive data.

And for each of the above steps, security monitoring (including logging) can be an effective mitigation. Of course, other mitigations such as incident response, stateful awareness, and threat intelligence are also necessary:

(1) Identify potential exploit indicators early.
(2) Efficiently investigate scenarios in which vulnerabilities are exploited.
(3) Take appropriate measures to mitigate the impact of vulnerability exploitation.
(4) Meet the requirements of relevant laws and regulations.

A typical system monitoring architecture is shown in Fig. 3.43.

The monitoring agent/client is used to collect logs from various devices and information sources, which may include internal system logs (such as operating system, firewall, and IDS) and external logs (such as cloud service logs and various service provider logs).

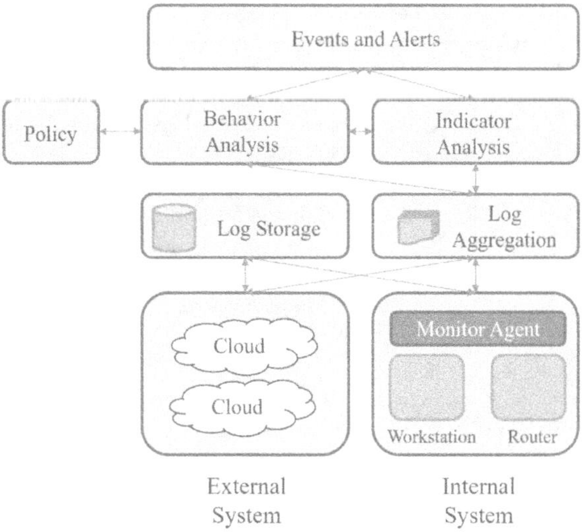

Fig. 3.43. Typical system monitoring architecture.

After the required logs are collected, the system passes them to the log storage and aggregation platform. These platforms can perform preliminary aggregation, correlation, and analysis. Some platforms also support metric-based analysis.

Correlation results from a single indicator or multiple indicators can correspond to similar behaviors. Based on predefined policies, the behavioral analysis platform analyzes the correlation results of a single indicator or multiple indicators to generate events that represent specific types of phenomena. Most security monitoring systems support the simultaneous generation of corresponding alarms for major network attacks, exploits, and data destruction events.

In earlier organizations, the activities described above were typically the responsibility of the organization's IT or information security department. In today's large organizations, a separate Security Operations Center (SOC) typically handles monitoring and alerting.

The four types of logs described in Table 3.8 are particularly necessary when gathering information from the organization's various IT systems to identify potential or existing network security incidents.

This section mainly describes the system log and system monitoring part, especially file and database monitoring in system monitoring.

Table 3.8. Log types.

Log Type	Description
System logs	System activity logs (such as administrator activity) Endpoint logs Authorization and authentication logs Physical security log
Network logs	VPN and NetFlow logs HTTP proxy logs DNS/DHCP/FTP logs
Server and middleware logs	Web server logs Database server logs Mail server logs
Network Security Logs	Malware detection tool logs Network-based IDS (NIDS) logs Network-based IPS (NIPS) logs Data Loss Prevention (DLP) logs Network firewall logs Other security system (e.g., honeypot and sandbox) logs

File Integrity Monitoring (FIM) refers to an IT security technology and process for testing and auditing operating systems (OS), databases, and applications for tampering or corruption. FIM acts as a change auditing mechanism to validate files by comparing their latest versions to a known, trusted "baseline". If FIM detects that a file has been changed, updated, or corrupted, it can generate an alert to ensure further investigation and, if necessary, repair of the file. FIM includes both reactive auditing (forensic scenarios) and proactive rules-based monitoring.

FIM solutions scan, analyze, and report on changes to critical files in IT environments. As a result, FIM can provide a critical layer of file, data, and application security while accelerating incident response.

FIM can be applied to the following four main scenarios:

(1) **Detection of illegal activity:** When hackers break into an IT environment, the defender needs to know if they are trying to modify files critical to the operating system or applications. Even if log files and other detection systems are compromised, FIM can

still detect changes to critical parts of the IT ecosystem. Files, applications, operating systems, and data can be monitored and secured with FIM.

(2) **Detect unexpected changes:** File changes are often made unintentionally by administrators or other employees. Sometimes, the impact of these changes is small and negligible, but other times, they can create vulnerabilities or backdoors, or cause business downtime or continuity issues. FIM can identify changes to files, which simplifies forensics, and can also roll back corresponding file changes or take other remediation actions.

(3) **Verify update status and monitor system operation:** Using the FIM solution's patch validation capability, installed versions can be scanned across multiple locations and computers to verify that files have been updated to the latest version.

(4) **Regulatory compliance:** Some regulations, such as GLBA, SOX, HIPAA, and PCI DSS, require organizations to be able to audit changes and monitor and report on certain types of activities.

The international standards and industry standards related to monitoring are typically combined with the logging requirements, such as Section 12.1 "Logging and Monitoring" of ISO 27002 and Requirement 10 of PCI DSS V3.1: "Track and monitor all access to network resources and cardholder data" and Requirement 11: "Regularly test security systems and processes". In the United States, based on the Federal Information Security Management Act (FISMA), NIST 800–137 "Information Security Continuous Monitoring of Federal Information Systems and Organizations" (Dempsey *et al.*, 2011) proposed "Information Security Continuous Monitoring (ISCM)" concepts and operational requirements.

File integrity monitoring exists because an organization's IT environment changes frequently: hardware assets change, software programs change, configuration state changes. Some modifications are normal operations during upgrade or maintenance cycles, while others require attention.

Organizations typically respond to this dynamic by investing in asset discovery and security configuration management. These controls allow organizations to track and monitor the configuration of

IT systems. However, a key challenge remains: coordinating changes to critical files. FIM is an effective way to address this scenario.

Five steps are required to complete file integrity monitoring, as follows:

(1) **Define the strategy:** This step involves determining which files on which systems the organization needs to monitor.
(2) **Establish a document baseline:** Before an organization can actively monitor files for changes, it needs a reference point against which to detect changes. Therefore, organizations should document a baseline, or known good state, for the files covered by the FIM policy. This baseline should include versions, creation dates, modification dates, and other data that can help IT professionals ensure that files are legitimate.
(3) **Monitor changes:** With detailed baselines in place, organizations can continue to monitor changes to all specified files. False positives are minimized by identifying and automatically adding expected changes.
(4) **Send an alert:** When a file integrity monitoring solution detects unauthorized changes, those responsible for the process should send an alert to someone who can resolve the problem.
(5) **Report results:** Sometimes organizations use FIM tools to ensure PCI DSS compliance. In this case, the organization may need to generate audit reports to demonstrate the use of its file integrity monitoring assessment program.

The above steps can establish a closed-loop process for file integrity monitoring.

3.7.8 *Database monitoring technologies*

According to a survey by Imperva, nearly 60% of the most sensitive data in various organizations is stored in the database (Kaur *et al.*, 2017). However, most organizations spend more on firewalls and other forms of perimeter security, which are not enough to prevent breaches. According to Verizon's 2018 Data Breach Investigations Report (DBIR) (Verizon Enterprise, 2018), of the 2,216 data breaches it investigated, the most affected asset was the database server.

The following two database security measures are commonly used:

One is a data activity monitoring (DAM) tool. This tool is used to detect suspicious activity against databases in a timely manner.

The other is data obfuscation techniques such as encryption and data masking. Data obfuscation techniques can render stolen data worthless.

This section focuses on data activity monitoring tools and techniques. Data activity monitoring tools do not work in isolation, and in most scenarios, they must be integrated with perimeter security tools such as firewalls, identity and access management (IAM) platforms, data leakage protection (DLP) tools, and tools or platforms, such as log management tools, security information and event management (SIEM), and so on, to provide the ultimate data security solution.

Mogull R. of Gartner Research notes that Database Activity Monitoring (DAM) (Mogull, 2006) "is ... functionality that identifies and reports fraudulent, illegal, or otherwise undesirable behavior with minimal impact on user operations and productivity". Data activity monitoring tools have evolved from basic account activity analysis to powerful "data-centric" security measures, such as data discovery and classification, user rights management, privileged user monitoring, data protection, and loss prevention.

A DAM solution should have at least some of the key features described in the following:

(1) Independent monitoring and auditing of all database activity, including administrator activity and SELECT query transactions. DAM tools can capture all SQL transactions: Data Manipulation Language (DML), Data Definition Language (DDL), Data Control Language (DCL), and sometimes Transaction Control Language (TCL). Ideally, this can be done without relying on local database logging, with minimal performance impact (less than 2%, depending on the data collection method).
(2) Securely store audit logs on a central server outside of the target audit database.
(3) Monitor, aggregate, and correlate activity from multiple heterogeneous database management systems (DBMS). DAM tools can be used with multiple DBMSs, such as Oracle Database, Microsoft SQL Server, and IBM DB2, and can standardize transactions from different DBMSs despite differences in SQL styles.

(4) Ensure that the service account accesses the database only from specified source IPs and executes only a set of authorized queries; source IP and authorized query limits are used to receive timely alerts when connections are made to unexpected systems or when service accounts are compromised.
(5) Ensure separation of duties by monitoring and recording the activities of database administrators.
(6) Generate alerts for policy violations based on rules or heuristics. For example, a rule could be created to generate an alert every time a privileged user executes a SELECT query (e.g., returns more than five results from a credit card column). The trigger will alert the application to the possibility of an SQL injection attack or other attack.

Some DAM tools can also perform the following functions:

(1) Discover and provide visibility into the location, volume, and context of data on-premises, in the cloud, and in legacy databases.
(2) Categorize discovered data based on the type of personally identifiable information (e.g., credit card numbers, email addresses, and medical records) and the level of security risk of the data found.
(3) Provides pre-defined policies for PCI DSS, SOX, and other general compliance requirements.
(4) Provides closed-loop integration with external change management tools to track approved database changes implemented in SQL. Leverages other tools to track administrator activity and provide change management reports for auditing when needed.

One of the key benefits of DAM is its ability to monitor multiple databases running on multiple database management systems (DBMS) on multiple operating systems (Windows, Unix, etc.) DAM tools aggregate information from multiple collectors to a central server. In some cases, the central server/management console also collects information, while in other cases, it simply acts as a repository for the collectors to send data to.

At the heart of a DAM solution are collectors that monitor database traffic and either store it locally or send it to a central server, depending on filtering rules and configuration. At a minimum, these collectors are capable of monitoring SQL traffic. This is one of the defining characteristics of DAM, and it's the key differentiator from log management, security information and event management, and other tools that provide some level of database monitoring. Collection techniques fall into three main categories: network monitoring, remote monitoring, and local agents.

3.7.8.1 *Network Monitoring*

This technology (see Fig. 3.44) monitors network traffic for SQL, parses the SQL, and stores it in the collector's internal database. Most tools can perform bi-directional monitoring, i.e., both inbound and outbound SQL network traffic can be monitored. The advantage of network monitoring is that there is no overhead on the monitored database, it can be monitored platform-independently, there is no need to modify the database, and it can monitor multiple heterogeneous database management systems simultaneously. The disadvantage is that it has no knowledge of the internal state of the database and misses database activity that does not access the network as SQL, such as direct login to the local database server and remote console connections. Due to the second drawback, network monitoring is only recommended in combination with other monitoring techniques that can capture local activity.

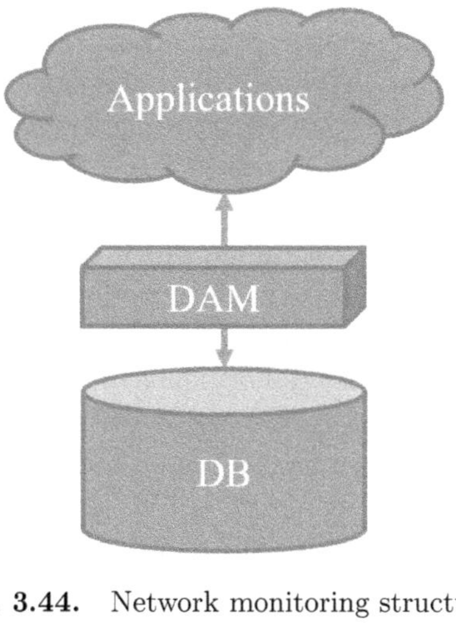

Fig. 3.44. Network monitoring structure.

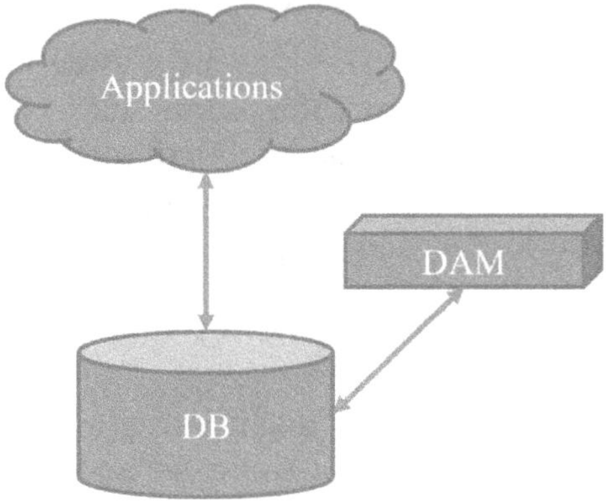

Fig. 3.45. Database remote monitoring structure.

3.7.8.2 *Remote monitoring*

The remote database monitoring architecture is shown in Fig. 3.45. Using this technique, the DAM collector is granted administrative access to the target database and the native database auditing feature is enabled. The DAM collector monitors the DBMS from the outside, collects database activity records, and aggregates the activity records submitted by the database. Therefore, the resource consumption of the monitored system is only the resource consumption of the logging and auditing modules of that system. Some database distributions, such as Microsoft SQL Server, have minimal to negligible resource consumption for remote database monitoring. However, some database distributions have huge resource consumption for database remote monitoring and high performance risk.

The advantages of database remote monitoring include the ability to monitor the performance of all database activity (including local activity), the ability to provide a close approximation of local logging and monitoring, and the ability to monitor all database activity (including local activity) regardless of client connection methods. Its main drawback is potential performance issues with some database platforms. Remote monitoring technology also requires administrative accounts to be enabled on the database platform and may require configuration changes.

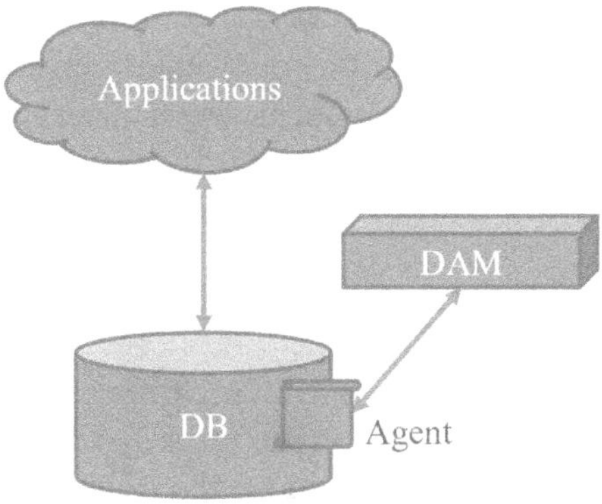

Fig. 3.46. Database local agent monitoring structure.

3.7.8.3 *Local proxy*

The local database agent monitoring structure is shown in Fig. 3.46. This technology requires software agents to be installed on the database server for collection activities. Due to the different support requirements for DBMSs and host platforms, the capabilities and technologies used by each agent vary widely. Some early proxies relied on local sniffing of the network loopback interface for monitoring, which lost certain types of client connections. Current agents can audit activity through platform kernel hooks without modifying the DBMS with minimal performance impact, or they can also take advantage of shared memory monitoring. Leading proxies typically impact performance by no more than 3%, which seems to be an acceptable range for DBAs. Advantages of the native agent include the ability to capture all activity without turning on native auditing and the ability to monitor internal database activity such as stored procedures and lower overhead. Disadvantages of native agents include limited platform support (a new agent must be written for each platform) and the requirement to install an agent on each monitored server.

In summary, with today's evolving security threats and the exponential growth in the volume and use of sensitive data, it is critical to implement data-centric security measures. These measures focus on securing data as it moves between networks, servers, applications, or endpoints. These measures come in two forms: native database auditing tools and database activity monitoring. Which form is right for an organization, or a combination of forms, must be fully considered based on specific scenarios, data types, actual threats, and compliance requirements.

3.8 Data Security Business Scenarios

There are many types of data, and many different business scenarios for data security. It is not only the user's data that needs to be fully protected. On the client side, which is used by consumers, ensuring that the copyright owner's data is not illegally copied, ensuring its confidentiality, etc., are applicable Digital Rights Management (DRM) scenarios. At the same time, when the organization's employees bring their own personal devices to work, ensuring that the organization's data is not leaked is also one of the typical data security scenarios, and it is also a key scenario of the MDM business. The following section selects several typical business scenarios and describes their key technologies and application solutions.

3.8.1 *DRM technologies and applicable scenarios*

DRM is a technology that uses a key to encrypt digital files. Keys are used to lock or unlock content. Typically, a user must obtain an authorization file that contains a key in order to access a target file. In most scenarios, the authorization file also contains restrictions on how the target file can be used.

DRM technology allows digital rights owners to distribute songs, videos, or other digital multimedia files over the Internet in a controlled and protected file format.

In the 12th issue of the Berkeley Technology Law Journal, Spring 1997, Mark Stefik published an article entitled "Shifting the possible: How trusted systems and digital property rights challenge us to rethink digital publishing" (Stefik, 1997). The paper states, "With the development of trusted system technology and usage rights languages with which to encode the rights associated with copyrighted material, authors and publishers can have more, not less, control over their work". This was one of the first descriptions of DRM technology.

In fact, the application of DRM technology is closely tied to the convenience of duplicating and distributing digital assets in the Internet age. In the era of physical books, even with the invention and popularization of photocopiers, the time and money required for an ordinary reader to copy an entire book is considerable. Therefore, physical books do not have a strong appeal for copyright protection technology. In the Internet age, the cost of copying and distributing a document in digital format is almost negligible. While the legal requirements for copyright protection of books in print and digital format are virtually identical, the law alone is not effective enough to prevent copying. Technological means are needed to protect digital works.

When it comes to technological means to protect digital works, one of the most intuitive considerations is "why not use encryption?" Encryption is necessary, but encryption alone is not a complete solution. First, encryption does not prevent files from being copied. Second, even encrypted files can be easily moved from one digital device to another, not to mention distributed via email, Internet downloads, and even P2P networks. This illustrates the difference between encryption and digital rights management: the point of control for DRM is the access and use of the content, not necessarily the act of copying the encrypted file itself. Even if hundreds or thousands of copies of an encrypted file are made and distributed, those copies are worthless without the key to access the file.

Digital Rights Management (DRM) has to face various scenarios, and simply using encryption technology cannot meet the requirements. Here are two ways to protect files: The first method is to provide the user with the encrypted file and a password to open the encrypted file. At this point, the user can give the file and password to anyone. The second, more complicated method is not to provide the user with the password to access the file but to hide the key to access the file in an authorization file (license) and provide it to the user along with the encrypted file. At this point, the user may not even know that the file is encrypted. Of course, there are obvious problems with this approach. A clever user could distribute both files to other unauthorized users, thereby harming the interests of the copyright owner.

A third approach is to bind the authorization file to the user's specific hardware device. This way, users can only open protected files on specific devices. In fact, this was the most popular DRM technology solution in the early days. The disadvantages of this method are also obvious: first, the lifecycle of common hardware devices is only three to five years at most, so it is affected by hardware replacement; second, the user's access to files is very obviously limited.

An ideal solution would be to associate a digital file with a "person" rather than a "device". In this way, a user can move a purchased digital file between devices in the same way that a user can take a book from home to the office.

As early as 2003, Microsoft supported users in its Microsoft Reader product to associate their account with a specific Microsoft Reader to protect the e-books they purchased from being illegally copied.

The previous discussions in this chapter have focused on protection measures when files are opened, but DRM technology covers much more than that. After the file is opened, additional controls are needed. For example, you might want to control whether the user can print the file, whether the user can copy and paste the contents of the file to the clipboard, or whether the file itself expires after a certain amount of time.

Using Acrobat Reader, a commonly used document reading software, as an example of digital rights management of documents, the digital rights management mechanism of documents can be vividly illustrated, as shown in Fig. 3.47. In the "Security" tab of the document properties, it can be seen that the creator controls various protection methods of the document, such as whether to allow copying of the document contents, whether to allow printing, and so on.

Both Acrobat Reader and Microsoft Reader implement some, but not all, features of DRM technology. From an abstract dimension, DRM technology systematically classifies the demands of copyright holders into functional restrictions (playing, printing, copying, etc.), usage restrictions (number of times restrictions, expiration time restrictions, geographical location-based restrictions, etc.), and user compliance obligations (payment, usage tracking, etc.). The functional classification of DRM technology is shown in Fig. 3.48. The design and implementation of the DRM system is also based on these three types.

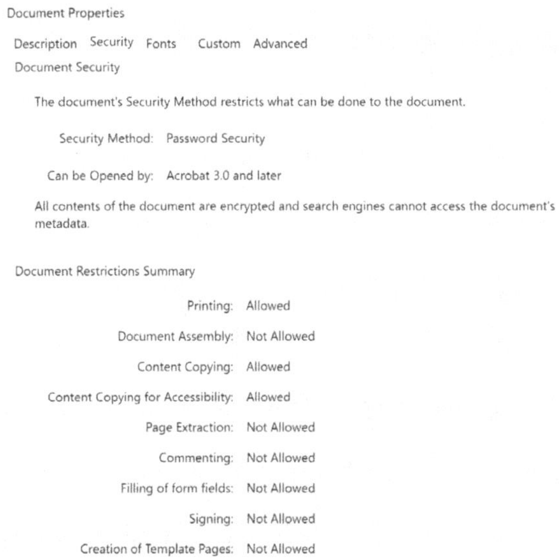

Fig. 3.47. Example of DRM for documents.

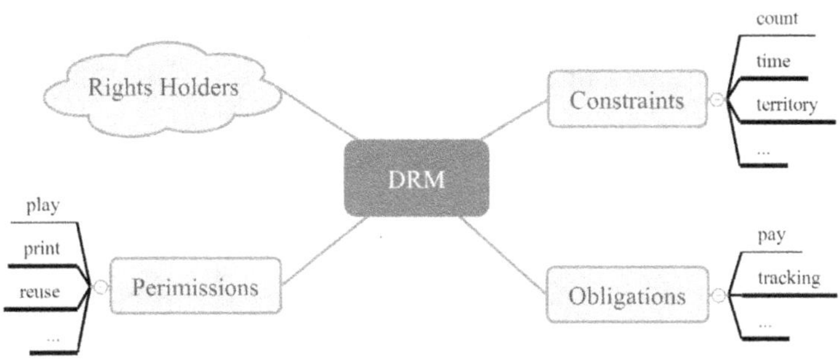

Fig. 3.48. DRM technical function classification.

3.8.2 *MDM technology and applicable scenarios*

With the gradual rise and popularity of telecommuting, home office, and mobile office, consumers expect to be able to more conveniently handle work email, access internal address books, use internal social platforms or collaboration tools (such as SharePoint), and use internal websites and applications. At the same time, users want to keep their personal data (such as photos, music, videos, and personal applications) out of corporate control. Especially in scenarios where personal devices are used to process work content, effective separation of work and personal data has become a common requirement.

Mobile Device Management (MDM) technology is the main application technology in the above-mentioned mobile office scenarios. Its purpose is to allow only authorized applications to access and process corporate assets and data in a secure and controllable manner.

From a technical perspective, MDM technology can isolate and protect an organization's data to reduce the attack surface for that data. Some MDM technologies can also provide capabilities, such as restricting application execution, discovering devices, wiping corporate data, and erasing all data on lost devices.

The working principle of MDM technology can be visualized in Fig. 3.49.

The device operating system provides support for MDM technology. MDM technology can virtualize an "isolated" space where the enterprise's applications run, separate from the consumer's personal applications.

It is worth noting that Gartner in its article "Four Architectural Approaches to Limit Business Risk on Consumer Smartphones and Tablets" (Girard and Dulaney, 2010) points out the following: "Management of these devices can help, but enterprises should

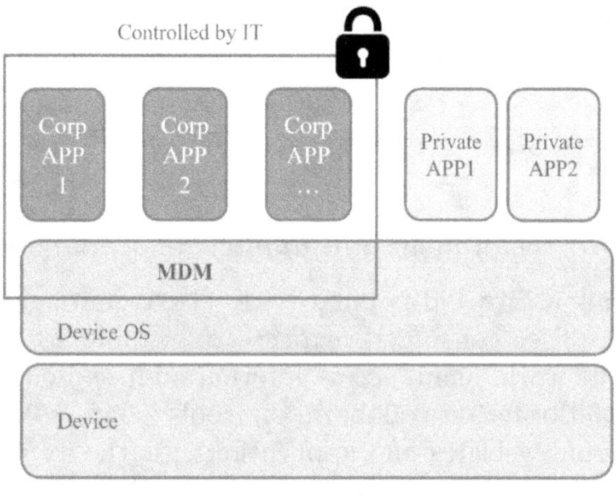

Fig. 3.49. Working principle of MDM technology.

also consider other technologies and application practices to reduce data exposure and leakage". In other words, relying solely on MDM technology solutions cannot effectively protect data and assets on devices. The main reason for this is that modern smartphones introduce multiple attack surfaces at the operating system level to enterprise assets, and there are relatively high risks, such as Bluetooth transmission, printing, clipboard, Wi-Fi, VPN, and so on.

Gartner offers the following recommendations:

(1) Prioritize self-protection applications from independent software vendors and internal development teams.
(2) Do not assume that device management policies will protect corporate interests and access control.

Data leakage prevention and management at the application level is required. All applications that need to access sensitive data should use similar mechanisms.

MDM technology focuses on device control and security. A popular view is that MDM solutions have been replaced by a more complete and comprehensive Enterprise Mobility Management (EMM) solution. EMM solutions broadly refer to the collection of technologies and services used to protect corporate data accessed and used on employee mobile devices. According to Forrester Research, EMM enables organizations to support the choice and autonomy of employees to use mobile devices for work while protecting the organization from the threats that arise from this activity.

In addition to the key capabilities of MDM technology, EMM technology generally includes the technical capabilities of Mobile Application Management (MAM) and Mobile Identity Management (MIM). MAM technology allows an organization's IT department to control applications and no longer be limited by devices. IT departments can set security policies for specific applications, limit the sharing of corporate data between applications, or selectively remove applications and their data without wiping the entire device. MIM technology includes technologies such as user and device certificates, application code signing, authentication, and single sign-on (SSO). The primary purpose of MIM technology is to ensure that only trusted devices and users have access to an organization's data or applications. In general, MIM technology also provides the ability to measure and aggregate application and device usage.

With the emergence of multiple device types and platforms, many software vendors of EMM technologies are repackaging their products and services with Unified Endpoint Management (UEM) technology solutions. Based on EMM technology, UEM technology enhances the management of different types of devices, such as iOS, Android, unified support for Windows, and Mac OS. UEM technology also provides a single point of access for users to their applications and files.

One of the main design goals of modern mobile devices is to make information processing and sharing faster and more convenient. To protect corporate assets and data, it is important to ensure the controllability and security of corporate data. MDM technology provides management of user devices. On this basis, it is necessary to examine the effectiveness of solutions and tools from the perspective of data rather than equipment.

It is also important to keep in mind that MDM technology and other device management solutions must comply with applicable privacy laws and regulations. In general, enabling sensitive operations such as taking photos or recordings on the device without the user's knowledge, or uploading all of the user's text messages and photos to the organization's server, may result in a violation of applicable privacy laws.

3.8.3 *DLP technology and applicable scenarios*

In terms of dynamic goals, data security targets important and sensitive data in the organization and is committed to discovering data (Discover), protecting data (Protect), and monitoring data (Monitor), and cycles through these three stages, as shown in Fig. 3.50.

In the Data Discovery phase, the first step is to search for repositories that contain sensitive or personal data (these may be files or databases or other types). After finding the protected data in the system, identify the business workflows that involve this data and their corresponding data flows. After that, it is necessary to identify the owner of the data and whether the data is involved in external transmission.

In the data protection stage, appropriate protection mechanisms, such as encryption mechanisms and access control mechanisms, should be formulated according to the type, storage methods, and

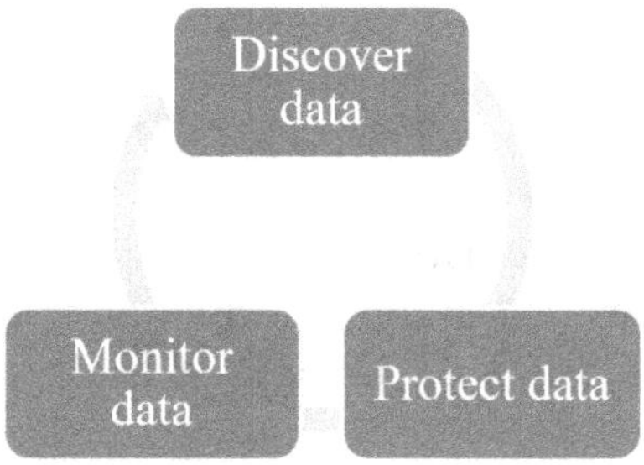

Fig. 3.50. Three stages of data security.

usage methods of the data obtained in the data discovery stage. For example, sensitive data that is no longer needed in the cloud platform should be deleted; sensitive data that is in an untrusted network should be moved to a trusted network or encrypted. In addition, appropriate access control should be formulated according to data access requirements and combined with the principle of privilege minimization.

At the data monitoring stage, automated means should be used to monitor the access of people and systems to data, especially the monitoring of data inbound/outbound transmission. Data scanning tools can be used to meet regulatory compliance requirements. Alerts, blocks, and logs can be recorded for abnormal data transfers.

Data Loss Prevention (DLP) or Data Leakage Prevention (DLP) refers to technologies and products that detect and prevent inappropriate data leakage through monitoring, detection, and blocking, etc. DLP technologies and products are in line with the characteristics of the three phases of "data discovery, data protection, and data monitoring" mentioned above and mainly focus on the implementation of the data monitoring phase.

The protected sensitive data can be divided into three categories: Data in Use, Data in Motion, and Data at Rest.

When designing and deploying DLP technologies and products, it is first necessary to classify and validate sensitive data and determine methods for identifying sensitive data. There are two main categories of commonly used techniques:

(1) **Data signature:** Define the identification rules of sensitive data according to the composition rules or content characteristics of

sensitive data. The most common form is regular expressions, which are used to identify sensitive data fragments in various fixed formats or components, such as ID numbers, bank accounts, phone numbers, and emails.

(2) **Data/file fingerprinting:** By analyzing all aspects of the information, such as the frame format, keyword composition, or content characteristics of the data or file, the unique characteristics of the data or file are extracted. Different vendors' DLP technologies and products use different data/file fingerprinting techniques, which are mainly used to identify complex data and files as a whole. Some vendors' technologies can still accurately identify data or files after certain changes.

The composition of the DLP system is shown in Fig. 3.51. The core of the DLP product is the inspection engine. The inspection engine uses configured or accumulated data characteristics and data/file fingerprints to analyze and identify incoming or captured data and handle corresponding sensitive data events according to the processing policies set by the organization. Common processing methods include denying (blocking) data dissemination, issuing warnings or reminders to relevant users or managers, encrypting sensitive data, or replacing/deleting relevant sensitive data.

In actual use by enterprises, there are two main deployment methods of DLP technology and products:

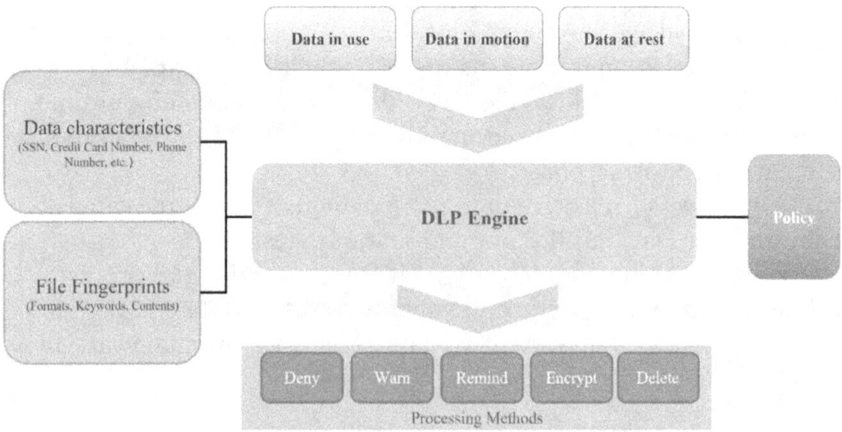

Fig. 3.51. DLP system composition.

(1) **Independent deployment:** Use independent hardware devices or install them on independent servers in the form of software to actively scan targets (hosts, file servers, etc.) or passively monitor network traffic, and block when sensitive data is found to be inappropriately used or alarm processing.
(2) **Integrated deployment:** Integrate with other products or devices, such as web security gateways, mail servers, IPS/IDS, Unified Threat Management (UTM), etc., and receive data through the integrated products or devices. When sensitive data is found to be used inappropriately, the operations will be blocked or alerted.

3.8.4 *Introduction of CASB and integration with DLP*

For nearly 20 years, organizations of all types have relied on the "walled garden" model for data security. In today's world, organizations are often faced with scenarios such as the proliferation of Bring Your Own Device (BYOD), increased employee mobility, and unauthorized employees accessing cloud services such as Office 365. Therefore, it is necessary to set a unified security policy across the multiple cloud services that an organization can access to protect the organization's assets and personal data.

The premise of establishing a unified security policy is that the organization's IT department has visibility into the cloud services that the organization is currently using, especially unapproved SaaS services and shadow IT (IT projects that the IT department does not control or even know about). According to the 2019 McAfee Cloud Adoption and Risk Report (McAfee, 2019), most IT professionals estimate that their organizations use about 30 cloud services, when in fact they use as many as 1,935 cloud services.

A Cloud Access Security Broker (CASB) is cloud-hosted software or local software or hardware that sits between the cloud service user and the cloud service provider, allowing organizations to extend the reach of their security policies from existing local infrastructure to cloud services and create new policies for cloud-specific contexts. In addition, CASBs provide visibility into an organization's security policies.

CASBs are becoming an integral part of enterprise-level security by enabling organizations to protect their sensitive data when using cloud services. According to a Datamation report (Anina Ot, 2022), the global CASB market is estimated to be $5.8 billion in 2020. It's expected to grow to $16.5 billion by 2027, at a compound annual growth rate (CAGR) of 16.2%.

The CASB acts as a policy enforcement hub, consolidating the enforcement of multiple types of security policies and applying them to everything an organization uses in the cloud, regardless of the device being used (including unmanaged smartphones, IoT devices, or personal laptops).

Through fine-grained data protection mechanisms and policies, CASB enables employees to protect the security of corporate data assets while using cloud services that can save time and improve work efficiency.

In general, a CASB solution provides the following basic functions, as shown in Fig. 3.52:

(1) **Visibility:** An organization may have many employees accessing multiple applications in many different cloud environments. When cloud usage is out of IT's sight, an organization's data is no longer subject to the organization's governance, risk, or compliance policies. To protect users, confidential data and intellectual property, CASB solutions provide complete visibility into cloud application usage, including user information, such as device and location. Cloud Discovery Analytics provides a risk assessment for each cloud service used, allowing an organization's security staff to decide whether to continue allowing access or block

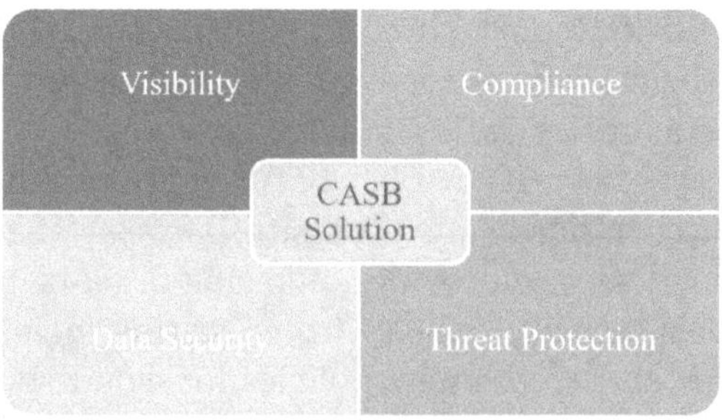

Fig. 3.52. Basic functions provided by the CASB solution.

applications. This information can also be used to implement more granular controls, such as granting different levels of access to applications and data based on an individual's device, location, and job function.

(2) **Compliance:** Whether enterprise data is stored in the enterprise data center or in the cloud, it must comply with applicable privacy and data security laws and regulations. CASB solutions can ensure compliance in the cloud by meeting various compliance laws such as HIPAA and regulatory or standard requirements such as ISO 27001 and PCI DSS. A CASB solution can identify areas of highest compliance risk and guide security teams on how to address them.

(3) **Data security:** The adoption of the cloud makes it possible to collaborate effectively over distance. While the seamless movement of data can have many benefits, it can also come at a significant cost to organizations that need to protect sensitive and confidential information. Traditional on-premises DLP solutions are designed to protect data, but their capabilities often do not extend to cloud services and lack cloud context. By combining a CASB solution with a modern DLP solution, IT can see when sensitive content is uploaded to the cloud, downloaded from the cloud, and moved from cloud to cloud. By deploying security features such as DLP, collaborative controls, access controls, information rights management (IRM), encryption, and tokenization, security operators can minimize the risk of corporate data breaches.

(4) **Threat protection:** Whether through negligence or malicious intent, employees and third parties with stolen credentials can leak or steal sensitive data from cloud services. Using machine learning-based User and Entity Behavior Analytics (UEBA) technology, CASB solutions can provide a holistic view of users' regular usage patterns to identify or pinpoint anomalous user behavior. CASB solutions detect and mitigate threats when someone attempts to steal data or gain unauthorized access. To defend against threats from cloud services, CASB solutions can block malware using features, such as adaptive access control, static and dynamic malware analysis, and threat intelligence.

In addition, CASB solutions can provide capabilities such as the following:

(1) Cloud threat analysis.
(2) Configuration auditing.
(3) Malware detection.
(4) Data encryption and key management.
(5) SSO and IAM integration.

The main purpose of CASB solutions is to provide visibility and control over data and threats to meet the data security needs of the organization. The typical mode of operation of the CASB solution is shown in Fig. 3.53.

(1) **Discovery:** CASB solutions provide auto-discovery capabilities to create an inventory of all third-party cloud services and who is using them.
(2) **Classification:** After obtaining a list of all cloud service usage in the discovery phase, the CASB solution can determine its risk level based on the type of application, what data the application accesses, and how the data is used and shared.
(3) **Remediation:** After identifying the risks associated with each application, the CASB solution can use this information to define and generate the organization's data policy and user access policy to meet security requirements.

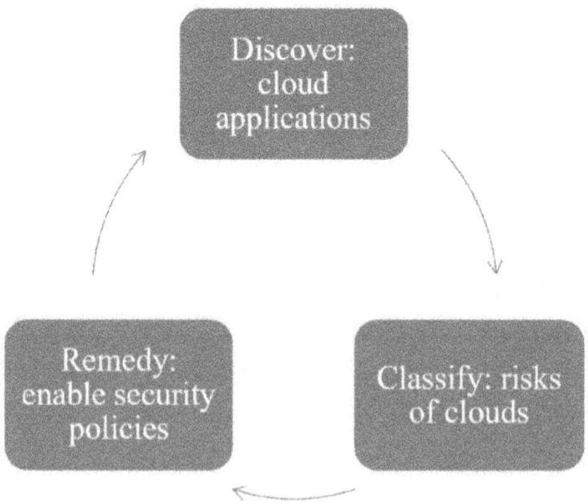

Fig. 3.53. CASB solution working mode.

CASB solutions provide an additional layer of protection on top of malware protection and data encryption mechanisms.

3.9 Chapter Summary

This chapter focuses on the data security architecture of organizations, products, and solutions and explains the requirements analysis, design, and presentation of the security architecture. Based on the data security architecture of a typical organization, this chapter introduces the organization's data protection model and its key components, and provides some typical examples of the security architecture.

In terms of the security design dimensions of systems, subsystems, and modules, this chapter focuses on the eight security design principles and cases in both positive and negative dimensions.

The Data security architecture is supported by a number of key technologies. To this end, this chapter introduces a number of basic key technologies such as file encryption, access control, and system monitoring, as well as application layer key technologies such as Data Leakage Protection (DLP).

The core goal of data security is to protect the confidentiality, integrity, and availability of data, and the key to achieving this goal is encryption and access control. Good encryption is based on the best practices of modern cryptography and key management.

Chapter 4

The Cornerstone of Data Security: Cryptography and Encryption

Protecting the confidentiality of information may be an issue that humans have pursued since the beginning of time. As early as BC, Caesar the Great of the Roman Empire (Gaius Julius Caesar, July 12, 100 BC–March 15, 44 BC, known in history as "Caesar") used some form of encryption to make military communications difficult to intercept and decipher. The term "cryptography" is used to describe the art of secret communication. The word comes from the Greek and means "secret writing".

Since its inception, cryptography has evolved with technology and grown in importance. History has shown that good cryptographic communication techniques can help win wars, both in ancient and modern times.

The "father of computer science" Alan Turing cracked the German Enigma cipher machine during World War II, giving the Allies a significant advantage, according to declassified files and documents.

In the modern information society, communication is becoming increasingly digital and remote. How to protect privacy and data security while enjoying the speed and convenience of digital communication is an inevitable issue in the development of communication services. In addition, with the rapid development of e-commerce, the security and credibility of digital transaction systems are becoming increasingly important. If the security of e-commerce cannot be guaranteed, global trade will be seriously threatened.

Table 4.1. The role of cryptography in data security.

Objectives of Data Security	Typical threats addressed	Related Cryptography Techniques
Confidentiality	Eavesdrop theft of information leakage of sensitive information	Symmetric and asymmetric encryption digital envelope
Integrity	Tampering replay attack	Hash functions and MAC data encryption digital signature
Authentication	Spoofing Identity	Passwords Digital certificates and digital signatures
Non-repudiation	Claiming to have not performed an action	digital signatures
Authorization	Access Control	Gain capabilities without proper authorization and Attribute certificate Access Control

Section 1.1.4 "Overall Objectives of Data Security" points out that the three core elements of data security are confidentiality, integrity, and availability. In more and more digital scenarios, these three core elements all rely to varying degrees on the basic services provided by cryptography, and cryptography relies on the engineering practice of key management.

Referring to the definition of the NIST SP 800–130 standard (Barker *et al.*, 2013), cryptography refers to the use of mathematical techniques to provide security services, and its data security objectives include confidentiality, data integrity, entity identity authentication, data source identity authentication, and anti-replay. The role of cryptography in data security is shown in Table 4.1. In a broader sense, cryptography refers to the technology or science related to the principles, methods, and procedures for converting plaintext to ciphertext and for recovering encrypted ciphertext to plaintext.

Cryptography is also the cornerstone of many cybersecurity and data security features and functions. The typical security scenario mapping table is shown in Table 4.2.

Table 4.2. Mapping between security features and cryptography services.

Scenario	Security features	Services provided by cryptography
HTTP connection	Transport Layer Security (TLS)	Symmetric encryption, hash algorithm, asymmetric encryption, DH key exchange, pseudo-random number generator
IP connection	IPSec	Symmetric encryption, digital signature, MAC, DH key exchange, pseudo-random number generator
Laptop data protection	Full Disk Encryption (FDE)	Symmetric encryption, hash algorithm
Database	Database Encryption (DBE)	Symmetric encryption, hash algorithm, asymmetric encryption

Of course, cryptography is not a castle in the air. In actual deployment scenarios, the security achieved by itself also needs the support of the security dimension. Its security is emphasized in the following two aspects:

(1) Access to confidential data such as keys and passwords requires authorization and authentication. This relies on the authentication mechanisms discussed in Section 1.3 "Identity Authentication and Data Access Control".
(2) Access to confidential data must support auditing, including monitoring, logging, and analysis functions.

When implementing a particular security service, a set of required **cryptographic schemes** is usually referred to as a cryptographic system or cryptosystem. Cryptographic schemes consist of four basic components:

(1) Plaintext information is the original message to be sent.
(2) Cryptographic system or cryptographic algorithm consists of mathematical encryption and decryption algorithms.
(3) Ciphertext is the result of applying an encryption algorithm to the original message before it is sent to the recipient.

(4) Key is the array of bits used by the algorithm during encryption and decryption.

The term "**cipher**" for cryptography is derived from the Arabic "sifr", which means "empty" or "zero". The encryption process refers to the conversion of plaintext information into ciphertext information using a cryptographic algorithm and a selected key. Ciphertext information is often presented in a form that is difficult to understand or read. The encrypted message can be transmitted through the communication channel to the intended recipient.

Encryption technology is one of the most commonly used security and confidentiality methods in the field of information security. Encryption technology uses technical means to convert important data into ciphertext (encrypted) before it is transmitted, and then restore it (decrypted) using the same or different means after it reaches its destination. Encryption technology is one of the most basic data security measures.

Encryption has two characteristics:

One feature is that only certain recipients can decrypt it, and others cannot see the plaintext. The schematic diagram of an encrypted transmission is shown in Fig. 4.1.

The other feature is the ability to prove who the sender is.

Encryption technology consists of two main elements: algorithm and key. According to different types of keys, cryptographic algorithms can be divided into the following three categories: symmetric key encryption, asymmetric key encryption, and hash functions. Each class of algorithm has its own role in the field of cryptography.

4.1 Origin and Evolution of Cryptography

Classical cryptography has a long history. From ancient times to the advent of computers, classical cryptography has mainly two basic methods:

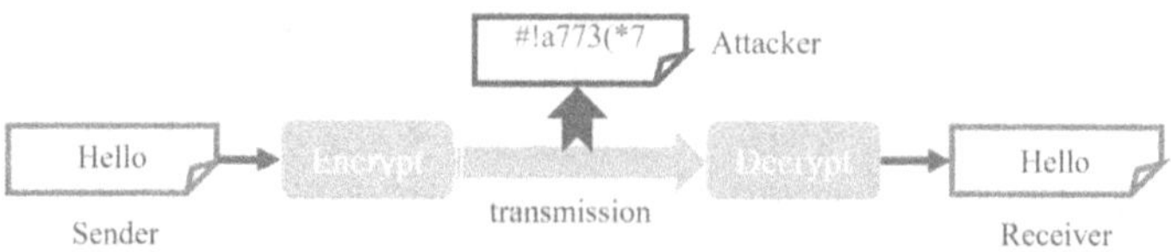

Fig. 4.1. Schematic diagram of encrypted transmission.

1. Transposition cipher

The letters of the plaintext remain the same, but the order is reversed. For example, "hello world" is transposed to "eholl owrdl".

2. Substitution cipher

A substitution cipher is the systematic substitution of one character of a plaintext with another character. For example, replacing the last letter with the previous letter in the Latin alphabet, as in the Caesar cipher.

4.1.1 *Caesar cipher*

Caesar cipher (Churchhouse, 2002), also known as Caesar's cipher, Caesar's code, and Shift cipher, is the simplest and most well-known encryption technique. It is named after Julius Caesar. According to records, Caesar used this method to communicate with his generals.

The Caesar cipher is a substitution cipher. All letters in the plaintext are shifted backwards (or forwards) according to a fixed number from 0 to 25 in the alphabet and then replaced with ciphertext. This number can be thought of as a kind of "encryption key".

The formula for the Caesar cipher is as follows:

Encryption formula: ciphertext = (plaintext + offset) mod 26
Decryption formula: plaintext = (ciphertext − offset) mod 26

Therefore, assuming that the offset is 13, at which point the name of the algorithm is noted as ROT13, the resulting correspondence between the plaintext and ciphertext alphabets is shown in Table 4.3.

Assuming that the plaintext is "hello", look up Table 4.3, and the corresponding ciphertext is "uryyb".

It is obvious that the Caesar cipher is a very insecure cipher. Even if we do not consider the attacks triggered by the probability of the distribution of plaintext letters, there are only 26 possible keys that an attacker only needs to be patient enough to enumerate.

Table 4.3. Mapping between plaintext and ciphertext letters.

Plaintext	a	b	c	d	e	f	G	h	i	j	k	l	m	n	o	p	q	r	s	t	u	v	w	x	y	z
Ciphertext	n	o	p	q	r	s	T	u	v	w	x	y	z	a	b	c	d	e	f	g	h	i	j	k	l	m

4.1.2 *Simple substitution ciphers*

The simple substitution cipher is a further development of the Caesar cipher. The program replaces a character with a predefined arrangement for another character.

The simplest solution is abcd...xyz in sequence and zyx...dcba in reverse order, see Table 4.4.

Thus, assuming the plaintext "hello" is looked up in Table 4.4, the corresponding ciphertext is "svool".

At this point, the mapping is the cipher key of this scheme.

Since there are 26 letters in English, there are 26! (factorial of 26). This number is approximately equal to $410^{2}6$, so the sample space for the key is large enough. Even with a modern computer, it is very difficult to crack the key using an exhaustive permutation method.

There are inherent flaws in the design of simple substitution ciphers, through chosen plaintext attacks, or through probabilistic attacks, it is easy to decrypt.

The substitution ciphers mentioned above are essentially single-letter ciphers. Their characteristic is that for a given key, the cipher letters corresponding to each fixed letter are fixed throughout the encryption process. For example, if "A" is encrypted to "D", "A" will always be encrypted to "D" for any scenario that occurs in that plaintext. These cryptographic schemes are extremely vulnerable to cryptanalysis.

Multi-letter ciphers are another type of substitution cipher. During the encryption process of a multi-letter password, the password letters corresponding to the plaintext letters can be different in different positions. The next two examples (the Playfair cipher and the Vigenère cipher) both use multi-letter ciphers.

The Playfair cipher treats the bigrams in the plaintext as a unit and converts these units into bigrams in the ciphertext. The frequency of occurrence of the ciphertext characters is balanced to some extent to resist cryptanalysis based on the probability of occurrence of the letters.

Table 4.4. Reverse order of alphabet.

Plaintext	a	b	c	d	e	f	g	h	i	j	k	l	m	n	o	p	q	r	s	t	u	v	w	x	y	z
Ciphertext	z	y	x	w	v	u	t	s	r	q	p	o	n	m	l	k	j	i	h	g	f	e	d	c	b	a

The core of the Playfair cipher is a 5×5 letter transformation matrix. There are 26 letters in English, but the letter J is rarely used in English, so it is agreed that wherever J appears, it is replaced by the letter I. In this way, both the plaintext and the ciphertext have a total of 25 letters.

Assuming that the key is "cipher", the remaining English letters are written in order into a 5×5 matrix, and the Playfair cipher table is shown in Table 4.5.

The encryption rule is to encrypt by letter pair. If the same letter occurs in a word, a separator (such as the letter "x") is added. If the word length is not even, add a suffix letter (usually "z") at the end.

For example, take the word "ballon". There are two "l"s, so a separator is added between them.

$$ballon \rightarrow balxloon$$

The calculation is a bit more complicated. If two letters of the plaintext are on the same line, the letters to the right of them are used instead.

A line search example of the Playfair cipher is shown in Fig. 4.2. The pair "ba" is replaced by "db".

If the two letters of the plaintext are in the same column, use the letters in the row below the letters to replace them (see Fig. 4.3).

Table 4.5. Playfair grid of letters.

c	i	p	h	e
r	a	b	d	f
g	k	l	m	n
o	q	s	t	u
v	w	x	y	z

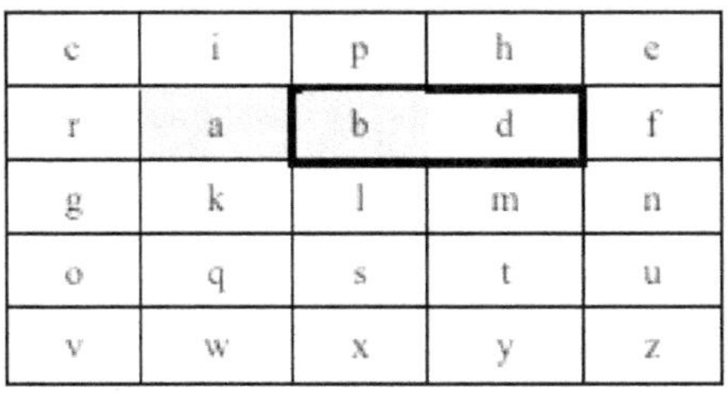

Fig. 4.2. Example of Playfair password peer lookup.

c	i	p	h	e
r	a	b	d	f
g	k	l	m	n
o	q	s	t	u
v	w	x	y	z

Fig. 4.3. Example of Playfair password same column lookup.

c	i	p	h	e
r	a	b	d	f
g	k	l	m	n
o	q	s	t	u
v	w	x	y	z

Fig. 4.4. Example of Playfair password rectangle search.

Pairs of letters "lx" in the same column are replaced by the letters below them. The letter "x" is in the last row, so wrap to the corresponding letter in the first row.

If the two letters of the plaintext are neither in the same row nor in the same column, draw a rectangle with the two letters as the two upper corners and use the letters at the other two upper corners of the rectangle instead, as shown in Fig. 4.4. The pair of letters "lo" can draw the rectangle in the figure, then the letters at the other two top corners are "gs".

So with "cipher" as the key, the word "ballon" is represented as "ba lx lo on" "db sp gs ug" "dbspgsug".

As long as the recipient of the message knows the key, he can construct the same 5×5 matrix and decrypt it in the same way.

The Playfair cipher is difficult to analyze. This is because the target of the analysis is no longer 26 English letters but $25 * 25 = 625$ possible pairs of letters. However, assuming the attacker has enough time and a certain amount of ciphertext samples, decryption is not impossible.

The Vigenère cipher is an extension of the Caesar cipher and introduces the concept of a key. The key can take the form of a word. The offset of the letters of the plaintext is determined based on the

Table 4.6. Vigenère cipher.

Plaintext	a	t	t	a	c	k	f	r	o	m	n	o	r	t	h
Key	c	i	p	h	e	r	c	i	p	h	e	r	c	i	p
Offset	3	9	16	8	5	18	3	9	16	8	5	18	3	9	16
Ciphertext	d	c	j	i	h	c	i	a	e	u	s	g	u	c	x

key, and then the corresponding replacement letters are determined as a way to combat word frequency statistics.

Suppose the plaintext is "attack from north". First, the key is repeated up to the length of the plaintext, and then the corresponding offset is calculated according to each letter of the key (assuming that the letter "a" corresponds to an offset of 1, and the letter "z" corresponds to an offset of 26); the Virginia cipher table is shown in Table 4.6.

Using the Vigenère cipher table, we can see that the letter "a" is shifted by 3 to get the letter "d", the letter "t" is shifted by 9 to get the letter "c", and so on. The resulting ciphertext is "dcjihciaeusgucx".

When decrypting, a similar table can be constructed using the same key so that the plaintext corresponding to the ciphertext can be obtained letter by letter.

The formula used is as follows:

Encryption formula: ciphertext = (plaintext + key) mod 26 – 1
Decryption formula: plaintext = [26 + (ciphertext – key)] mod 26 + 1

The Vigenère cipher is much more secure than the Caesar cipher by reducing the efficiency of ciphertext analysis. Historically, some sensitive military information has been preserved using the Vigenère cipher.

There are two variants of the Vigenère cipher:

(1) The length of the key is the same as the length of the plaintext, which is called the Vernam cipher. The Vernam cipher provides better security than the Vigenère cipher.
(2) Under certain conditions, the Vigenère cipher has the properties of "one-time pad" and "perfect security".

The specific conditions mentioned above are as follows:

(1) The length of the key must be equal to the length of the plaintext.
(2) Each character in the key must be selected from the character table using true random numbers.
(3) The key is used only once.

Other commonly used ciphers include transposition ciphers, another type of cipher in which the order of letters in the plaintext is reorganized to form the ciphertext. Individual letters in the plaintext are not replaced one by one.

An example is the "simple column transposition" cipher, in which the plaintext is written horizontally with a certain letter width (the value of the key) and the ciphertext is read vertically, sometimes called a rail fence cipher or a columnar transposition cipher.

Take the plaintext "THE LONGEST DAY MUST HAVE AN END" as an example. Assume the selected key is 6. Obtain the transposition password table horizontally, see Table 4.7.

The ciphertext obtained in the vertical way is as follows:

TGYANHEMVDESUELTSAODTNAHE
Other ciphers worth mentioning include the Hill cipher.

The Hill cipher was invented in 1929 by Lester S. Hill. It is a polygraphic substitution cipher based on linear algebra. The Hill cipher uses matrices and matrix multiplication to mix plaintext.

To counter accusations that his system was too complex for everyday use, Hill built a cipher machine for his system that used a series of gears and chains. However, the machine was never actually sold.

Hill's major contribution was the use of mathematics to design and analyze cryptographic systems. It should be noted that the analysis of this cipher requires a branch of mathematics called

Table 4.7. Transposition cipher table.

T	H	E	L	O	N
G	E	S	T	D	A
Y	M	U	S	T	H
A	V	E	A	N	E
N	D				

number theory. Many textbooks on elementary number theory cover the theory behind the Hill cipher.

The Hill cipher uses the basic principles of matrix theory and uses matrix multiplication to replace each letter. Each character is treated as a hexadecimal number: A = 0, B = 1, C = 2, ... A string of letters is treated as an n-dimensional vector, multiplied by an $n \times n$ key matrix, and then the result is modulo 26. The advantage of the Hill cipher is that it completely hides the frequency information of the characters, but its weakness is that it is easily broken by known plaintext attacks.

Suppose the key used is "DSECURITY". Convert to a 3∗3 matrix:

$$\begin{bmatrix} 3 & 8 & 4 \\ 2 & 10 & 17 \\ 8 & 19 & 24 \end{bmatrix}$$

The plaintext message is "ATTACK AT DAWN", and every 3 letters must be converted into a one-dimensional matrix. For example, according to the representation method agreed above, the first 3 letters "ATT" are transformed into a $3 * 1$ matrix:

$$\begin{bmatrix} 0 \\ 19 \\ 19 \end{bmatrix}$$

Multiply that matrix with the key matrix:

$$\begin{bmatrix} 3 & 8 & 4 \\ 2 & 10 & 17 \\ 8 & 19 & 24 \end{bmatrix} \times \begin{bmatrix} 0 \\ 19 \\ 19 \end{bmatrix} = \begin{bmatrix} 418 \\ 703 \\ 817 \end{bmatrix} (mod\ 26) = \begin{bmatrix} 2 \\ 1 \\ 11 \end{bmatrix}$$

The corresponding ciphertext is "CBL".

To decrypt the ciphertext, simply multiply the ciphertext by the corresponding inverse matrix. The calculation of inverse matrices involves more complex mathematical knowledge, such as modular multiplicative inverse, determinants of matrices, and adjugate matrices, which will not be discussed here.

4.2 The Birth of Modern Cryptography

Classical cryptography is mainly based on linguistics and relies on human intelligence, skill, and creativity for both invention and decryption. In modern times, due to the rapid development of computers and the exponential growth of computing power, classical cryptography is no longer applicable and its cracking cost is insignificant in front of computers. The computers with their powerful computing power eliminated classical cryptography and at the same time gave birth to modern cryptography.

Kerckhoffs' principle of cryptography (Petitcolas, 2023) (also known as Kerckhoffs' desideratum, assumption, axiom, or law) is one of the fundamental principles of cryptography. It was formulated by the Dutch cryptographer Auguste Kerckhoffs in the late 19th century. The principle can be expressed as follows: even if the entire design and algorithm of the cryptosystem is publicly known, if only the key is kept secret, the cryptosystem should be secure.

In 1949, Claude Shannon published the paper "Communication Theory of Secrecy Systems" (Shannon, 1949), which was the first qualitative leap in cryptography and marked the real beginning of modern cryptography. Claude Shannon reformulated the Kerckhoffs' principle as "the enemy understands the system", a phrase known as "'s maxim".

In 1976, Diffie and Hellman published the article "New Directions in Cryptography" (Hellman, 1976), marking the birth of the public key cryptosystem, which can be considered as the second qualitative leap in cryptography.

Three cryptography experts, Ron Rivest, Adi Shamir, and Len Adleman, described the first widely used public key algorithm in 1977 (Rivest *et al.*, 1978) and named it the RSA encryption algorithm after the initials of each person's name. The algorithm is based on the mathematical treatment of two large prime numbers and their product. Its strength is thought to be related to the difficulty of factoring large integers. Taking advantage of the current and foreseeable speed of modern digital computers, choosing long enough prime numbers when generating RSA keys should make the algorithm infinitely secure. However, this idea is not mathematically proven, and either a fast factoring algorithm or an entirely different way to break RSA encryption is possible. Moreover, when a practical quantum computer is developed, factoring large prime numbers will no longer be an intractable problem.

Nevertheless, the RSA public-key cryptosystem represents a new stage in the development of cryptography.

Table 4.8. XOR operation.

$\oplus$	0	1
0	0	1
1	1	1

In addition, modern encryption algorithms no longer focus on characters or words but are based on byte or bit encryption.

Among bit operations, the XOR operation (see Table 4.8) is widely used in encryption algorithms.

For example, $0011 \oplus 0101 = 0110$.

The specificity of the XOR operation is as follows:

$x \oplus x = 000\ldots$ (XOR itself results in 0)
$x \oplus 000\ldots = x$ (XOR 0 results in itself)
$x \oplus 111\ldots = \bar{x}$ (XOR 1 results in inversion of all bits)
$x \oplus y = y \oplus x$ (XOR is symmetrical)
$(x \oplus y) \oplus z = x \oplus (y \oplus z)$ (XOR is associative)

From the associability of XOR, it can be deduced that if x and y are XORed twice, the result will still be x (y = z scenario). Due to this property, a simple encryption method was developed: one-time password (OTP).

The security of various cryptographic algorithms in classical cryptography and even in modern cryptography is based on "wisdom" and "creation", so it may be difficult to discover their security flaws, but it is difficult to prove that they are secure. One day in the future, this cryptographic algorithm may be completely broken. In order to break out of the vicious cycle of "creating solutions and breaking solutions", modern cryptography is based on a more rigorous and scientific foundation. According to the views of cryptography expert J. Katz in the book "Introduction to Modern Cryptography" (Katz and Lindell, 2014), the three main principles of modern cryptography are as follows:

Principle 1 — Formulation of Precise Definitions: The first step in solving any cryptographic problem is to formulate a rigorous and precise definition of security.

Principle 2 — Reliance on Precise Assumptions: If the security of a cryptographic construction relies on an unproven assumption, that assumption must be stated precisely. Furthermore, the assumption should be as minimal as possible.

Principle 3 — Rigorous Proofs of Security: Cryptographic constructions should be accompanied by a rigorous proof of security with respect to a definition formulated according to Principle 1 and with respect to an assumption stated according to Principle 2 (if an assumption is needed at all).

Based on these three principles, modern cryptography has evolved from an art to a science, gradually moving toward standardization and formalization.

Principle 1: Formulate Precise Definitions.

One of the most important intellectual contributions of modern cryptography has been the realization that formal definitions of security are essential prerequisites for the design, use, or study of any cryptographic primitive or protocol:

(1) **Importance for design:** In short, without a clearly defined goal, there is no way to know if (or when) the goal will be achieved. Having a clear and precise definition clarifies the effort of designing cryptographic algorithms, and the quality of the construction scheme can be better evaluated and thus improved.
(2) **Importance for use:** Suppose you want to use an encryption scheme in some larger systems. How do you know which encryption scheme to use? If there are several candidate ciphers, how do you know which cipher is sufficient for the scenario? These questions can be answered by having a precise security definition for a given encryption scheme, as well as a security proof associated with formal security assumptions (in accordance with Principles 2 and 3). In particular, the user can first define the security required by the system and then verify that the given encryption scheme satisfies it. Note that it may be unwise to choose the "most secure" encryption scheme. A weaker security definition may be sufficient for the system's use case and may be more efficient.
(3) **Importance for research:** Given two construction schemes, how do we compare them? Comparing efficiency is the most intuitive and observable, but it is definitely not enough to consider only efficiency. A highly efficient scheme may be

completely insecure and unusable. The precise specification of the level of security achieved by a scheme provides another point of comparison. If two schemes are equally efficient, but the first satisfies a stronger definition of security than the second, then the first is preferable. Alternatively, there may be a trade-off between security and efficiency (see the previous two points), but at least with precise definitions, we can understand what this trade-off entails.

Principle 2: Reliance on Precise Assumptions.

Most modern cryptographic constructions cannot currently be proven to be secure in an unconditional sense. Instead, security is often based on some widely believed (though unproven) assumption. The modern cryptographic approach requires that such assumptions be clearly and unambiguously defined. This allows not only an objective evaluation of the assumption but also, more importantly, a rigorous proof of security:

(1) **Verification of assumptions:** An assumption is a statement that has not been proven but is assumed to be true. It is necessary to examine the credibility of the assumption on which cryptography is based. The more carefully an assumption is examined and no counterexamples are found after testing, the more credible the assumption is. In addition, the study of an assumption may provide supporting evidence for its validity that can be derived from some other widely believed assumptions. Therefore, a prerequisite for increasing the credibility of an assumption is to have a precise statement of what exactly is being assumed.
(2) **Comparison of schemes:** Often in cryptography we are presented with two schemes, both of which can be shown to satisfy some definition but each with respect to a different assumption. Assuming that both schemes are equally efficient, which scheme should be preferred? If the assumption on which one scheme is based is weaker than the assumption on which the second scheme is based (i.e., the second assumption implies the first), then the first scheme is to be preferred, since it may turn out that the second assumption is false while the first assumption is true. If the assumptions used by the two schemes are incomparable, then the general rule is to prefer the scheme based

on the better-studied assumption (for the reasons emphasized in the previous paragraphs).

(3) **Ease of proof:** Modern cryptographic schemes are presented along with security proofs. If the security of the scheme cannot be proved unconditionally and must rely on some assumption, then a mathematical proof that "the construction is secure if the assumption is true" can only be provided if there is a precise statement of what the assumption is.

Principle 3: Rigorous Proofs of Security.

The first two principles discussed above lead naturally to this principle. Modern cryptography emphasizes the importance of rigorous proofs of security for proposed schemes. The fact that exact definitions and precise assumptions are used means that such a proof of security is possible. But why is proof necessary? The main reason is that the security of a design or protocol cannot be verified in the same way that software is typically verified. For example, the fact that encryption and decryption "work" and that the ciphertext looks garbled does not mean that a sophisticated adversary cannot break the scheme. Without proof that no adversary of the specified strength can break the scheme, we must rely on our intuition that this is the case. Of course, intuition in general is very problematic. In fact, experience has shown intuition to be disastrous in cryptography and computer security. There are countless examples of unproven schemes that were broken (sometimes immediately and sometimes years after they were presented or even deployed).

Most modern cryptographic proofs use the so-called reductionist approach. Given a theorem of the form:

"If the given assumptions X are correct, then the construction Y is secure according to the given definition".

Proof usually shows how to reduce the assumptions X to defeat the construction scheme Y. In particular, proofs typically show (via a constructive argument) how an adversary's method of defeating construction Y leads to a conflict with assumption X.

The above three principles constitute a rigorous modern cryptographic approach that differs from the *ad hoc* approach of classical cryptography. The *ad hoc* approach may lack one of the above three principles and often ignores all three. Unfortunately, *ad hoc* approaches are still developed and used by those who want a “fast and direct” construction scheme (or who are unaware of rigorous proofs). Only cryptographic algorithms produced by strict adherence to the above principles should be widely accepted and used.

4.3 Key-based Encryption

The study of cryptography has long focused on the confidentiality of information — first converting it from an intelligible form into an unintelligible form and then converting it again on the other side so that it cannot be read by an interceptor or eavesdropper who does not have the secret knowledge (the key needed to decrypt that information). In fact, the terms cryptography and encryption can be used interchangeably.

A good cryptographic algorithm ensures that once the ciphertext is transmitted over a protected or unprotected channel, it is very difficult to crack the ciphertext and get the plaintext. A strong encryption algorithm can also support the transmission of ciphertext in open channels without worrying about the interception and recovery of the plaintext. The decryption process is to recover the plaintext from the ciphertext using the key and the decryption algorithm. A hallmark of a good cryptosystem is that the security of the system does not depend on the secrecy of the encryption or decryption algorithm but on the secrecy of the key. This means that encryption algorithms can be understood and used by many people as long as the key is kept secret. In fact, before the 1970s, key-based encryption algorithms were basically symmetric algorithms. Symmetric means that both the encryption and decryption algorithms use the same key.

4.3.1 *One-time password*

One-time password (OTP) was first invented in 1917 by Major Joseph Mauborgne and Gilbert Vernam of AT&T.

The one-time cipher uses only the XOR operation, i.e., XORing each bit of the plaintext data stream and the random key stream to obtain the ciphertext stream. The one-time password diagram is shown in Fig. 4.5. Theoretically, this encryption method provides the strongest security guarantee. If every bit of the keystream is truly random, i.e., the attacker cannot predict it in any way, and the key is used only once, the attacker will not be able to obtain any useful

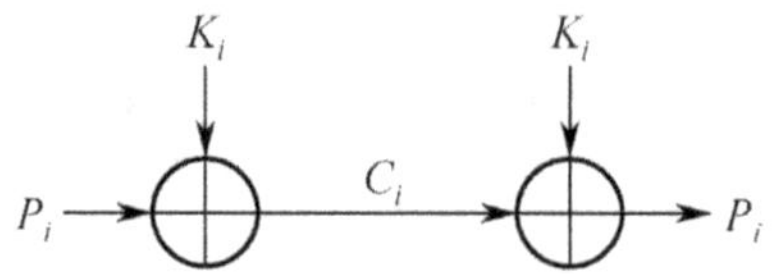

Fig. 4.5. Schematic diagram of a one-time password.

information in the plaintext, even if he can obtain the ciphertext. Of course, "any useful information" here is somewhat absolute; in fact, the attacker can still sense the existence of the message and know the length of the message. Some modern cryptosystems are even designed to hide the existence of the message and try to hide its length.

Assuming the attacker gets C_i without knowing K_i, he still cannot get any information about P_i. According to Shannon's theorem, this can be defined as zero information entropy. This property is called "perfect security".

Of course, the perfect security of one-time passwords depends on their correct use. The first condition for correct use is that the keystream is truly random, and the second condition is that the keystream is used only once. It is worth noting that many commercial products that claim to support OTP have difficulty meeting either or both of these conditions. In practice, perfect security is unrealistic.

Pseudorandom keystreams undermine "perfect security" but do not necessarily pose fatal security risks, and reused keystreams are more destructive. Suppose the attacker obtains two ciphertexts C_1 and C_2 using the same keystream K:

$$C_1 \oplus C_2 = (P_1 \oplus K) \oplus (P_2 \oplus K)$$

According to the characteristics of XOR operation, the above form can be transformed into

$$C_1 \oplus C_2 = P_1 \oplus K \oplus P_2 \oplus K = P_1 \oplus P_2 \oplus K \oplus K$$

And the result of K is 0 after its own XOR. So the XOR result of the ciphertext is equal to the XOR result of the plaintext:

$$C_1 \oplus C_2 = P_1 \oplus P_2$$

At first glance, this doesn't seem like a problem. But the real problem is that the XOR of two plaintext messages reveals important information about the plaintext itself.

OTP attacks also include crib-dragging attacks.

As mentioned above, the correct use of one-time passwords provides the strongest security guarantee of "zero information entropy" in the information-theoretic dimension. If one-time passwords can be successfully applied, most problems in cryptography can be solved, but it is obviously not realistic in terms of technical implementation.

First, OTP requires that the length of the key be equal to the length of the message; second, these keys must be securely exchanged between the two communicating parties or even more parties, and the exchange must be completed before the information is transmitted. If the recipient of the communication is unknown, the key exchange is more difficult. Finally, OTP requires keys to be truly random, key generation is likely to be hardware dependent, and performance is a key risk.

Clearly, to solve the communication security problem from an engineering perspective, it is necessary to have a key of appropriate length, maintain the confidentiality of the key, and negotiate the communication key over an unreliable transmission channel.

4.3.2 *Symmetric encryption*

Symmetric encryption refers to an encryption method that uses the same key (same encryption and decryption credentials) for both encryption and decryption. The sender uses a key to encrypt plaintext data into ciphertext and then sends it out. After receiving the ciphertext, the receiver uses the same key to decrypt the ciphertext into plaintext and read it. The schematic diagram of symmetric encryption is shown in Fig. 4.6.

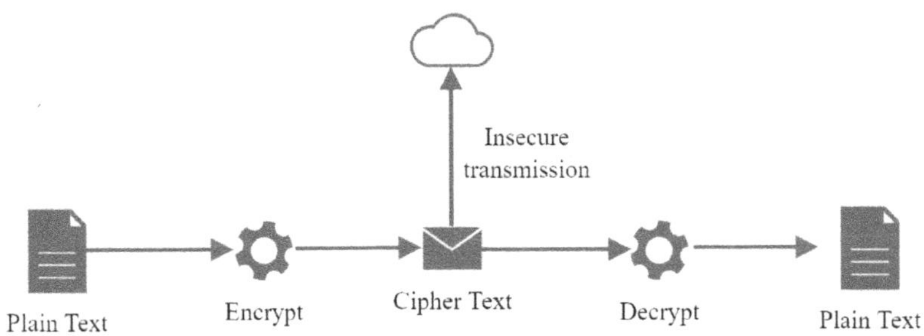

Fig. 4.6. Schematic diagram of symmetric encryption.

Symmetric encryption has low computation and high speed and is suitable for scenarios where a large amount of data is encrypted.

However, symmetric encryption has two shortcomings:

One is the issue of key transfer (or key distribution). Since symmetric encryption uses the same key for encryption and decryption, the security of symmetric encryption depends not only on the strength of the encryption algorithm itself but also on whether the key is securely transmitted and stored. How the sender securely transmits the key to the receiver has become a key issue in symmetric encryption.

The second is key management. As the number of keys increases, key management issues will gradually emerge. For example, when encrypting user information, it is necessary to use a different encryption/decryption key for each user. Otherwise, once the key is leaked, it is equivalent to leaking all the user information. Therefore, the cost of key management is very high.

Of course, in the communication application scenario, the above problem is also manifested in the fact that each sender and receiver must share a key in pairs. Symmetric encryption requires a prior relationship between the sender or receiver of a communication to establish a connection and obtain a key. In the uncertain communication environment of the Internet, it's impossible to predict everyone's communication goals and distribute keys in advance. In a distributed environment with complex communication topologies, it is difficult for both the sender and receiver to maintain enough keys to support all communications.

In addition to the issues discussed above, the use of symmetric encryption involves the following issues:

(1) The integrity of the data may be compromised because the recipient cannot confirm whether the message has been altered before receiving it.
(2) It is possible for the sender to deny the message because there is no mechanism to ensure that the message was sent by the purported sender.
(3) If the encryption process is compromised, the attacker is likely to observe the key in plaintext, and confidentiality cannot be guaranteed at that time.
(4) If the key is not changed frequently enough, confidentiality may be compromised.

Therefore, modern cryptosystems usually do not use symmetric encryption alone but rather a combination of symmetric and asymmetric encryption to achieve encryption.

Symmetric ciphers are implemented as block ciphers or stream ciphers. Typically, block ciphers encrypt input in the form of blocks of plaintext, while stream ciphers encrypt input in the form of individual characters.

4.3.3 *Block ciphers*

Block cipher is a mode of symmetric encryption.

Described by the formula, a block cipher provides a cryptographic function E that takes a plaintext block P and a key K as input and finally produces a ciphertext block C as output:

$$C = E(P, K)$$

A block is a sequence of bits of fixed length. The fixed length of a block is related to the specific packet cipher algorithm and is called the block size.

The decryption function corresponding to the encryption function E is denoted as D, and the decryption process can be described as follows:

$$P = D(C, K)$$

The difference between a block cipher and a stream cipher is that a stream cipher is based on a specific algorithm that generates a randomized stream of keys, which are then combined with the plaintext on a bit-by-bit or byte-by-byte basis to generate a stream of ciphertext; the plaintext of a block cipher is divided into blocks according to the block lengths specified by the block cipher algorithm, and each block is encrypted independently.

Common block ciphers include 3DES, AES, IDEA, Blowfish, and RC5. Correspondingly, the most common stream cipher is RC4. In theory, the longer the key, the longer the encryption time and the stronger the security. The strengths of common cryptographic algorithms are shown in Table 4.9.

Block cipher algorithms continue to evolve with modern cryptographic systems. The DES algorithm was standardized in 1977 and is one of the first widely used block cipher algorithms. With the rapid

Table 4.9. The security strength of cryptography algorithms.

Algorithm	Strength	Key size/bit	Algorithm	Strength	Key size/bit
3DES	High	64, 112, 168	Blowfish	Low	32 448
AES	High	128, 192, 256	RC4	Low	40
IDEA	High	64, 128	RC5	High	32,64,128

development of computing power, the 56-bit key length of the DES algorithm has become difficult to resist brute force attacks. In January 1999, distributed.net, in cooperation with the Electronic Frontier Foundation (EFF), publicly cracked the DES key in 22 hours and 15 minutes, proving that the DES algorithm is no longer secure. Some theoretical studies of DES algorithms also prove that DES algorithms are vulnerable to linear cryptanalysis. The Triple DES (3DES) algorithm is an evolution of the DES algorithm. From a technical point of view, the triple DES algorithm is safe for the time being. However, the triple DES algorithm has the shortcomings of slow processing speed and difficult software implementation, and is no longer the most reasonable choice for a block cipher algorithm. However, from the point of view of the evolution of cryptography, it is valuable to explain the DES algorithm.

4.3.3.1 *DES*

The Data Encryption Standard (DES) is a 16-rounds Feistel-structure cipher, with a 64-bit block length and a 56-bit key. The basic structure of DES is shown in Fig. 4.7. Due to the short key length of only 56 bits, the DES algorithm is too insecure for modern applications, but it has been very influential in the development of modern cryptography.

As shown in Fig. 4.7, during the encryption process, DES first divides the plaintext data into 64-bit blocks of information. In subsequent work, each 64-bit plaintext information block is independently encrypted into 64-bit ciphertext. All DES encryption processes consist of 19 steps, including 1 round of initialization permutation (IP) performed on the plaintext data block, 1 round of key block permutation and left circular shift operations on the key, 16 rounds of key permutation, and the final round of inverse initial permutation performed.

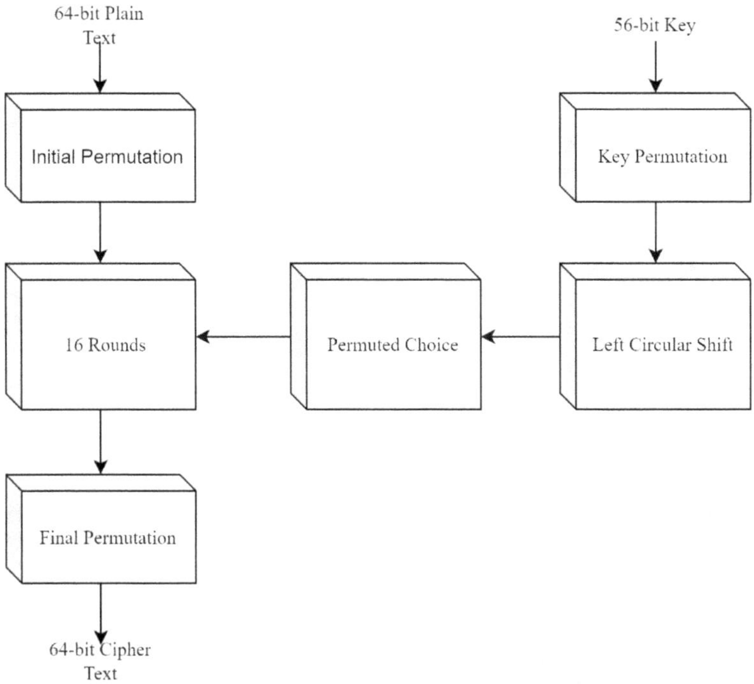

Fig. 4.7. DES basic structure.

Many design elements of DES were quite advanced at the beginning of its development, which influenced the development path of symmetric encryption algorithms. DES uses the S-box, which is now an integral part of almost all block ciphers. The round function structure of DES is the Feistel network, which is now one of the classic round function structures.

In the early days, the 56-bit key made DES extremely vulnerable to "brute force attacks". A well-known improvement measure is the Triple Data Encryption Standard (Triple-DES, 3DES), also known as the Triple DES algorithm. 3DES is an improvement for the security problem that the key of the DES algorithm is too short. It achieves increased key length and security by simply performing DES 3 times. The operation of the Triple DES algorithm is divided into three steps: encryption-decryption-encryption (Encrypt-Decrypt-Encrypt, EDE). It works by taking three 56-bit keys (K_1, K_2, and K_3), first encrypting with K_1, then decrypting with K_2, and finally encrypting with K_3. There are two versions of the 3DES algorithm: dual-key and

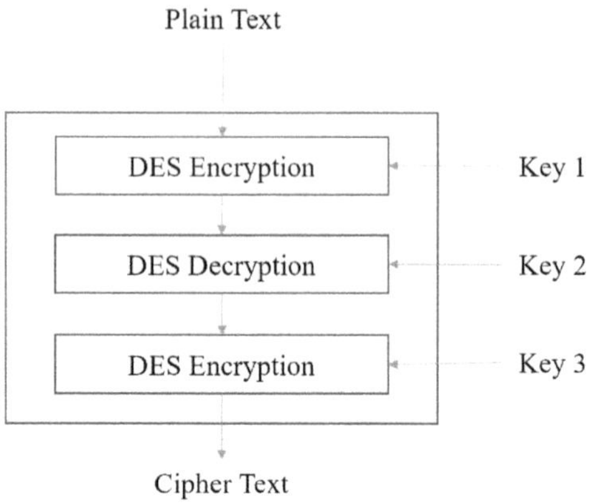

Fig. 4.8. 3DES algorithm encryption process.

triple-key. The dual-key version runs the same algorithm three times but uses K_1 for the first and last pass. In other words, $K_1 = K_3$. If $K_1 = K_2 = K_3$, then the triple DES algorithm is actually DES. The encryption process of the 3DES algorithm is shown in Fig. 4.8. When encrypting the plaintext, three encryption processes are used, of which the first and third encryption algorithms are used for DES, and the decryption algorithm is used for the second time. This produces the final ciphertext.

4.3.3.2 *AES*

As the security of the DES algorithm gradually weakened, in January 1997, the U.S. National Institute of Standards and Technology (NIST) issued a worldwide call for the Advanced Encryption Standard (AES) to replace the DES algorithm, with the requirements that AES be no less secure and more efficient. The AES call received responses from many cryptographers around the world. Many people submitted their own designs, and ultimately five algorithms, namely Rijndael, Serpent, Twofish, RC6, and MARS, were selected for the final round of selection. After security analysis and performance evaluation, the Rijndael algorithm became the Advanced Encryption Standard (AES). The Rijndael algorithm is a block cipher. Its block lengths are 128 bits, 160 bits, 192 bits, 224 bits, and 256 bits. The key lengths also include these five types. However, in AES, only the block length is chosen to be 128 bits, and the key length is any one of 128 bits, 192 bits, and 256 bits. In this sense, the AES algorithm can be considered a subset of the Rijndael algorithm.

The security of the AES algorithm is higher than similar algorithms such as DES, it uses the S-box as the only non-linear component, the structure is simple, and its security is easy to analyze. There is currently no known way to break the AES algorithm. As a typical block cipher, the AES algorithm is more flexible than the DES algorithm, with faster processing speed and higher efficiency, and can be quickly encrypted using software and hardware. Under certain conditions, the AES algorithm has a lower memory requirement compared to the DES algorithm and is very suitable for performing encryption and decryption operations in a limited space environment. These characteristics strongly support the AES algorithm to replace the DES algorithm, and it is gradually widely accepted and used.

Due to the above origin, the AES algorithm is also called Rijndael cipher in cryptography. The AES algorithm includes four operations: byte substitution (SubBytes), row shifting (ShiftRows), mixed columns (MixColumns), and round key addition (AddRoundKey). Figure 4.9 shows the flow of AES encryption and decryption, from which we can see the following:

(1) Each step of the decryption algorithm corresponds to the inverse operation of the encryption algorithm.
(2) The order of all encryption and decryption operations is exactly reversed.

The above points, together with the reciprocal operation of each step of the encryption algorithm and the decryption algorithm, ensure the correctness of the algorithm. Each round key in encryption and decryption is obtained from the seed key by the key expansion algorithm, such as $W[0,3]$ in Fig. 4.9. The 16-byte plaintext, ciphertext, and round key in the algorithm are represented by a 4×4 matrix.

4.3.4 *Modes of operation of block ciphers*

Block cipher is a deterministic encryption algorithm in which an encryptor operates on a fixed-length group of bits using a key. The fixed-length group of bits is called a block. Block ciphers are more secure than stream ciphers, another form of symmetric encryption.

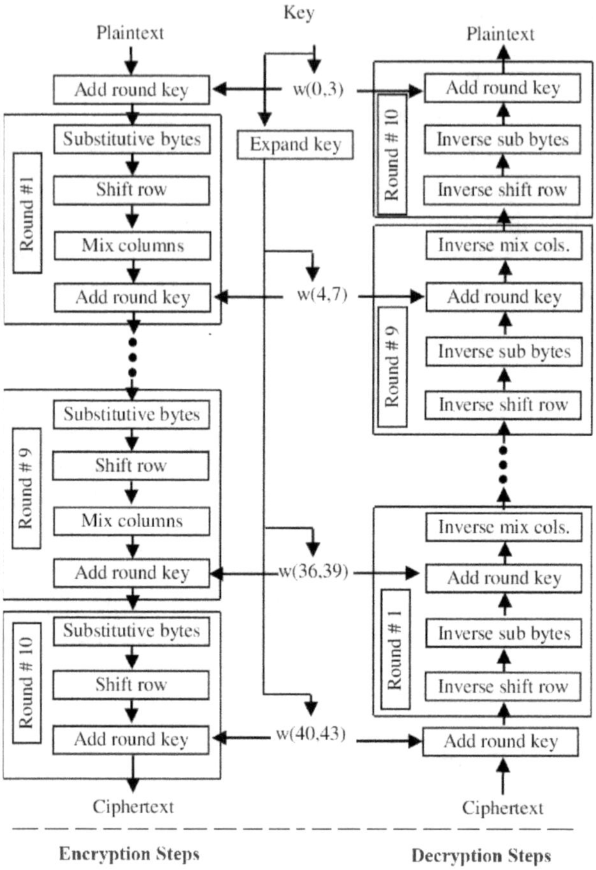

Fig. 4.9. AES encryption and decryption process.

Block ciphers are an important basic component in the design of many cryptographic protocols and are widely used in bulk data encryption scenarios.

A block cipher is a secure cryptographic transformation (encryption or decryption) of a fixed-length group of bits called a block. The operation of a block cipher describes how the block operations of the cipher can be repeatedly applied to securely encrypt or decrypt amounts of data larger than a block.

When performing simple block cipher operations, the same plaintext block will produce the same ciphertext block, which is vulnerable to cryptanalysis or replay attacks. Therefore, some operating modes introduce correlation between ciphertext blocks to solve this problem. Similarly, to prevent two encryptions of the same message from producing the same ciphertext, the concept of an initialization vector (IV) is introduced. The combination of a randomly generated initialization vector with the first plaintext block and the key ensures that the ciphertext for all blocks does not match the previously encrypted ciphertext.

There are five common block cipher modes of operation: ECB, CBC, OFB, CTR, and GCM. They are described in detail in the following.

4.3.4.1 *ECB mode*

The simplest encryption mode is the Electronic Code Book (ECB) mode. The message to be encrypted is divided into several blocks (P_1, P_2, P_3, ...) according to the block size of the block cipher, and each block is encrypted independently, and the corresponding ciphertext C_1, C_2, C_3 is output. The ECB mode is shown in Fig. 4.10. If the plaintext is not a complete packet and the last segment does not

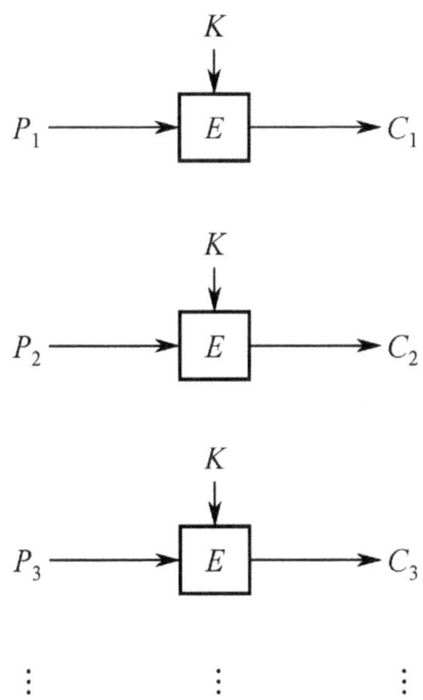

Fig. 4.10. ECB encryption mode.

match the length of the data block, the missing bytes are filled or added as needed for the calculation.

As you can see, the encryption operation in ECB mode is simple, and since the encryption of each data block is independent, both encryption and decryption can be computed in parallel, which is more efficient. This independent encryption mode also makes each piece of data independent of the others, ensuring that errors in encryption operations are not transmitted.

However, it is precisely because of this simple encryption mode that ECB mode encryption is vulnerable to the cryptanalytic attack of pattern recognition. If there are two identical blocks of data in the plaintext, there must also be two identical blocks of data in the ciphertext. This feature provides an opportunity for subsequent attacks.

To visually demonstrate the extent to which ECB mode reveals plaintext information in ciphertext, the "ECB penguin" image (see Fig. 4.11) is widely used (Valsorda, 2013). A bitmap version of this image (see the left image in Fig. 4.11) can be encrypted into the right image in Fig. 4.11 after being encrypted in ECB mode.

It can be seen that the ciphertext encrypted with ECB mode can still see the plaintext information, which makes ECB mode very insecure. Similarly, ECB mode causes the protocol using it to lack data integrity protection and to be vulnerable to replay attacks. Since each block of data is encrypted in the same way, an attacker can benefit from replaying the ciphertext even if he does not know the corresponding plaintext. An intuitive but impractical example is replaying a ciphertext packet containing the transfer object and the transfer

Fig. 4.11. Comparison between the original image (left) and the encrypted image in ECB mode (right).

amount. Therefore, more secure encryption modes are needed for plaintext encryption.

4.3.4.2 *CBC mode*

In 1976, IBM invented the Cipher Block Chaining (CBC) encryption mode and filed a U.S. patent (Ehrsam *et al.*, 1978) in 1978. In the CBC mode, each plaintext block P_i is first XORed with the previous ciphertext block C_{i-1} and then encrypted. In this encryption mode, each ciphertext block depends on all preceding plaintext blocks. At the same time, an initialization vector (IV) must be used in the first block to ensure the uniqueness of each message. CBC mode encryption is shown in Fig. 4.12.

If the index of the first plaintext block is 1, then the CBC mode encryption process can be summarized by the following formula:

$$C_i = E_k(P_i \oplus C_{i-1}) \tag{4.1}$$

$$C_0 = IV \tag{4.2}$$

And the decryption process is

$$P_i = D_k(C_i) \oplus C_{i-1} \tag{4.3}$$

$$C_0 = IV \tag{4.4}$$

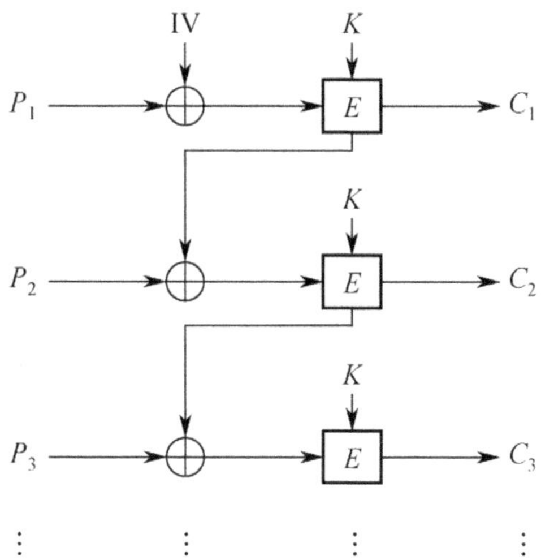

Fig. 4.12. CBC encryption mode.

CBC mode is one of the most commonly used encryption modes. Each block of ciphertext (except the first) is dependent on the previous block, making it less vulnerable to attack, more secure than ECB mode, and suitable for transmitting long messages. CBC mode is also the standard for secure transmission protocols, such as SSL, IPsec, and so on. However, since the encryption of each data block depends on the encryption result of the previous data block, the encryption process can only be done serially and cannot be parallelized. During decryption, a plaintext block can be obtained from two adjacent ciphertext blocks. Therefore, the decryption process can be parallelized.

It should be noted that this encryption mode causes each ciphertext block to depend on all information blocks. During encryption, small changes in each plaintext block will cause changes in all subsequent ciphertext blocks, while during decryption, changing one bit in the ciphertext will only cause the corresponding plaintext block to be completely changed and the corresponding bit in the next plaintext block to be changed and will not affect the contents of other plaintext blocks. At the same time, encryption and decryption can only be performed correctly if both the sender and the receiver know the initialization vector.

Also, like ECB mode, CBC mode requires the plaintext to be padded to an integer multiple of the block size. Due to the padding process, the decrypted result is not necessarily the original plaintext but may also contain padding bits that must be removed to restore the plaintext.

As you can see from the original image (left) and the result (right) of the original image shown in Fig. 4.13, the result of CBC encryption is similar to random noise compared to ECB mode, making it difficult to see information in plaintext. Other cryptographic algorithms or other cryptographic modes described in the following may also produce this randomization result. However, it should be noted that the "random result" shown on the right in Fig. 4.13 requires rigorous proof to show that the image has been securely encrypted.

4.3.4.3 *OFB mode*

The Output Feedback (OFB) mode is shown in Fig. 4.14, which can turn a block cipher into a synchronous stream cipher. The keystream

Fig. 4.13. Comparison between the original image (left) and the encrypted result in CBC mode (right).

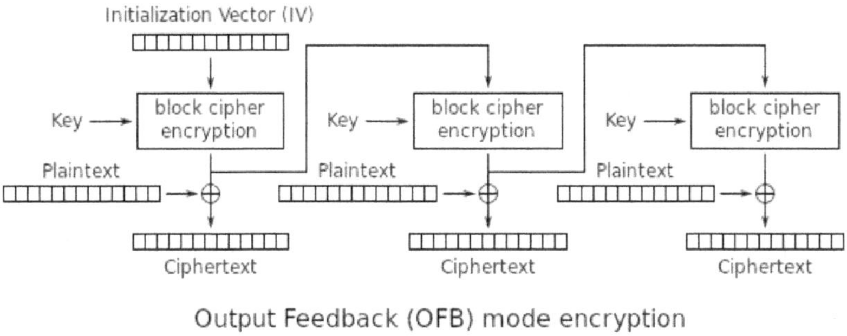

Fig. 4.14. OFB mode.

blocks are generated and then XORed with the plaintext block to obtain the ciphertext. As with other stream ciphers, flipping a bit in the ciphertext flips the same bit in the plaintext. This property allows many error correction codes to function normally, even when applied before encryption.

Based on the symmetry of the XOR operation, the encryption and decryption operations are exactly the same:

$$C_i = P_i \oplus O_i \tag{4.5}$$

$$P_i = C_i \oplus O_i \tag{4.6}$$

$$O_i = E_k(O_{i-1}) \tag{4.7}$$

$$O_0 = IV \tag{4.8}$$

In the above formula, O_i represents the i-th output block. Formula 4.7 represents that O_i is the result of E_k encryption of $O_i - 1$. Each output feedback block cipher operation depends on all previous ones and so cannot be performed in parallel.

It is possible to obtain an OFB mode keystream by using the CBC mode with a constant string of zeros as input. This can be useful because it allows fast hardware implementations of CBC mode to be used for OFB mode encryption.

In the case of CFB, a single bit error in one block is propagated to all subsequent blocks. OFB solves this problem because it is free of bit errors in the plaintext block. The disadvantage of OFB is that it is more susceptible to a message stream modification attack than CFB due to its operating modes.

4.3.4.4 *CTR mode*

Counter mode (CTR mode), as shown in Fig. 4.15, is also called ICM mode (Integer Counter Mode) and SIC mode (Segmented Integer Counter). The CTR mode converts a block cipher into a stream cipher. It generates a continuous key stream by incrementing an encryption counter, where the counter can be any function that guarantees no repeated output for a long time, but using an ordinary accumulating counter is the simplest and most common approach. The CTR mode no longer encrypts the ciphertext, but instead encrypts an incrementing counter, and then XORs the encrypted bit sequence with the plaintext group to obtain the ciphertext.

In CTR mode, the bit sequence XORed with the plaintext packet is different each time. Therefore, CTR mode solves the problem in ECB mode that the same plaintext gets the same ciphertext.

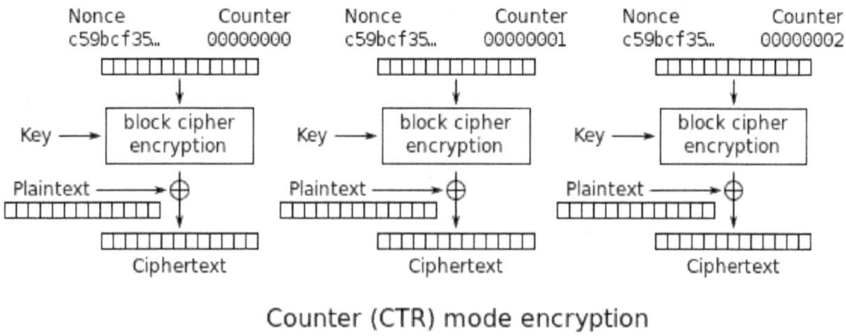

Fig. 4.15. CTR mode.

CBC mode can also solve this problem, but the other two advantages of CTR mode encryption are as follows:

(1) Supports parallel computation of encryption and decryption.
(2) An incorrect 1 bit in the ciphertext will only affect the corresponding 1 bit in the plaintext.

Note that the "one-time random number" (nonce) in Fig. 4.15 is similar to the IV in CBC mode. Random IVs, one-time random numbers, and counters can generate different ciphertexts from the same plaintext using reversible operations such as concatenation, addition, or XOR.

Although the CTR mode solves the problem of error propagation and serial encryption in the CBC mode, the CTR mode still does not provide the integrity checking function of the ciphertext message. If integrity protection of the ciphertext message is required, it is recommended to use the Block cipher mode of the Message Authentication Code.

4.3.4.5 *GCM mode*

In general, a Message Authentication Code (MAC), usually in the form of a keyed hash function, is used to verify the authenticity and integrity of a message.

Informally, a Message Authentication Code system consists of three algorithms:

(1) A key generation algorithm selects a key from the key space uniformly at random.
(2) A signing algorithm efficiently returns a tag given the key and the message.
(3) A verifying algorithm efficiently verifies the authenticity of the message given the same key and the tag. That is, return accepted when the message and tag are not tampered with or forged, and otherwise return rejected.

Galois/Counter Mode (GCM) combines the well-known counter mode of encryption with the new Galois mode of authentication. The key feature is the ease of parallel computation of the Galois field multiplication used for authentication. This feature permits higher throughput than encryption algorithms.

Galois Message Authentication Code (GMAC) (Dworkin, 2007) is an authentication-only variant of the GCM that can form an incremental message authentication code. Both GCM and GMAC can accept IVs of arbitrary length. GCM can take full advantage of parallel processing and implementing GCM can make efficient use of an instruction pipeline or a hardware pipeline. The CBC mode of operation introduces pipeline congestion that hampers its efficiency and performance.

The **GCM** mode (shown in Fig. 4.16) uses a counter that is incremented for each block and calculates a message **authentication tag** (MAC code) after each block is processed. The final authentication tag is calculated from the last block. Like all counter modes, GCM works as a stream cipher, and so it is essential that a different IV is used at the beginning for each stream that is encrypted.

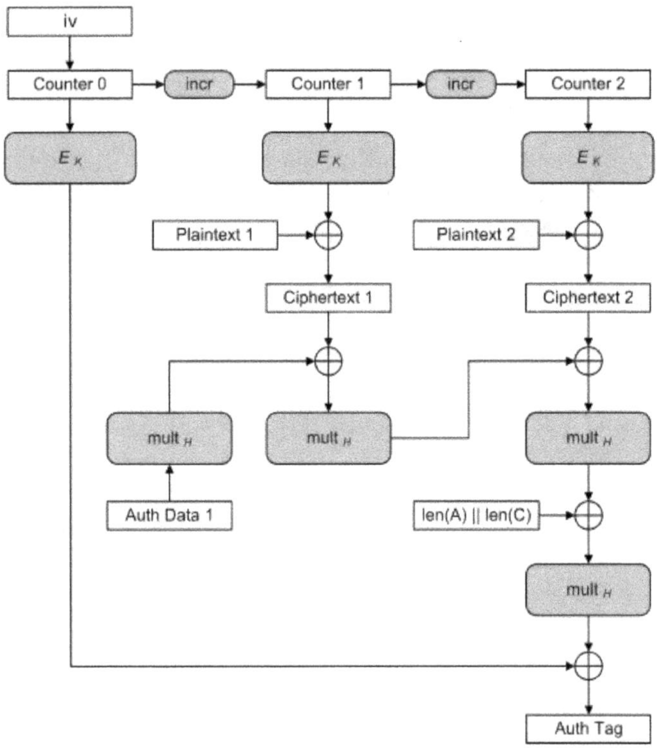

Fig. 4.16. Process of GMAC mode.

As shown in Fig. 4.16, A represents the data used for authentication, and C represents the encrypted ciphertext, $mult_H$ is to multiply the input authentication data 1 and the key H on the finite field $GF(2^128)$, and A is additional data for authentication.

Galois/Counter Mode (GCM) can solve the problem of error transmission while parallelizing encryption/decryption and can also perform message integrity verification. The design motivation of the GCM mode is to realize an authentication and encryption mode with efficient hardware implementation, good software performance, provable security, and no patent, which is suitable for network transmission and application environments that require high speed.

4.3.4.6 *Summary of modes of operation for block ciphers*

Block cipher modes (such as CBC, CFB, OFB, CTR, and GCM) are the ability to repeatedly apply single-block encryption/decryption of the cipher to securely encrypt/decrypt amounts of data larger than the block length.

Some grouping modes (such as CBC) require the input to be split into blocks and a padding algorithm (such as adding special padding characters) is used to pad the last block to the standard block length. Other block-mode ciphers, such as CTR, CFB, OFB, and GCM, don't require padding at all because they XOR byte-by-byte or bit-by-bit.

Different block cipher modes are suitable for different scenarios and must be selected according to specific requirements and constraints.

In general, the most commonly used secure grouping modes are CBC, CTR, and GCM, which require a random (unpredictable) initialization vector (IV) to generate a one-time random number at the beginning.

In most cases, the CTR mode is the appropriate choice. It has strong security and parallel processing capabilities and can accept arbitrary data length (no need to pad). Note that it does not provide authentication and integrity protection, only encryption.

GCM mode has all the advantages of CTR mode and adds message authentication (generates an encrypted message authentication tag). GCM mode is a fast and efficient way to implement authenticated encryption in symmetric ciphers and is generally highly recommended.

CBC mode works in fixed size blocks and therefore requires padding; most applications use the PKCS7 padding scheme. Under certain conditions, CBC mode is vulnerable to padding oracle attacks. Therefore, it's best to avoid using CBC mode for encryption.

A notoriously insecure mode is ECB mode, which encrypts equal input blocks into equal output blocks.

In summary, when designing a data security solution based on a block cipher algorithm, it is necessary to consider not only the security and performance of the algorithm itself but also the appropriate block cipher mode of operation.

4.3.5 *Stream ciphers*

Modern cryptography can be divided into two categories: symmetric cryptography (single-key cryptography) and asymmetric cryptography (public-private key cryptography). Stream cipher is a type of symmetric cipher, also known as sequence cipher. It combines the plaintext data stream and the pseudorandom cipher data stream (key stream) to form a ciphertext data stream. In terms of implementation, the two data streams are generally combined bit by bit, and the combined operation is an exclusive OR (XOR). Ideally, the length of the plaintext and ciphertext data streams is unlimited. In practice, streams of any length required by the business are generally supported.

The keystream is typically generated serially from a random seed using a digital shift register. At this point, the random seed is the encryption and decryption key for the ciphertext stream. Stream ciphers differ from block ciphers, which perform fixed, invariant transform operations on blocks. The distinction is not clear-cut. In certain modes of operation, block cipher primitives are used in ways that effectively implement stream ciphers. Stream ciphers generally perform faster and have less hardware complexity than block ciphers. However, stream ciphers are prone to serious security problems if used improperly. Crucially, the same initial state (random seed) should never be used twice.

Streaming ciphers can be seen as an approximation of one-time passwords (OTPs), which have proven to be unbreakable. OTPs use a completely random key stream that is combined verbatim with the plaintext stream to form a ciphertext stream. In 1949, Shannon proved that one-time passwords were secure. However, OTP requires that the key stream be completely random, at least as long as the plaintext, and cannot be used more than once, which makes the key extremely difficult to generate, distribute, and manage.

Stream ciphers use a smaller, more convenient key, such as 128 bits. Based on this key, stream ciphers produce a pseudorandom keystream that can be combined with plaintext, similar to an OTP implementation. However, this type of keystream is pseudorandom, not truly random. As a result, the security proofs associated with one-time pads are no longer valid, and it is quite possible for a stream cipher to be completely insecure.

4.3.5.1 *Encryption process of stream cipher*

The idea of stream cipher encryption and decryption is very simple. Divide the plaintext stream m into its basic unit (such as bit), then use the key k_1 (seed) to generate a key stream k, and encrypt bit by bit with the plaintext stream m to obtain the ciphertext. Decryption is performed by decrypting the ciphertext stream c bit by bit with the same synchronously generated key stream k to recover the plaintext stream m. The stream cipher encryption and decryption process are shown in Fig. 4.17.

Let the plaintext stream be

$$m = m_1 m_2 \cdots m_i \ldots, m \in M$$

The key stream is

$$K = k_1 k_2 \cdots k_i \ldots, k \in K$$

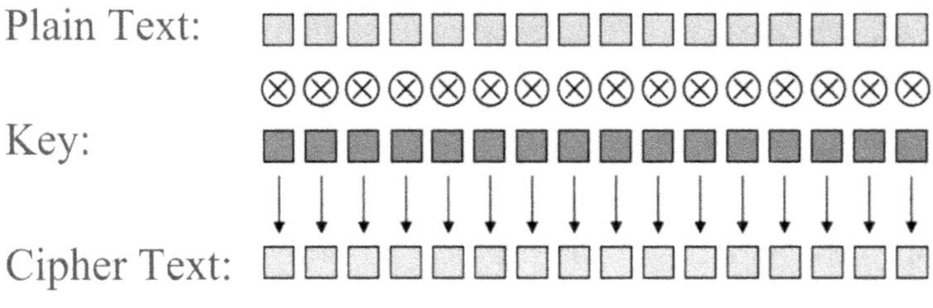

Fig. 4.17. Stream cipher encryption and decryption process.

The ciphertext stream is

$$c = c_1 c_2 \cdots_i \ldots, c \in C$$

Then the encryption process can be expressed as

$$c = c_1 c_2 \cdots c_i \cdots = E_{k_1}(m_1) E_{k_2}(m_2) \cdots E_{k_i}(m_i) \cdots$$

The decryption process can be expressed as

$$m = m_1 m_2 \cdots m_i \cdots = D_{k_1}(c_1) D_{k_2}(c_2) \cdots D_{k_i}(c_i) \cdots$$

4.3.5.2 *Typical stream cipher algorithms*

The inherent characteristics of stream ciphers determine that their application scenarios are generally not as rich as block ciphers. Common stream cipher algorithms include the A5/1 algorithm used in the GSM (Global System for Mobile communication, also known as 2G) network and the RC4 algorithm.

The RC4 algorithm is a stream cipher developed in 1987 by Ron Rivest of the RSA Security Company. Its official name is "Rivest Cipher 4", but the acronym RC can also be understood as "Ron's Code". The RC4 algorithm has a variable key length and uses byte-oriented rather than bit-oriented operations.

The RC4 algorithm has fast execution speed, small memory footprint, simple software and hardware implementation, and wide application. It uses a 256-byte state vector $S_{[0]}$ $S_{[255]}$ and a variable-length key of k bytes (the value ranges from 1 to 256) to initialize the state vector. At all times, S contains permutations and combinations of 8-bit unsigned numbers from 0 to 255. During encryption and decryption, a byte k is generated from S by systematically selecting one of the 255 entries, and then the entries in S are permuted again.

The RC4 algorithm was originally developed as a trade secret. However, it was posted anonymously to the Cypherpunks mailing list in September 1994 and then posted to the sci.crypt newsgroup, which attracted a lot of attention. The RC4 algorithm used to be a widely used stream cipher system, especially in transmission security protocols and standard dimensions (such as the Secure Sockets Layer SSL protocol and the WEP protocol in wireless communications). Since then, cryptanalysis and attacks on the weaknesses of the RC4 algorithm and implementation have appeared many times, leading to the prohibition of the use of the RC4 algorithm in all versions of TLS (Transport Layer Security) in 2015 in RFC 7465 (Popov, 2015).

Similar to the RC4 algorithm, the A5/1 algorithm used to secure wireless communications in the 3GPP standard for the GSM network is also a stream cipher algorithm. The principle is also to encrypt the plaintext with the key stream at the sender and then decrypt the ciphertext with the same key stream at the receiver.

The A5/1 algorithm is one of the seven security algorithms specified by GSM. The A5/1 algorithm was also kept secret in the early stages but became public knowledge through document leaks and reverse engineering. Some serious weaknesses in the algorithm have been discovered.

The A5/1 algorithm was developed in 1987, and the slightly weakened A5/2 algorithm in 1989. Although both were initially kept secret, the overall design was leaked in 1994. In 1999, Marc Briceno reverse engineered the two algorithms from the software of a GSM mobile phone (Briceno *et al.*, 1999).

GSM radio waves are transmitted in burst sequences. On a typical channel, a burst containing 114 bits of useful information is transmitted every 4.615 ms. The A5/1 algorithm is used to generate a 114-bit key stream sequence for each burst, which is XORed with 114 bits before modulation. The A5/1 algorithm is initialized with a 64-bit key and a 22-bit public frame number.

The A5/1 algorithm uses three irregularly clocked linear feedback shift registers (LFSR). The lengths of these three LFSRs are 19 bits, 22 bits, and 23 bits, respectively, for a total of 64 bits. The key K of the A5/1 algorithm is also 64 bits. In GSM, each time a call is made, the base station generates a 64-bit random number that is used in conjunction with the key (referred to as K) in the mobile phone's SIM card, and an encryption algorithm is used to generate the master key used during the call. This master key is used for the initial population of the three LFSRs. Once the three LFSRs are populated with keys, keystream generation can begin. The lifecycle of a call key is from the beginning to the end of a call. Once the call is complete, the call key for that call is discarded.

Over the decades, there have been numerous cryptanalytic and practical attacks on the security of the A5/1 algorithm.

The first attack on the A5/1 algorithm was proposed by Ross Anderson in 1994 (Anderson, 1994). Anderson's basic idea was to guess the entire contents of the LFSR R1 and R2, and about half the contents of the linear feedback shift register R3. This way, all three LFSRs can be clocked and the contents of the second half of R3 can be calculated.

In 1997, Golić JD *et al.* proposed an attack method based on solving a system of linear equations with a time complexity of 240.16 (the unit of time complexity is the number of solutions to the system of linear equations required) (Golić, 1997).

Since then, there have been actual attacks against the A5/1 algorithm (Biryukov *et al.*, 2001) in GSM. In 2003, Barkan *et al.* published several attacks on GSM encryption (Barkan *et al.*, 2008). The first type of attack is an active downgrade attack. Attackers attempt to temporarily cause targeted GSM phones to use the much weaker A5/2 cipher. The A5/2 algorithm is easily broken. The second attack is a time, memory, and tradeoff attack on the ciphertext, which requires a lot of precomputation.

Nevertheless, the A5/1 algorithm is still a well-designed and widely used stream cipher algorithm. Since its invention in 1987, although GSM has gradually faded from the historical stage, there is still no publicly available technical cracking method for the A5/1 algorithm.

4.3.5.3 *Salsa20 stream cipher algorithm*

Salsa20 (Bernstein, 2008) is a new stream cipher developed by cryptographer Dan J. Bernstein. Bernstein is known for designing several modern cryptographic algorithms and making their software implementations available as open source. There are two smaller variants of Salsa20, Salsa20/12, and Salsa20/8, which are the same algorithm but have 12 and 8 fewer rounds, respectively, than the original 20 rounds.

ChaCha (Bernstein *et al.*, 2008) is another orthogonal adaptation of the Salsa20 cipher that attempts to increase the number of diffusions per round while maintaining or improving performance. There is no "20" after ChaCha, but there is a number after the particular algorithm (ChaCha8, ChaCha12, and ChaCha20), which indicates the number of rounds. Salsa20 and ChaCha represent the state of the art in modern stream cipher technology. There are currently no attacks against Salsa20, ChaCha, or even their proposed reduced-round variants, so their actual security cannot be compromised. Both cipher families are also very fast. For longer data streams, on modern Intel and AMD processors, Salsa20 needs about 4 CPU cycles per byte for the full version, about 3 CPU cycles per byte for the 12-round version, and about 3 CPU cycles per byte for the 8-round version. ChaCha is slightly faster than Salsa20 on most platforms. However, ChaCha is more than three times faster than RC4 and three times faster than AES-CTR.

Salsa20 has two interesting features. First, it is possible to "jump" to a particular point in the keystream without computing all the previous bits. This is especially useful in scenarios where large files are being encrypted since random reads can be done in the middle of the file. Many encryption schemes require the entire file to be decrypted, but with Salsa20 it is possible to select only the desired portion of the encrypted file to decrypt.

This ability to "jump" also means that Salsa20 blocks can be computed independently of each other, allowing encryption or decryption to occur in parallel, which can improve performance on multi-core CPUs. At the same time, it is resistant to many types of side-channel attacks. This is achieved by ensuring that no key is used to choose between different code paths in the cipher and by ensuring that each round consists of a fixed number of constant-time operations. The result is that each block is generated with exactly the same operands, regardless of the key. Both the Salsa20 and Chacha stream ciphers are based on the ARX cipher design (Mouha and Preneel, 2013). The ARX cipher is characterized by modular addition (A), rotation with fixed amounts of rotation (R), and XOR (X). The advantage is that the operation is essentially constant time and resistant to differential cryptanalysis. These cryptographic algorithms also perform well on modern CPU architectures without cryptographic-specific optimizations. They work by taking advantage of general-purpose vector instructions, which means that the CPU performs related operations on multiple pieces of data in a single instruction. Thus, even though AES has dedicated hardware, ChaCha20's performance is competitive with AES on modern Intel CPUs.

4.3.5.4 *Stream cipher summary*

The key sequence in the stream cipher is a pseudo-random sequence obtained by a key stream generator and a deterministic algorithm, which is the fundamental difference between it and the "one-time password". The system of stream cipher does not have the perfect security of "one-time password", but it improves the practicality. If the algorithm is properly designed, the security of stream cipher can meet the requirements of practical applications. The security strength of a stream cipher is determined by the key stream. Therefore, the analysis of the security and reliability of pseudorandom sequences and the method of generating them has become a very important issue in the field of stream cipher design and analysis.

4.4 Public Key Encryption

Symmetric key cryptography uses the same key to encrypt and decrypt data. A major drawback of this method is that it is very difficult to manage the key securely. Ideally, each pair of different communicating parties must share a different key. Even each individual message must be encrypted with a different key to achieve perfect forward secrecy. As a result, the number of keys required increases exponentially, requiring complex key management schemes to maintain key consistency and confidentiality for both sender and receiver.

Due to the aforementioned problems with symmetric ciphers, a more modern encryption scheme is needed to address these shortcomings.

4.4.1 *Key exchange*

The goal of key exchange is to solve a seemingly impossible problem. Suppose the two people communicating (Alice and Bob) have not met but need to negotiate a shared secret on the communication channel. Suppose also that the channel is insecure so that the content transmitted over the channel can be eavesdropped on.

In 1976, W. Diffie and M. Hellman made a seminal breakthrough in the field of key exchange when they first publicly introduced the concept of public key cryptography in their published article "New Directions in Cryptography" (Hellman, 1976).

The DH key exchange algorithm, named after the initials of the two people's names, depends on the difficulty or complexity of the mathematics. It can be visualized as the problem that it is easy to compute from the "right" direction but difficult to compute from the "wrong" direction.

Since W. Diffie and M. Hellman proposed the idea of public key cryptography in 1976, a variety of public key encryption algorithms based on the same or different mathematical foundations have appeared in the world. For example, the RSA algorithm is based on the factorization problem of large prime numbers, the ElGamal public key encryption algorithm (ElGamal, 1985) is based on the discrete logarithm problem on finite field multiplicative groups, and the Elliptic Curve Cryptography (ECC) is based on the discrete logarithm problem on elliptic curves. Some new public key encryption algorithms, such as NTRUEncrypt, are based on the shortest vector problem in a lattice, in an attempt to defend against the impact of possible quantum computing on public key cryptography.

The mathematical formula underlying the discrete logarithm problem can be stated as follows:

$$y \equiv g^x (mod\ p)$$

The symbol "≡" here means "modular congruence". Given the integers x, g, and p, computing y is easy (at least for a computer). Given y, g, and p, it is very difficult to compute x. This is called the discrete logarithm problem. The reason for the term "discrete logarithm" is that if there is no modulo operation, and y and g are known, computing x is called a logarithmic problem.

Now, suppose Alice and Bob want to communicate over an unreliable channel. The constant information they both know is a large prime number p and the base number g.

Each person chooses a random number to be kept secret, which will be recorded as r_a and r_b. Perform the following calculations:

$$m_a \equiv g^{r_a} (mod\ p) \tag{4.9}$$

$$m_b \equiv g^{r_b} (mod\ p) \tag{4.10}$$

m_a and m_b can be transmitted on the network. According to the characteristics of discrete logarithm, it is very difficult to calculate r_a and r_b.

After the two got each other's calculation results, they added their own confidential random numbers. Alice calculates the following:

$$s \equiv (g^{r_a})^{r_b} (mod\ p)$$

Table 4.10. Comparable strengths.

Security Strength/bit	IFC Key Length/bit	ECC Key Length/bit
56	512	112
80	1024	160
112	2048	224
128	3072	256
256	15360	512

Bob's calculation is slightly different in form. But according to mathematical principles,

$$(g^{r_a})^{r_b} = (g^{r_b})^{r_a}$$

Both will get the common result s. This is the shared secret negotiated by the two.

Suppose there is an adversary Eve who observes on the communication channel, can see m_a and m_b, and knows g and p, but it is difficult to compute r_a and r_b, and it is also difficult to compute s.

The above DH key exchange algorithm is based on the computational complexity problem of discrete logarithms. The DH key exchange algorithm can also be based on the computational complexity of elliptic curves. Since the mathematical formulas involved are too complex, they are not discussed here.

It is worth noting that elliptic curve problems are mathematically harder to solve than ordinary discrete logarithm problems. Therefore, based on the same security level and algorithm strength requirements, the elliptic curve algorithm has a smaller key length. Refer to the key length table of different algorithms shown in Table 4.10.

4.4.2 *Public-key cryptography*

Public key cryptography (often referred to as asymmetric encryption) uses two different keys: a public key that is known to everyone, and the other is the private key, which is kept by the sender and receiver, respectively.

The application scenarios of public-key cryptography algorithms can be divided into four types:

(1) **Key pair generation:** Generate a random key pair of "private key + corresponding public key".
(2) **Key exchange:** The communication parties exchange session keys.
(3) **Encryption/decryption:** The sender encrypts the data with the public key, and the receiver decrypts the data with its own private key. The application of public key encryption algorithm is shown in Fig. 4.18.
(4) **Digital signature/message authentication:** The sender signs the message with his own private key, and anyone can verify the signature with the public key.

As shown in Fig. 4.18, to encrypt a message from sender A to receiver B, both sender A and receiver B must create their own key pair. Sender A and receiver B publish their public key — anyone can obtain the public key. When sender A sends message M to receiver B, sender A encrypts M with receiver B's public key. Once M is received, receiver B decrypts message M using its own private key. Simply ensuring that only receiver B has receiver B's own private key ensures that only receiver B can decrypt the message. This ensures not only data confidentiality but also data integrity since an attacker would need recipient B's private key to modify the data. Therefore, public key cryptography can guarantee data confidentiality and integrity.

Since the public key can be made public and has a mathematical correspondence to the private key, all public key schemes are

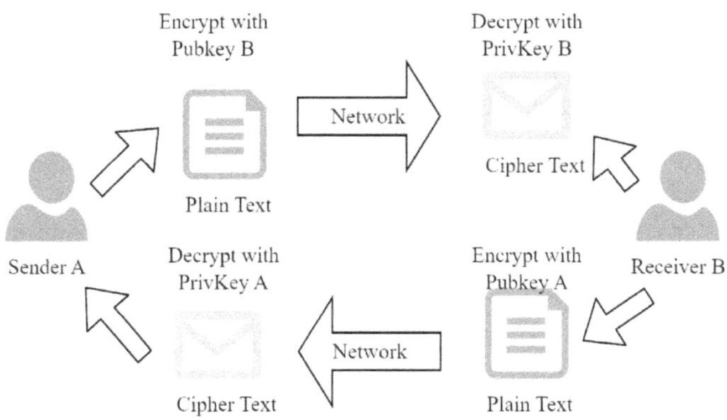

Fig. 4.18. Application of public key encryption algorithm.

theoretically vulnerable to exhaustive key search attacks. However, such an attack is impractical if the amount of computation required to succeed is beyond the capabilities of all potential attackers. This amount of computation was called the "work factor" by Claude Shannon. In many cases, the work factor can be increased simply by increasing the key length. Specific attacks against RSA and ElGamal encryption algorithms have been described in several research papers, and the decryption speed is much faster than exhaustive search but has not yet reached a level that can be used in engineering.

Public key cryptographic algorithms depend on the difficulty of solving a particular mathematical problem. This specific mathematical problem is usually represented by a one-way function or a more specific trapdoor function. In computer science, a one-way function is a function that is easy to compute given any input but difficult to infer the input given any output. The "easy" and "difficult" here should be understood in the theoretical sense of computational complexity, especially the theory of polynomial time problems. Not all many-to-one functions are one-way functions.

A trapdoor function is a function that is easy to compute in one direction but difficult to compute (find its inverse) in the opposite direction without special information. Trapdoor functions are often used in cryptography.

Mathematically, if f is a trapdoor function, then there exists some secret information t such that it is easy to compute x when $f(x)$ and t are given. The metaphor is a padlock and its key. To change a padlock from open to closed without using a key, you simply push the striker into the lock. To simply open a padlock, however, a key is required. The padlock fulfills the key function of the trapdoor: the padlock is easy to close, but a key is required to open it, otherwise it is very difficult to open.

A simple example of a math trap door would be "4,367,653 is the product of two prime numbers. What are these two primes?" A typical "exhaustive" solution is to try every prime number smaller than that number until you find the answer. However, if you tell people that 1979 is one of the numbers, they can find the answer by simply typing "4367653 ÷ 1979" into a calculator. This example is not a practical trapdoor function — modern computers can find all possible answers in 1 second. In fact, some public-key cryptography algorithms still rely on the factorization problem of large prime numbers, but the above product numbers have grown to astronomical lengths to withstand exhaustive search attacks.

4.4.3 *Man-in-the-middle attack and protection*

Using the DH key exchange algorithm, it is possible to negotiate a shared secret between the two communicating parties over the insecure Internet, a secret that is reasonably resistant to eavesdropping. Although an attacker may not be able to simply obtain secrets by eavesdropping, it is still possible to compromise the confidentiality of the system. The attacker (usually called Mina, meaning the man-in-the-middle) runs the DH key exchange protocol twice between Alice and Bob: once with Alice (Mina impersonating Bob) and once with Bob (Mina impersonating Alice).

There are two shared secrets: one between Alice and Mina, and the other between Mina and Bob. The attacker (Mina) can then simply send all the messages received from one person to another person, and view, delete, or manipulate the messages as she wishes. Figure 4.19 shows a schematic diagram of a man-in-the-middle (MITM) attack.

This attack is called a man-in-the-middle attack because the attacker (Mina) is located between two peer-to-peer participants (Alice and Bob). Given that the network infrastructure is operated by many different operators, this type of attack is very realistic, and therefore a secure cryptosystem must address the attack in some way. Although the DH protocol succeeds in generating a shared secret between the two peers, it clearly still does not solve the problem of authenticating Alice and Bob, so some other means of authenticating both parties must be used. In addition, message integrity tools must be used so that the receiver can verify that the incoming message is indeed the one that the sender intended to send.

Of course, it is not enough to simply establish a secure communication channel by exchanging keys; it is also necessary to

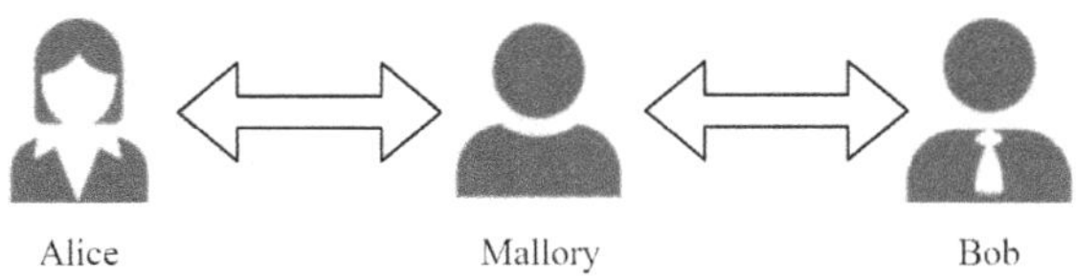

Fig. 4.19. Schematic diagram of a man-in-the-middle attack.

identify the identities of both communication parties and to prevent tampering with the content of the communication.

Similar to a certificate in the physical world that represents a person's identity or capability, a "digital certificate" is used in the digital world to identify the identity of any entity. The digital certificate system is based on the Public Key Infrastructure (PKI), which can effectively resist the MITM attack, an important cornerstone of network security. PKI enables two parties in a network communication, such as a client and a server, to exchange certificates that are issued and verified by a trusted third party called a Certificate Authority (CA). If the CA has not been MITM attacked, then the certificate issued by the CA can be used to verify the message sent by the owner of the certificate. When mutual authentication is used, both the server and the client authenticate each other's communications, and if the identity of the server or client is not authenticated or is deemed invalid, the session is terminated. This method can prevent MITM attacks. However, the default behavior for most connections is for the client to verify the server's identity, which means that mutual authentication is not always used, so MITM attacks can still occur.

4.4.4 *Replay attack and defense*

A replay attack is a type of network attack. In this attack, an attacker maliciously or fraudulently repeats or delays valid data transmissions. The premise of this attack is that the attacker can intercept and retransmit the data. Replay attacks can also be performed by malicious senders of transmissions. Replay attacks can be used as a spoofing attack scenario for IP packet substitution. Replay attacks are typically passive and are a low-level version of man-in-the-middle attacks.

An intuitive example is where an attacker intercepts a data packet for a user to place a shopping order on a website and replays the data packet in an attempt to generate a malicious order or cause economic loss to the user or the website. Of course, encryption alone does not prevent replay attacks. Figure 4.20 shows the replay attack scenario.

To deal with replay attacks, the first intuitive way is to add flags related to that session, such as adding a one-time effect of random numbers. Before each session, a one-time random number (called a nonce) is generated and sent with the message. One-time nonces

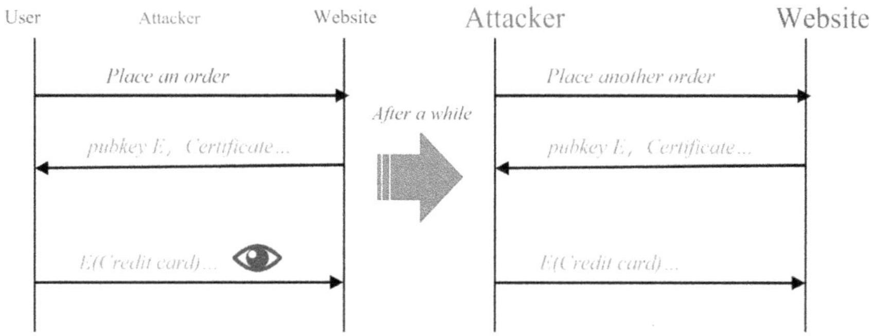

Fig. 4.20. Replay attack scenario.

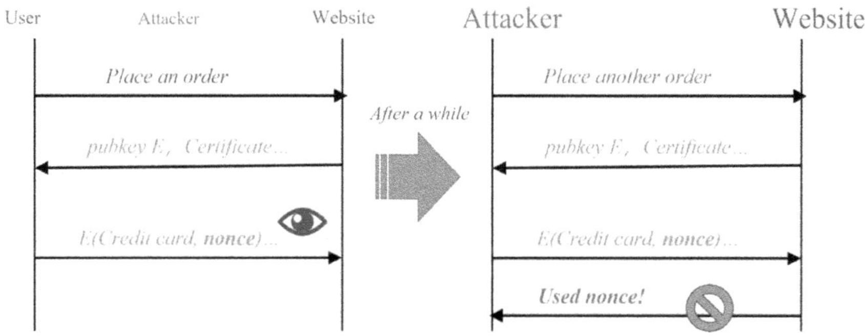

Fig. 4.21. The method of adding a one-time random number.

also require a tamper-resistant mechanism. Since the one-time nonce changes with each communication, replay attacks cannot be performed. The method of adding a one-time random number is shown in Fig. 4.21.

The second way to deal with replay attacks is to add timestamps. The two sides of the timestamp-based verification need an accurate synchronization time. Synchronization should be implemented using a secure protocol. For example, a receiver periodically broadcasts the time on its clock. When a sender wants to send a message to a receiver, the message contains its best estimate of the time on the receiver's clock. The receiver will only accept messages with timestamps within a reasonable tolerance. Timestamps can also be used in the mutual authentication process. Both the sender and receiver use unique session IDs to authenticate each other to prevent replay attacks. The advantage of this approach is that no (pseudo) random number generation and exchange is required. In networks that are unidirectional or nearly unidirectional, this can be an advantage. A remaining risk of this approach is that if the replay attack is performed fast enough, within a "reasonable" time limit, the attack may be successful.

Based on the session ID and one-time pad, other measures can be taken to prevent replay attacks.

4.4.5 *Elliptic curve cryptography*

What the public key cryptography system needs is a set of algorithms that are easy to compute in one direction but hard to compute in the opposite direction. For example, RSA relies on the difficulty of factoring prime numbers. That is, it is very easy to obtain the product of two very large prime numbers, but it is very difficult to factor the product into two prime numbers using the appropriate factorization algorithm. Such an algorithm, which is easy to compute in one direction but hard to compute in the opposite direction, is called a trapdoor one-way function. Finding a good trapdoor function is very important for constructing a secure public key cryptosystem. Simply put, the greater the difference in difficulty between the one-way calculation and the reverse calculation in the trapdoor function, the more secure the system based on the key pair.

After the advent of the RSA and DH key exchange algorithms, researchers focused on algorithms better than prime factorization to construct better trapdoor functions. In 1985, researchers proposed an encryption algorithm based on elliptic curves (Lopez and Dahab, 2000).

From the usage scenario, Elliptic Curve Cryptography (ECC) realizes asymmetric encryption. All major functions of cryptographic systems are encryption, signing, and key exchange.

ECC is considered the successor to the RSA cryptosystem. At the same level of security, ECC uses smaller keys and signatures than RSA and provides very fast key generation, key exchange, and signatures.

ECC provides cryptographic algorithms based on the mathematical principles of elliptic curves on finite fields:

(1) ECC digital signature algorithms, such as ECDSA (for classical curves) and EdDSA (for twisted Edwards curve).

(2) ECC encryption algorithms and hybrid encryption schemes, such as ECIES integrated encryption scheme and ECC-based ElGamal encryption scheme (EEECC).
(3) ECC key exchange algorithms, such as Elliptic Curve Diffie-Hellman (ECDH), X25519, and FHMQV.

All of these algorithms perform computations on some type of elliptic curve (such as secp256k1, curve25519, or p521) and rely on the difficulty of the Elliptic Curve Discrete Logarithm Problem (ECDLP). All of these algorithms use a public/private key pair, where the private key is an integer and the public key is a point on the elliptic curve (EC point).

The following is a detailed introduction to elliptic curves over finite fields. Elliptic curves are not a full part of the middle school curriculum like the prime factorization on which the RSA algorithm is based, and some of the content is relatively difficult to understand.

In mathematics, elliptic curves belong to plane geometric curves and consist of all $\{x, y\}$ satisfying the following binary equation:

$$Ax^3 + Bx^2 + Cxy^2 + Dy^3 + Ex^2 + Fxy + Gy^2 + Hx + Iy + J = 0$$

Elliptic curve cryptography uses the following simplified form:

$$y^2 = x^3 + ax + b$$

Among them, the different values of the two parameters a and b represent different elliptic curves.

For example, the secp256k1 curve (Mayer, 2016) (the curve used by Bitcoin) is based on the following elliptic curve instance:

$$y^2 = x^3 + 7$$

Obviously, the values of the two parameters are $a = 0$ and $b = 7$. An example of the drawn elliptic curve is shown in Fig. 4.22.

The elliptic curves described above have some interesting properties that are more useful in cryptography.

The first property is horizontal symmetry. That is, any point on an elliptic curve has a symmetrical point based on the X-axis that is also on the curve.

The second property is that any straight line through a curve has up to 3 points of intersection. Depending on the position of the line

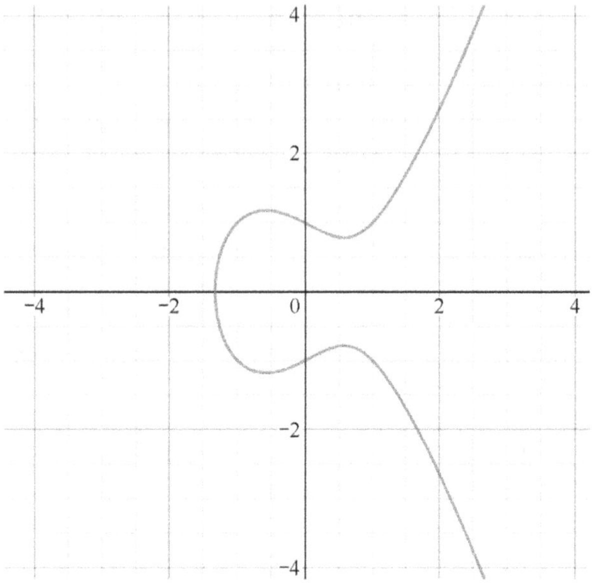

Fig. 4.22. Elliptic curve sample.

and the curve, it is obvious that there can also be 0 to 2 intersection points. Only scenarios where there are 3 intersection points occur can be exploited by elliptic curve cryptography.

An example of an elliptic curve intersection is shown in Fig. 4.23. For example, select an appropriate straight line:

$$y = bx + c$$

There are 3 intersection points (A, B, C) with the elliptic curve. Define the following series of operations:

(1) Let C be the intersection of the straight line defined by points A and B and the elliptic curve.
(2) Find the mirror intersection D of the intersection point C relative to the X-axis.

This series of operations is written as a **dot**. Expressed using the operation symbol dot, it can be recorded as

$$A \ dot \ B = D$$

This series of operations can be continued like $A \ dot \ D = E$, $A \ dot \ E = F, \ldots.$

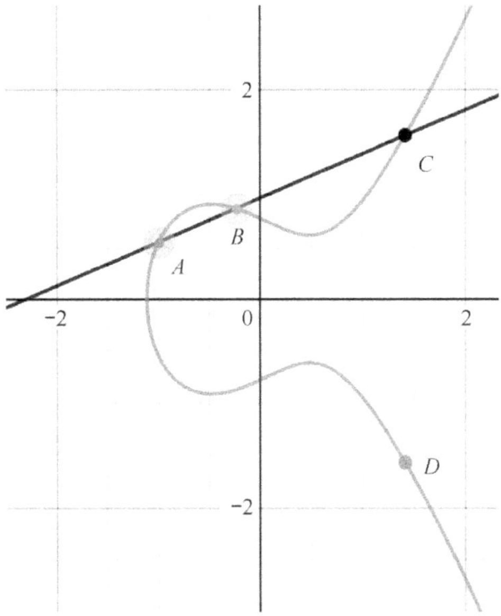

Fig. 4.23. Example of elliptic curve intersection.

Mathematically, it is very difficult to determine the number of dots (n times) given the starting point A and the end point Z of the dot. However, it is very easy to calculate each dot if n dots are known. If the observer does not observe each dotting action but only sees the starting point A and the end point Z, the number n of dot cannot be calculated. Obviously, this is a very good trapdoor function.

If the first operation (the intersection of the straight line and the elliptic curve defined by the intersection A and the intersection B) calculates a very large coordinate $\{x, y\}$ of the intersection C, it may be difficult to use in cryptography. Therefore, the geometric position of the intersection point should be limited to a certain range. Assuming that a large prime number is chosen as the maximum value, the elliptic curve is then called a "prime curve" and has excellent cryptographic properties.

Select the curve $y^2 = x^3 + 7$ and choose the prime number 97, the resulting prime curve is shown in Fig. 4.24.

Although this discrete scattering matrix does not look like an "elliptic curve" at all, it is in fact an elliptic curve used in cryptography. The X-axis-based symmetry and even point operations described above are still valid for this elliptic curve.

Fig. 4.24. Simple example of elliptic curve.

Note that the above example is for ease of understanding, and the key length it provides is only 6–7 bits. For the real curve, the key length is usually 256 bits or more.

Given the above prime number curve, it is very easy to determine whether a certain point $\{x, y\}$ is on the curve; just calculate

$$x^3 - x + 1 - y^2 \equiv 0(mod97)$$

After the modulo operation, the prime curve actually contains finite points. In mathematical language, elliptic curves over finite fields can form a finite cyclic algebraic group (called a cyclic group). This cyclic group consists of all points of the curve. In this cyclic group, adding two points or multiplying a point by an integer yields another point from the same cyclic group (on the same curve). The **order of a curve** is the **total number** of all points on the curve. The total number of points includes a special point called "point of infinity", which is obtained by multiplying the point by 0.

Some prime curves form a single **cyclic group** (containing all the points), while others form several non-overlapping **cyclic subgroups** (each subgroup containing a subset of the points of the curve). In the second case, the points on the curve are divided into h cyclic subgroups (partitions), each of order r (each subgroup has the same number of points). The **order** of the whole group is $n = h \times r$ (number of subgroups multiplied by the number of points in each subgroup). The number of subgroups **h** containing the EC points is called the **cofactor**. A schematic diagram of a cyclic subgroup is shown in Fig. 4.25.

The cofactor h can be expressed as

$$h = n \div r$$

In the formula, n is the order of the curve, that is, the number of points the curve contains. h is the cofactor of the curve, that is, the number of non-overlapping subgroups. r is the order of the subgroup, that is, the number of points contained in each subgroup. Each subgroup needs to contain the point at infinity".

In other words, the points on the elliptic curve lie in one or a few non-overlapping subsets, i.e., cyclic subgroups. The number of subgroups is called a cofactor. The total number of points in all subgroups is called the order of the curve, usually denoted by n. If the curve contains only one cyclic subgroup, its cofactor $h = 1$. If the curve contains multiple subgroups, its cofactor is greater than 1.

The number of cofactors in different commonly used elliptic curve algorithms is shown in Table 4.11.

For elliptic curves over finite fields, elliptic curve cryptography defines a special predefined (constant) point, called the **generator point G** (base point), which can generate any other point in its subgroup over the elliptic curve by **multiplying G by some integer**

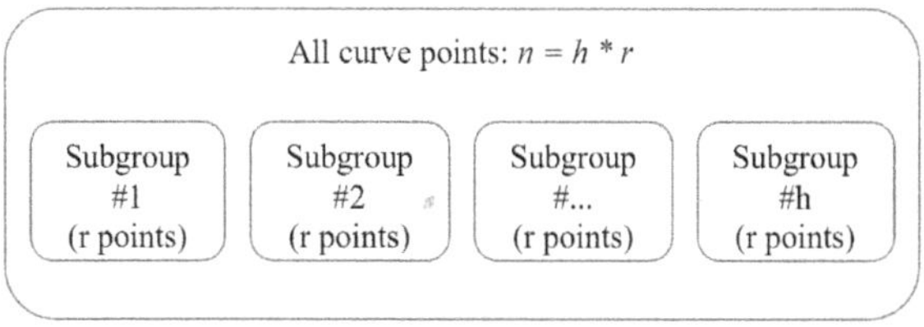

Fig. 4.25. Schematic diagram of cyclic subgroup.

Table 4.11. Cofactor of elliptic curves.

Elliptic Curve	Cofactor
Secp256k1	1
Curve448	4
Curve25519	8

in the range $[0...r]$. The number r is called the "order" of the cyclic subgroup (the total number of all points in the subgroup).

A curve with cofactor $h = 1$ has only one subgroup whose order n (the total number of distinct points on the curve, including the infinity) is equal to the order r of the subgroup.

If G and n are chosen carefully and the cofactor $h = 1$, the generator point G can be multiplied by an integer number in the range [1, n] to generate all possible points on the curve (including the special point infinity) from the generator point G. The integer n is the "order of the curve".

Importantly, the **order r of a subgroup** obtained from some generator point G (which may be different from the order of the curve) defines the total number of all possible private keys for this curve: $r = n \div h$ (order of the curve n divided by the curve cofactor h). Cryptographers carefully choose the parameters of the elliptic curve domain (curve equation, generator point, cofactors, etc.) to ensure that the **key space is large enough** to guarantee a certain cryptographic strength.

In elliptic curve cryptography, multiplying a fixed generator point G by an integer k yields another point P. Among them, k can be used as the private key, and P is the public key corresponding to k.

From this, key concepts of elliptic curve cryptography are defined:

(1) Elliptic curve EC on the finite field $\mathbb{F}p$.
(2) Generator point G (fixed constant, base point on the elliptic curve).
(3) k: Private key (integer).
(4) P: Public key (a specific point on the elliptic curve).

The computational difficulty that elliptic curve cryptography relies on is that computing $P = k \times G$ is very fast, but computing $k = P \div G$ is very hard. This asymmetry (fast multiplication and infeasible inverse operations) is the basis for the security strength of elliptic curve cryptography (also known as the "Elliptic-Curve Discrete Logarithm Problem", ECDLP).

According to the difficulty of the Elliptic Curve Discrete Logarithm Problem, the security strength of different elliptic curve cryptographic algorithms can be obtained. It can be roughly estimated that a well-designed elliptic curve algorithm of n bits provides a key security strength of $n \div 2$. Therefore, the secp256k1 elliptic curve algorithm ($p = 256$) provides approximately 128 bits of strength (the exact number is 127.8). The Curve448 elliptic curve algorithm ($p = 448$) provides an approximate strength of 224 bits (the exact number is 222.8).

4.5 Key Management System

As mentioned earlier, encryption is extremely important for data security. For data lifecycle security, there are different encryption techniques for different lifecycle stages:

(1) **Storage:** In-App Encryption, Database Encryption, File Encryption, Disk Encryption are used.
(2) **Usage:** Access control technology and memory encryption are commonly used, and suitable encryption algorithms are still under research (such as homomorphic encryption).
(3) **Transmission:** Application layer encryption, channel encryption, and asymmetric encryption are used.

Of course, for different data classifications, encryption must also meet the appropriate security, regulatory, and forensic requirements.

For encryption and decryption, modern cryptography adheres to the openness of algorithms and relies on the confidentiality of keys to ensure the confidentiality of data. Therefore, the lifetime security of the data depends on the lifetime security of the key.

Key management refers to the management of the entire lifecycle of the key. In the practice of data security, key management is more important and more difficult than data encryption algorithms.

Key management faces the following challenges:

(1) **Key system scalability:** The explosive growth of key types has caused the key management system to face scalability problems.
(2) **Key system security:** Key management systems store and process a large number of keys and are a target of great interest to both external hackers and malicious insiders.
(3) **Key availability:** How to securely distribute keys to authorized users to ensure data availability to authorized users; how to stop authorization by revoking users' keys; how to recover data in the event of key loss (e.g., a user has left an organization and

their archived documents are accidentally or intentionally left encrypted).

(4) **Heterogeneous key system support:** In the IT system of modern organizations, there are many heterogeneous databases, operating systems, applications, and network protocols. Key management systems need to provide various forms of support for them.

(5) **Key system governance:** How to comply with relevant legal and regulatory requirements for data security and privacy protection.

Common regulatory requirements involving key management systems are as follows:

(1) Keys and encrypted data should be separated.
(2) Regular key rotation.
(3) Monitor and audit keys.
(4) Long-term retention of keys and encrypted data.

4.5.1 *The importance and principles of key management*

To implement a good key management framework, the following must be considered:

(1) **Key storage mechanism:** Secure key storage is achieved by supporting physical devices, such as HSM. It is generally required that the root key can only be stored in physical hardware and cannot be read from the outside, thus ensuring high reliability and high availability.

(2) **Provide encryption services:** Provide encryption of sensitive data, symmetric key storage, random number generation, asymmetric key pair generation, common encryption algorithms and other functions.

(3) **Key attribute and metadata processing mechanism:** Record the attributes of the key, such as version, one-time random number or IV, authorization records, etc., to facilitate operations such as auditing and key rotation.

(4) **Key management strategy:** Support key authority control, business-based grouping and authorization, etc.

(5) **Support auditing and reporting functions:** Logs of all key system operations must be generated and recorded to a centralized logging service for auditing and reporting.
(6) **Key access authentication:** Ensure that only users who have passed authentication and authorization checks can access and operate the key management system through IAM, role-based authentication, and other means.
(7) **Compliance with Laws and Standards:** Comply with applicable laws, regulations, and standards. For scenarios where the key management system crosses administrative regions (countries and regions), special attention must be paid to possible conflicts of laws and regulations.
(8) **Key lifecycle management:** Implement secure key management at all stages of the key lifecycle (before operation, during operation, after operation, and deletion).

The key management framework is shown in Fig. 4.26.

Fig. 4.26. Key management framework.

If encryption technology and key management are compared to a delicious pizza, encryption technology is like the filling on a pizza that attracts people's interest and attention, and key management is like the leavened dough of a pizza, and although it is indispensable, few people want to know more about its practices.

Data security depends on key security. However, this does not mean that the level of data security is equal to the level of key security. In practice, keys must provide a higher level of protection, especially in the dimensions of generation, storage, and transmission. Keys have their own security protection requirements.

After encryption, key management must be done carefully. The crypto key is like the key to the door lock at home. Locking the door is important, but managing the key is more important. Simply putting the key under the carpet in front of the door, no matter how strong the lock (encryption algorithm) is, cannot guarantee security.

4.5.2 *Key lifecycle management*

Whether it is a symmetric cryptographic algorithm or an asymmetric cryptographic algorithm, its security actually depends on the security protection of the key. The strength of data encryption protection provided by modern cryptography depends entirely on the key system, so key security management is an important factor in ensuring the security of cryptosystems.

The main stages of the key lifecycle include key generation, storage, distribution, injection, backup, application, archiving, replacement, and destruction, etc. The confidentiality, integrity, and availability of keys must be maintained throughout their lifecycle.

Key management refers to the secure management of the key lifecycle. The purpose of key management is to maintain the key relationship between the system and various entities to resist various possible threats (key disclosure and unauthorized use, etc.). All work in key management revolves around one purpose — to ensure that the keys in use are secure. Key management relies on techniques, such as encryption, authentication, signing, agreements, and notarization.

Following the definition of the American National Standards Institute ANSI X9.79 standard [97], the key lifecycle is divided into seven stages: generation, distribution, use, backup, revocation/rotation, destruction, and archiving. Figure 4.27 shows

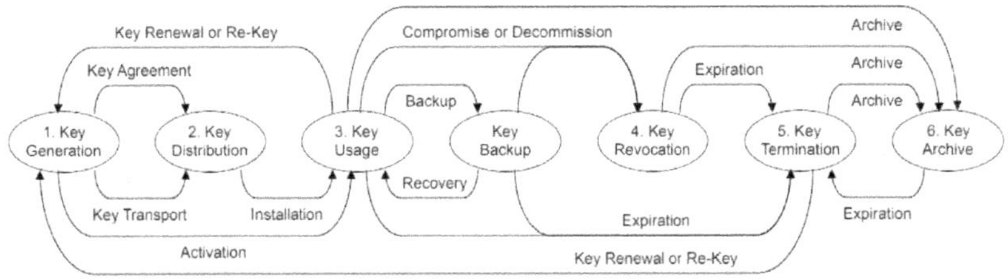

Fig. 4.27. Security management process of the key lifecycle.

the security management process of the key lifecycle. Note that the content of the following seven lifecycle stages will be slightly different in different key usage scenarios.

4.5.2.1 *Key generation*

The first phase of the key management lifecycle is key generation. Key generation must take place in a secure environment and may need to consider separation of duties (user and administrator) requirements. In most cases, the key generated will be a symmetric key (also known as a "shared key"). Keys should be of sufficient strength and rely in part on the basic ability to generate random numbers. Key strength (usually measured in bits) is usually based on the effective lifetime of the protected data, taking into account the time required to crack the key. For example, if the data is valid for 5 years, a key is needed that can withstand a brute force attack for more than 5 years. In addition to choosing an appropriate key strength, engineers responsible for designing and implementing the key management system should also strive to choose a standard encryption algorithm that has been peer-reviewed and scientifically validated. Proprietary encryption algorithms should be avoided because the algorithm itself has a high probability of failing peer review and academic scrutiny if it needs to be kept secret.

The key generation mechanism should use a vetted library of cryptographic algorithms of known good quality. Symmetric key protection mechanisms should be implemented as soon as possible after key generation, especially in environments where encryption keys need to be distributed (especially over a network) to other systems. A common scheme is to encrypt using the public key of an asymmetric key pair (also known as public key cryptography). From a segregation of duties perspective, the team responsible for key generation and management (the administrator) should be separated from the team that needs to access the encrypted system (the consumer).

Additional factors to consider when generating keys include the value of the data, the attacker's resource situation, the overall cost of the system, computational overhead, etc., which must be weighed in engineering practice, but the goal of data confidentiality should not be compromised.

4.5.2.2 *Key distribution*

Key distribution is an extremely complex issue in key management systems. When the system distributes the key, the confidentiality and integrity of the key itself should be considered first, and the key should be distributed to the appropriate users. Other considerations include segregation of duties between key managers and consumers, logging, and auditability requirements.

Key distribution refers to the deployment of new keys in production or encrypted environments. Secure deletion of old keys is not considered at this stage. Note that the key distribution process can cause data or business disruption and requires rigorous testing and deployment. It is also necessary to avoid losing the key during the key distribution process to avoid losing access to important data. Errors in cryptosystems can be very costly.

In terms of implementation, most modern systems use dynamic key exchange protocols. From the perspective of key distribution technology, it can be divided into key distribution based on symmetric cryptosystem and key distribution based on public key cryptosystem; from the perspective of key exchange method, it can be divided into manual key distribution, center-based encryption key distribution, and certificate-based key distribution. Currently, for symmetric cryptography, an authoritative third party such as a key distribution center (KDC) is often required for distribution. For public key cryptography, asymmetric key management is mainly realized through the PKI architecture.

4.5.2.3 *Key usage*

In the key usage phase, attention should be paid to the validity of the key usage and the security of the key itself. From the perspective of effectiveness, a single key can usually be used for only one purpose (such as encryption, integrity authentication, key wrapping, random number generation, or digital signature). In particular, there are three reasons:

(1) Using the same key for two different encryption algorithms can weaken the security of the encryption provided by one or both of the encryption algorithms.
(2) It can limit the damage that can be caused by key theft.
(3) Certain uses of the keys interfere with each other.

For example, the private key is used both to decrypt transmitted data encrypted with the public key and to digitally sign the application. To ensure that the data can be decrypted, the private key must be kept for a longer period of time. The signing private key, on the other hand, should be destroyed when it expires to prevent its disclosure. In this example, the two scenarios have conflicting lifecycle requirements, and a longer retention period cannot simply be chosen.

In addition, it should be noted that the security of the key itself is mainly confidentiality. Depending on the scenario, integrity and availability may also need to be considered.

In a production environment, key security is primarily protected by access control mechanisms.

4.5.2.4 *Key backup*

Before using the key, it is best to back it up. Optional backup schemes include physical backup and logical backup. Physical backup refers to writing keys to external media (such as CDs, DVDs, and USB drives) and storing them in a physical media vault. Logical backup refers to traditional network-based or local-based backup.

Regardless of the backup scheme chosen, a separate asymmetric key pair should be used to protect the symmetric key. This asymmetric key pair is commonly referred to as an "escrow key". It is critical to encrypt the key to be protected with the public key of the asymmetric key pair and to ensure that the private key is stored securely so that the protected key can be decrypted when needed.

Key backup mechanisms must be considered in conjunction with business continuity or disaster recovery plans. The location of the key must meet business continuity and disaster recovery requirements, and it must be recoverable within the specified time in the event of the above scenario. Once the encryption system has been seriously compromised, the emergency response team should implement the recovery plan according to the established requirements and procedures for key backup and recovery operations.

4.5.2.5 *Key revocation/rotation*

The concept of key revocation/rotation is related to key distribution, but they are used in different scenarios. In the key revocation/rotation phase, the goal is to make the new encryption key effectively used by the cryptosystem and to convert all stored encrypted data to the new key. This process can be very time consuming and consumes a lot of CPU resources. However, assuming that best practices have been followed for all previous phases of the key lifecycle, the key action in this phase is the conversion of keys and data, not the activation of new keys for future encryption requests.

During the key revocation/rotation phase, do not remove old keys from production systems unless you can prove that business data was not encrypted with the old keys. Failure to find the old key in the required phase can result in data loss or service interruption if a query using the old key ID returns a failure.

It is also important to note that there may be good reasons not to batch convert stored data from the old key to the new key. For example, if the data is highly transient (accessed, written, or rewritten very frequently), it may be appropriate to perform compatibility processing in the application so that the new key is automatically used when the data is read/written. Using such mechanisms can reduce some of the system load associated with key revocation/rotation.

4.5.2.6 *Key destruction*

The life of the key ends when it is destroyed. Key destruction should follow secure deletion procedures to ensure that keys are properly deleted. Note that key destruction is only possible after a sufficiently long period of archiving and after at least double-checking to ensure that key loss does not equate to data loss.

4.5.2.7 *Key archive*

As mentioned above, destroying keys associated with valuable data can have serious consequences. Therefore, correctly identifying whether a key needs to be destroyed or archived is a very important task.

Archiving expired or disabled keys should be based on determining whether data encrypted with the key to be archived still exists in the production system. Not only the currently running production system but also scenarios such as disaster recovery backup sites, offline backups, and cloud backups should be considered. If any data is required to be recoverable, the key must be archived with that data.

When archiving keys, the following points should be considered:

(1) If archived keys must be used to recover data, the keys and their associated data should be recorded and indexed to make data recovery as efficient as possible. Archived information for keys should also include the length of time the key was used to help narrow the search for an appropriate key in a recovery scenario.
(2) Ensure that the archived copy of the key itself is protected. The symmetric key is encrypted with the public key of the asymmetric key pair, as suggested in the generation and distribution phase.
(3) During business continuity/disaster recovery exercises or tests, try to recover encrypted data using the archived keys.

It's worth noting that some encryption appliances automatically archive expired keys in a secure manner and may never enter the final stage, the key destruction phase. Overall, the presence or absence of a key destruction phase should be considered in terms of business risk. Furthermore, the risk of moving encryption keys from the archiving phase to the destruction phase and the risk of permanent loss of associated data may even outweigh the risk of exposing the archived key. Therefore, the transition from the archiving phase to the destruction phase should be done more cautiously, unless some regulations specify a time limit for archiving.

Key lifecycle management may also include key storage. Key storage should comprehensively consider the following factors:

(1) Determine the storage location and usage method of the key in the application. Identify the storage devices on which keys can be stored.
(2) Keys should never be stored in clear text.
(3) The key must be protected on both volatile and permanent storage. Keys should preferably be handled in a separate secure cryptographic module.
(4) Ensure that keys are stored with integrity protection applied. Dual-purpose algorithms that support encryption and message authentication codes (MACs) should be considered.
(5) If you plan to store the key in an offline device/database, you must encrypt the key with a key encryption key (KEK) before exporting the key information. The strength of the KEK (and the algorithm) should be equal to or greater than the strength of the protected key.
(6) If the key is provided in the form of a service or a dynamic library, it is not recommended to provide an API interface for reading/writing the key in plain text.
(7) It is recommended to store the key in a hardware security module (HSM) or in an isolated area of the software.
(8) It is recommended to perform key access, encryption, decryption, and signature operations within the isolated area of the hardware or software.

Another requirement not included in the above lifecycle phases, but also very important, is the monitoring of the entire key lifecycle phase. Monitoring should address three key aspects:

(1) Monitor all unauthorized administrative access to cryptographic systems to ensure that no unauthorized key management operations are being performed. Any form of unauthorized operation can be catastrophic to systems and data.
(2) Monitor the performance of the cryptographic system. The performance of cryptographic computations tends to be CPU-intensive, which means that the system may be under heavy load. Overloaded encryption services or overloaded business systems, or even the mutual influence of the two, can cause data to be corrupted or unavailable.

(3) Monitor the normal operation of the encryption system. If the encryption system fails, the service may be interrupted, which will have a huge negative impact on the business.

4.5.3 *Regulatory requirements for key management*

Key management faces not only technical challenges but also regulatory requirements. For example, the processing of credit card data in cross-border financial services requires compliance with the requirements of the PCI DSS (US Payment Card Industry Data Security Standard) (FIPS Publication, 2000).

Relevant regulatory requirements mainly include clause 3.5 and clause 3.6. Clause 3.5 requires that key protection processes be documented and implemented to securely store cardholder data from disclosure and misuse. Clause 3.6 requires documenting and implementing a process for the management of the keys that encrypt cardholder data.

In particular, Clause 3.5.1 is an additional requirement for service providers. This article requires the documentation of the cryptographic architecture, which must include the following content:

(1) Details of the algorithms, protocols, and keys used to protect cardholder data, including key length and expiration date.
(2) A description of how each key is used.
(3) An inventory of all Hardware Security Modules (HSMs) or Secure Cryptographic Devices (SCDs) used for key management.

Clause 3.5.2 requires that access to cryptographic keys be restricted to business administrators.

Clause 3.5.3 is a requirement for secure storage of keys. When the data encryption key is stored, it must be stored encrypted with the key encryption key or stored in a separate hardware cryptographic module.

Clause 3.5.4 requires that keys should be stored in as few locations as possible. Too many unnecessary storage locations make it difficult to protect and update the key.

For global companies dealing with health data, such as blood pressure and medical records of consumer users, it is necessary to assess compliance with the "Health Insurance Portability and

Accountability Act" (HIPAA) and the "Health Information Technology for Economic and Clinical Health Act" (HITECH Act) compliance.

Based on the requirements of HIPAA, the U.S. Department of Health and Human Services (HHS) formulated and issued what are commonly known as the HIPAA Privacy Rule and the HIPAA Security Rule. The HIPAA Privacy Rule defines "protected health information", and the HIPAA Security Rule establishes security standards for the storage and transmission of health information. The HIPAA Security Rule provides basic key management requirements to protect the security of the storage and transmission of health information.

Relevant regulatory requirements are reflected in Section 164.312, such as 164.312(a): "(iv) Encryption and decryption (Addressable). Implement a mechanism to encrypt and decrypt electronic protected health information".

In 164.312(e), there are also requirements for the protection of confidentiality and integrity using encryption and decryption during transmission.

In addition, compliance with the EU's General Data Protection Regulation (GDPR) is a challenge for companies doing business in the EU or processing EU consumer data:

(1) **Article 32:** Encryption and Pseudonymization: "...the controller and the processor shall implement appropriate technical and organisational measures to ensure a level of security appropriate to the risk, including inter alia as appropriate: (a) the pseudonymisation and encryption of personal data; ..."
(2) **Article 30:** Each controller and, where applicable, the controller's representative shall maintain a record of processing activities under its responsibility.

The relevant technologies are described in detail in the following.

4.5.4 *Regulatory compliance for key management*

Key management has the following classifications:

(1) **Key lifecycle management:** This includes key generation, distribution, usage, backup, revocation/rotation, destruction, and archiving, etc.

(2) **Key storage:** This includes file/database storage, key encryption key storage, and hardware storage (Hardware Security Module, HSM).
(3) **Key attributes/metadata:** This supports multiple attributes of keys and types of key attributes.
(4) **Access authentication:** This includes user access authentication, device access authentication, and administrator access authentication.

Based on the key lifecycle perspective, key management has the following best practices:

When generating keys, the key generator should comply with FIPS 140-2 or other applicable standards. The standard also includes requirements for secure random number generators. In general, hardware-based cryptographic modules are superior to software-based cryptographic modules for key generation, storage, and protection.

When distributing keys, the generated keys must be transmitted over secure channels, unless the business scenario cannot be met at all. The user of the key, i.e., the executor that uses the key to perform operations such as encryption and decryption, should also be a software module that complies with the FIPS 140–2 standard.

When storing keys, developers and implementers need to understand where encryption keys are stored in applications and storage devices. Keys must be protected when stored in both volatile and permanent memory. If the deployment environment allows, it is best to use an independent security encryption module for processing, such as a hardware security module (HSM). Keys should never be stored in clear text.

Key escrow and backup are also very important. If the encryption key is lost, the data it encrypts may be difficult or impossible to recover. Key escrow and backup are especially important if the application's business involves long-term storage of encrypted data at rest. When backing up keys, at a minimum, the database that stores the keys should be encrypted using a FIPS 140–2 validated cryptographic module. Keys used to perform digital signatures need not and should not be escrowed. Escrow refers primarily to the support of encrypted keys. Typically, escrow is provided by a certificate authority or a key management system for certificates and keys. Both types of systems may also provide separate APIs for hosting application execution.

The key lifecycle process also includes accountability and auditing. Accountability addresses who has access to or control over the key throughout its lifecycle. Accountability is an effective tool for preventing key breaches and reducing the impact of key breaches when they are discovered. It's best not to give anyone permission to view the plaintext key. If viewing plaintext keys is a must, the key management system should be able to identify all individuals who can view plaintext cryptographic keys and log and audit their actions. In addition, more complex key management systems must identify all individuals who are authorized to access or control any encrypted plaintext or ciphertext keys.

Key accountability and auditing, and more generally the design, implementation, and maintenance of key management systems, must be consistent with ergonomic theories and best practices. It is necessary to prevent inappropriate human behavior from damaging the key management system due to overly loose rules, as well as to avoid overly strict rules and restrictions that may lead to excessive psychological pressure on the personnel using, operating, and maintaining the system and increase the likelihood of making mistakes or failing to follow established procedures.

Key disclosure and recovery are a branch of the key lifecycle process. Unauthorized disclosure of a key used for encryption means that any data encrypted with that key may lose its confidentiality protection. Most keys themselves also have integrity and availability requirements. Tampering with or replacing a key affects not only the integrity of the key but also the integrity of the data it protects. Deleting the key affects not only the availability of the key but also the availability of the associated data.

A compromise recovery plan is critical to restoring cryptographically secure services in the event of a key compromise. The compromise recovery plan should be documented and easily accessible.

A compromise recovery plan should include the following:

(1) The identity and contact information of the person to be notified.
(2) The identity and contact information of the person who will perform the recovery.
(3) The identity and contact information of all persons needed to assist in the recovery process.

(4) The training of all relevant personnel on the recovery procedures.
(5) How to re-enter the key.
(6) A list of all encryption keys and their uses (such as the location of all certificates on the system).
(7) A policy for enforcing key revocation checks (to minimize the impact of violations).
(8) Monitoring of key rotation operations (to ensure that required actions are performed on all affected keys).

In conclusion, from a key lifecycle management perspective, the risks at each stage of the lifecycle can be systematically analyzed and mitigation measures can be implemented in a targeted manner to ensure the confidentiality, integrity, and availability of keys.

4.5.5 *Problems with key exchange*

If the key exchange is done only in an encrypted system, it may not be the most serious problem. But because of human involvement, the problem becomes complicated. People are very unpredictable, what they might do, what they will do, why they will do it... are all unpredictable.

In modern large-scale networks supporting many-to-one, many-to-many, and one-to-many communication topologies, the generation and distribution of millions of keys and their security always becomes a nightmare.

Although symmetric encryption is widely used due to its historical place in cryptography and its speed, it still faces a serious problem of how to securely transfer the key from the sender to the receiver. This problem is the basis of the key exchange problem. The key exchange problem includes the following:

(1) Ensuring that keys are exchanged so that both the sender and receiver can perform encryption and decryption.
(2) Ensuring that eavesdroppers or external parties cannot compromise the key.
(3) Ensure that the message received by the recipient has been encrypted by the sender.

The strength of an encryption algorithm depends on the key distribution technique. Poor key distribution techniques create an ideal environment for man-in-the-middle attacks. Thus, the key exchange problem emphasizes the need for strong key distribution techniques. Although the key exchange problem is more pronounced in the symmetric encryption cryptography method and has basically been solved by the public key cryptography method, there are still some key exchange problems in the public key cryptography method. Symmetric key encryption requires the communicating parties to agree on their secret key in advance before communicating, while public key encryption has the difficulty of how to securely obtain the recipient's public key. One of the ways to address both of these issues is to use a trusted third party or intermediary. For symmetric key encryption, the trusted intermediary is called a Key Distribution Center (KDC). For public key cryptography, a trusted and scalable intermediary is called a Certificate Authority (CA).

Another solution is to rely on users to distribute and track each other's keys and trust in an informal, distributed manner. PGP software and PGP-based email encryption is an implementation of this solution, known as the "web of trust" model.

4.5.6 *Key distribution center*

The Key Distribution Center (KDC) is a single trusted network entity through which all network communication parties must establish a shared secret key. It requires that all communicating parties have a shared secret key with which they can secretly communicate with the KDC. The tricky part, however, is how to distribute the shared secret. In general, the KDC does not generate keys for communicating parties; it is only responsible for storing and distributing keys. Key generation must be done elsewhere. The Diffie–Hellman key exchange algorithm is a common key generation algorithm that provides a means to distribute keys between two communicating parties. However, because Diffie–Hellman key exchange is vulnerable to man-in-the-middle attacks, it is best used in conjunction with public-key cryptography to ensure authentication and key integrity.

Each network communication party shares a unique secret key with the KDC. Therefore, the KDC secretly distributes keys to the appropriate communication parties upon request. Any network entity that wants to communicate with any other entity on the network

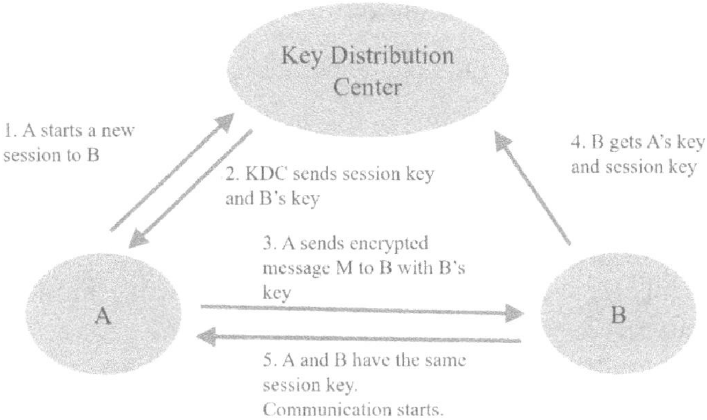

Fig. 4.28. Key distribution center.

using a symmetric encryption scheme can use the KDC to obtain the shared secret key needed for that communication.

The Key Distribution Center is shown in Fig. 4.28. The following uses a good scenario to describe the main work of the KDC:

First, both message sender A and message receiver B must have their own private keys that are shared with the KDC.

The interaction steps between the communication parties (A, B) and the KDC are as follows:

(1) A initiates the communication process by sending a request to the KDC for a session key and B's secret key.
(2) The KDC responds to this request by sending a two-part packet to A. The first part to be sent to A consists of A's request to the KDC, B's secret key, and a session key. The second part, to be sent to B, consists of A's identity and a copy of the session key given to A. Since the packet is to be sent to A, it is encrypted by the secret key the KDC shares with A.
(3) When A receives the packet, A retrieves B's secret key and encrypts the message along with B's part of the packet with B's secret key and sends it to B.
(4) Upon receipt, B uses the secret key that B shares with the KDC to decrypt the packet from A to recover the session key.
(5) Now the session key has been distributed to both A and B. After a few operations such as authentication handshake, communication can begin.

KDC has the following drawbacks:

(1) Both parties in the network communication must belong to the same KDC.
(2) Centralized security becomes a risk.
The KDC is naturally a very valuable target for attackers because it is the center for centralized key management. Since all parties involved in the communication trust the KDC unconditionally, a security breach on the KDC will compromise the security of the entire system.
(3) In a large network with many communicating parties, since each pair of users who need to use the key must visit the central node at least once, the KDC is likely to become a performance bottleneck.

In large-scale networks with different communication topologies, where network elements do not usually belong to the same KDC, key distribution can become a real problem. Such problems are solved by Public Key Infrastructure (PKI).

4.5.7 *Public key management*

There are authenticity and integrity issues with public key distribution. For example, a hacker may use his own public key to replace the public key of a website or service to create a forgery. The goal of public key management is to find ways to solve these problems.

In fact, there are two levels of problems in public key management: the distribution of public keys and the use of public key encryption to distribute secret keys. For public key distribution, there are several solutions:

(1) Public announcement, any user can broadcast their public key or send it to selected individuals.
(2) A public directory maintained by a trusted authority. The public directory must support dynamic additions and deletions.
(3) A Certificate Authority (CA) that distributes certificates to each communicating party. Each communicating party in the network or system securely communicates with the CA to register its

public key with the CA. Since the public key does not need to be kept secret, registration can be accomplished by a variety of methods.

Certificate Authorities are discussed further in Section 4.6.2 "Certificate Authorities".

4.5.8 *Key escrow*

Key escrow refers to entrusting a copy of the keys to a third party, similar to entrusting a spare key for a house or car to a trusted friend. This isn't a bad idea and can be useful in situations where you lose your keys or lock them in your house or car. However, in a public communication network like the Internet, this idea doesn't work very well.

According to cryptography expert Bruce Schneier, the need for key escrow will result in significant sacrifices in security and greatly increase costs for end users.

Designing a secure system that supports key escrow must address at least the following challenges:

First, there are new risks to the basic functionality of the system.

Potential risks include the addition of a potentially illegal data access channel, possible misuse by internal maintenance personnel, key escrow destroys the forward encryption scheme, etc. and also include the risks of storing and transmitting keys during the implementation process.

The second risk is complexity. Such systems can be extremely difficult to design. The US Escrowed Encryption Standard, which was based on the Clipper chip during the Bush administration, had various problems such as the length of the hash algorithm and the authentication mechanism.

In addition, there are complex risks such as increased complexity of operations and maintenance, key recovery authorization, and technical difficulties.

Third is cost. Design costs, operational costs, access control costs, regulatory compliance costs, and certification costs must all be considered, and they all affect the cost to the user.

4.6 Public Key Infrastructure

Section 4.5.6 mentions that in large networks such as the Internet, it is difficult to build a central key distribution center node to achieve secure communication at the entire Internet level. How to distribute keys to both parties in the communication becomes a complex problem. Utilizing a Public Key Infrastructure (PKI) instead of a KDC to provide trusted and efficient key and certificate management can solve this problem.

The purpose of PKI is to facilitate the transmission of secure electronic information for a range of network activities such as e-commerce, Internet banking, and confidential email. This is necessary in scenarios where a simple password is not sufficient for authentication and stronger evidence is required to confirm the identity of the parties involved in the communication and to authenticate the information being transmitted.

The ITU Telecommunication Standardization Sector (ITU-T) X.509 standard states that PKI is a set of hardware, software, personnel, policies and procedures required to create, manage, store, distribute, and revoke certificates based on public key cryptography. PKI automates all certificate management activities. PKI can generate digital certificates on a large scale and distribute them to a large number of users in a trusted manner.

PKI consists of four main components:

(1) A certificate that represents an authentication token.
(2) The CA, which has the final decision-making authority for the subject certificate.
(3) A Registration Authority (RA) that receives and processes certificate signing requests on behalf of end users.
(4) A Lightweight Directory Access Protocol (LDAP) directory that stores public certificate information.

4.6.1 *PKI system*

In cryptography, PKI is a protocol that binds public keys to the respective identities of entities, such as people and organizations. The binding is established through the certificate enrollment and issuance process of a Certificate Authority (CA). Depending on the level of assurance of the binding, the binding can be performed by an automated process or under human supervision.

The PKI role that can be delegated by a CA to ensure valid and correct registrations is called a Registration Authority (RA). Essentially, RAs are responsible for receiving digital certificate requests and authenticating the requesting entity.

An entity must be unique and identifiable within each CA domain based on information about the entity from which the request originated. A third-party Verification Authority (VA) can provide this entity information on behalf of the CA.

The PKI system business process is shown in Fig. 4.29, which represents a typical PKI system business process:

(1) Sender requests registration with RA.
(2) RA verifies the information provided by the sender (certificate applicant) and the identity of the certificate applicant.
(3) After passing the verification (if the registration is successful), the RA forwards the information to the Certificate Authority (CA).
(4) The CA issues a certificate to the applicant.
(5) Sender uses the private key of the certificate to sign and deliver the data to recipient B.

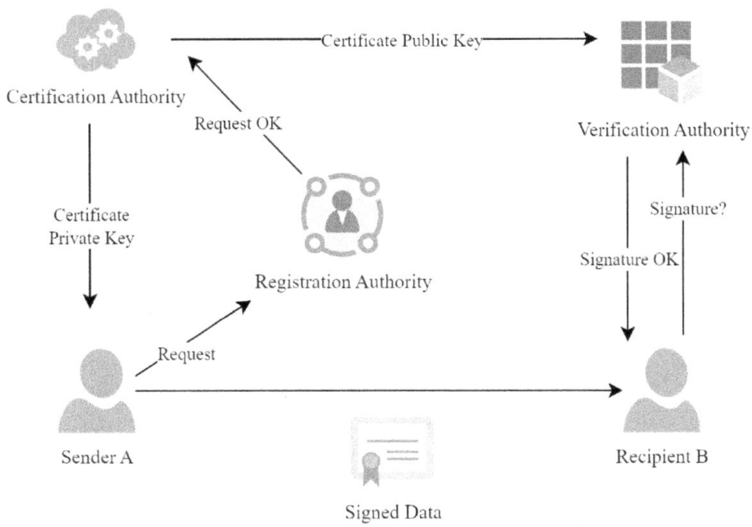

Fig. 4.29. PKI system business process.

(6) Recipient B requests signature verification from the Verification Authority (VA) to confirm the validity of sender A's certificate.
(7) The VA checks the Certificate Revocation List and queries the CA (optional); the CA responds to confirm the validity of the public key.
(8) VA returns that the signature is valid.

Note that the above network entities are functional concepts. In the actual network structure, the CA often performs all of the functions of the VA.

4.6.2 *Certificate authority*

The concept of the public key and the difficulties of public key management have been explained in Section 4.5.7. A Certificate Authority (CA) can certify that a public key belongs to a particular entity. This entity may be a person or a server on a network. If the entity trusts the CA that certified the public key, then the certified public key can be used with confidence. CA can be thought of binding keys to verified network elements or servers. The above description is rather abstract. In fact, CA is similar to the passport issuing authority in the digital world. In the digital world, these authorities issue digital certificates and verify the identity and authority of the holder. Just as the information about the passport holder is embedded in the passport in the real world, the digital certificate issued by the CA has the public key of the individual or organization and other identification information embedded in it, and the CA uses a cryptographic algorithm to record the timestamp on the digital certificate, and it is signed and sealed to prevent tampering. The most important function of a digital certificate is to prove that the user listed in the certificate is the legitimate owner of the public key listed in the certificate. CAs use digital signatures to make it impossible for attackers to forge and tamper with certificates.

CAs have the following roles:

(1) Verify to one entity in a communication that the other entity in the communication is who it claims to be. Assuming, of course, that the entity trusts the CA.
(2) Once the entity is verified, the public key and the entity are bound by a digital certificate. A digital certificate contains a

public key and other identifying information about the owner of the public key (such as the name of a natural person or an IP address).

Since the CA must verify the validity of the communicating party's certificate, it is also responsible for registering, issuing, and revoking certificates. Certificates are issued by many different CAs, so many certificate formats are defined to ensure scheme validity, manageability, and consistency. To reduce CA response time and improve CA performance, users who obtain certificates are responsible for managing their own certificates. To do this, each user initiating a communication must provide its certificate and other identifying information (such as date and nonce) and send it to the recipient along with a request for the recipient's certificate. Upon receipt of these files, the recipient sends its certificate. Both parties then verify each other's certificates and, upon approval by either party, begin communication. During the verification process, any user can periodically check the CA's list of certificates that may have expired prior to their original expiration date due to key compromise or administrative reasons. Since this may require online access to the CA's central facilities, it can sometimes be a bottleneck.

In terms of implementation, the CA certification authority is a tree structure. The root CA certification authority can authorize multiple second-level CA certification authorities, and the second-level CA certification authority can also authorize multiple third-level CA certification authorities. Digital certificate applicants can apply for digital certificates from the root CA certification authority or from the second-level and third-level CA certification authorities. Theoretically, there is no limit. After the application is successful, the applicant has a digital certificate, so its public key can be trusted, and its trust level is the same as that of the issuing CA. In the HTTPS protocol, the client confirms that the website is trustworthy by verifying the website's digital certificate and confirming that it is not subject to man-in-the-middle attacks.

4.6.3 *Registration authority*

The Registration Authority (RA) receives and processes the user's certificate signing request and can be thought of the "front end" of the CA. The RA is responsible for receiving registration applications from users (organizations, individuals, servers, applications, and other entities), authenticating the identity of certificate applicants, and deciding whether to allow the CA to issue digital certificates to them.

The IETF's RFC 3647 defines an RA as "an entity that is responsible for one or more of the following functions: the identification and authentication of certificate applicants, the approval or rejection of certificate applications, initiating certificate revocations or suspensions under certain circumstances, processing subscriber requests to revoke or suspend their certificates, and approving or rejecting requests by subscribers to renew or re-key their certificates. RAs, however, do not sign or issue certificates (i.e., an RA is delegated certain tasks on behalf of a CA)". RAs do not have the signing authority of the CA and only manage the validation and provisioning of certificates.

For small PKI systems, the functions of the RA may be integrated into the CA rather than existing as a separate entity.

4.6.4 *Digital certificates*

A digital certificate is a digitally signed message that proves the validity of the public keys of the communicating parties. To maintain the versatility and validity of digital certificates, the format of digital certificates must have a consistent standard. Most digital certificates follow the ITU Telecommunication Standardization Sector (ITU-T) X.509 standard. According to RFC 1422, the fields of X.509 digital certificates are listed in Table 4.12.

A sample of digital certificate attributes is shown in Fig. 4.30.

Many organizations offer digital certificate services. They provide most, if not all, of the functions of a CA. Typical organizations include Comodo, Symantec, DigiCert, Verisign, etc. There are also several CA providers in China. In addition, the non-profit CA "Let's Encrypt" also issues millions of digital certificates for website authentication.

In modern communications, the use of digital certificates has become commonplace and essential to the security of communications. For example, in a network environment, to encrypt a

Table 4.12. Elements of a X.509[AW2] certificate

Element	Description
Version number	The version number of the certificate
Serial number	The unique number for each certificate issued by a certificate authority (CA)
Issuer name	The X.500 name of the CA which signed and created the certificate
Subject name	The name of the user to whom this certificate has been issued
Period of validity	The period for which the certificate is valid
Signature algorithm Identifier	The algorithm that is used for signing the certificate
Signature	The hash code of all other fields which is encrypted by the CA private key
Subject's public key information [AW3]	The subject's public key along with an identifier of the algorithm for which this key is supposed to be used

Fig. 4.30. An example of the properties of a digital certificate.

server's transmission, the client must use the server's public key. The integrity of this key is critical to the security of subsequent sessions. For example, if a third party (a man-in-the-middle) intercepts the communication and replaces the legitimate key with its own public key, that man-in-the-middle can view all traffic and even modify the data in transit without either the client or the server detecting the intrusion. To prevent this, the client first makes a request to the server, which then sends the public key of a certificate signed by a certificate authority. The client verifies this digital signature. If the signature is valid, the client knows that the server has proven that the certificate is authentic and not a forged certificate from an intermediary. To provide meaningful authentication, it is important that the CA be a trusted third party.

4.6.5 *Lightweight directory access protocol*

The Lightweight Directory Access Protocol (LDAP), developed by the University of Michigan, is an open, vendor-neutral, industry-standard application protocol for accessing and maintaining distributed directory information services over an Internet Protocol (IP) network. Through this protocol, a PKI system can support the distribution of certificates.

The CA can use the LDAP standard directory service to store certificates. The role of directory service is the same as a traditional database, the advantage is that in the case of fewer modification operations, for access to the efficiency than traditional databases. In addition, it can also provide support for Certificate Revocation List (CRL) based on LDAP protocol.

It is worth noting that in the certificate status query scenario, Online Certificate Status Protocol (OCSP) has been more widely used in recent years. The protocol is based on the TLS protocol and extended to provide feedback on whether the certificate provided by the CA has been revoked.

4.6.6 *New PKI system*

With the evolution of distributed and blockchain technologies, concepts and prototypes of some new PKI systems have been continuously proposed, including Blockchain-based PKI and distributed PKI (DPKI).

Since blockchain technology is designed to provide a distributed and immutable ledger of information, it is considered well suited for the storage and management of public keys. Some cryptocurrencies support storing different public key types (SSH, GPG, RFC 2230, etc.) and provide open source software that directly supports the PKI of the OpenSSH server. Blockchain technology provides trustworthiness based on proof of work or other proof methods, which can generally enhance trust in PKI, but there are still issues such as policy consistency, operational security, and software implementation quality. The Certificate Authority paradigm has these problems regardless of the underlying encryption methods and algorithms it uses, and a PKI that seeks to add trustworthy attributes to certificates must also address these issues.

Decentralized ID (DID) eliminates the centralized registration of identifiers in the hierarchical PKI standard and the reliance on a centralized Certificate Authority for key management. When the DID registry uses a distributed ledger, each entity can act as its own root certificate authority, an architecture called distributed PKI.

4.7 Hash Algorithm

The hash algorithm is a basic algorithm in cryptography and is widely used in various security protocols. The input of this type of algorithm is a bit string of arbitrary length, and the output is a bit string of fixed length, often referred to as a hash, hash value, or message digest. This type of algorithm is often referred to as a digest algorithm. For a given message, the digest or hash can be thought of as the fingerprint of the message, the unique representation of the message. Unlike other encryption algorithms, hash algorithms have no keys themselves.

In general, a hash algorithm can be thought of as a procedure that applies a hash function to a set of data elements. In this sense, a hash function is a mathematical abstraction, and a hash algorithm is an implementation combined with data processing methods. In many cases, however, the terms hash algorithm and hash function are used interchangeably. For example, the NIST FIPS 180–4 standard begins as follows:

"This standard defines a family of secure hash algorithms, SHA-1, SHA-224...all of which are iterative one-way hash functions..."

Note that there are two types of hash functions. For example, a hash table, which is common in computer data structures, generally relies on a simple algorithm for its hash function. This type of algorithm is fast and guarantees to produce the same output for the same input. However, it does not fully guarantee that two identical outputs must correspond to two identical inputs. This section focuses on another type of hash function, the cryptographically secure hash function, sometimes called a cryptographic hash function.

Cryptographic hash functions have the following important properties:

(1) It is very difficult to modify the message and leave the hash value unchanged.
(2) It is very difficult to obtain the message corresponding to the given hash value.
(3) It is very difficult to find two different messages with the same hash value.

The first property means that cryptographic hash functions exhibit changes similar to the “avalanche effect”. Changing even a single bit in the input will produce a large number of changes in the entire digest: each bit of the digest will have about a 50% chance of flipping. This doesn’t mean that every change will necessarily cause about half of the bits to flip, but the cryptographic hash function ensures a very high probability. Importantly, this makes it difficult to find collisions or near-collisions.

The second property means that it is difficult to find a message m with a given hash value h. This property is called preimage resistance and is sometimes called irreversibility or stealth. This makes the hash function a “one-way” function: it is very easy to compute a hash for a given message but very difficult to compute a message for a given hash.

The third property has to do with finding messages with the same hash, and this property comes in two forms. In the first form, given a message m, it is difficult to find another message m' with the same hash value h: this is called Second Preimage Resistance. The second form of secrecy is stronger and requires that it should be very difficult to find any two messages m and m’ with the same hash value, which is called collision resistance. Since collision resistance is the stronger form of the second preimage resistance, these two forms can also be referred to as weak collision resistance and strong collision resistance, respectively.

The uses of hash algorithms in cryptography are diverse: hash algorithms are a fundamental building block of digital signature schemes and message authentication codes. Hashing algorithms are also widely used in other cryptographic applications, such as password storage or key derivation. Many solutions also make heavy use of cryptographic hash functions, such as blockchains.

In implementation, a hash algorithm takes an input message M of a given length and produces a unique output message of fixed length. The fixed length of the output message is typically 128, 160, or 256 bits. One-way hashing is a variant of hashing that is used to create a signature or fingerprint of a message that is unique, like a human fingerprint.

The usage scenario is that to ensure the integrity and authenticity of the data, both the sender and receiver use the same hash algorithm to perform the same hash calculation on the message before it is sent and after it is received. If two calculations on the same message produce the same value, the message has not been tampered with in transit.

There are different standards for the message digest length of the hash algorithm, including 160 bits (SHA-1 and MD5) and 128 bits (MD2 and MD4). The message digest hash algorithms MD2, MD4, and MD5 were invented by Ronald Rivest, while the Secure Hash Algorithm (SHA) was developed by NIST. The most popular hash algorithms are SHA and MD5. Table 4.13 lists more details about these algorithms.

Table 4.13. Comparison of cryptographic hash functions.

Function	Output size/bit	Block size/bit
SHA-1	160	512
MD5	160	512
HMAC-MD5	160 (Same as MD5)	512 (Same as MD5)
HMAC-SHA-1	160 (Same as SHA-1)	512 (Same as SHA-1)
RIPEMD	160	128

4.7.1 *MD5*

MD5 is a hash function designed by Ronald Rivest in 1991 and is an extension of MD4 (Rivest, 1992). This hash function outputs a 128 bit digest. Over the years, the cryptography community has repeatedly discovered weaknesses in MD5. In 1993, Bert Den Boer and Antoon Bosselaers published a paper demonstrating a "pseudo-collision" of MD5's compression capabilities (Den Boer and Bosselaers, 1993). Dobbertin extends on the above work and is able to generate collisions for compression functions (Dobbertin, 1996). Building on Dobbertin's work, Xiaoyun Wang, Dengguo Feng, Xuejia Lai, and Hongbo Yu demonstrated in 2004 that MD5 is vulnerable to real collision attacks (Wang *et al.*, 2004). The real attack on MD5 used as a digital signature scenario was proposed by Wang Xiaoyun *et al.* in 2005. In their published paper, they described that two different X.509 certificates have the same digital signature, thus proving that the MD5 algorithm is no longer applicable in digital signature scenarios (Lenstra *et al.*, 2005).

Therefore, it is *not* recommended to use MD5 to generate digital signatures. When designing and implementing new cryptographic systems, it is also necessary to avoid using MD5 as the basic hash function. Theoretically, however, HMAC-MD5 is still secure when used in message authentication scenarios.

The algorithm libraries of most programming languages contain packages of common cryptographic algorithms. For example, in Python 3, the following code can be used to compute MD5:

```
import hashlib
hashval = hashlib.md5(b"datasecurity").hexdigest()
print(hashval)
```

4.7.2 *SHA-1*

SHA-1 is another hash function in the MD4 family designed by the National Security Agency (NSA) that produces a 160 bit digest. Consistent with the MD5 algorithm, SHA-1 is no longer considered a security algorithm for digital signatures. Many software and browsers, including Google Chrome, have begun to drop support for the SHA-1 signature algorithm. On February 23, 2017, researchers from CWI Amsterdam and Google had a collision on the full SHA-1 function.

In Python 3, SHA-1 can be calculated with the following code:

```
import hashlib
hashval = hashlib.sha1(b"datasecurity").hexdigest()
print(hashval)
```

4.7.3 *SHA-2*

SHA-2 refers to a family of hash functions that includes SHA-224, SHA256, SHA-384, SHA-512, SHA-512/224, and SHA-512/256. The summary lengths are 224 bits, 256 bits, 384 bits, 512 bits, 224 bits, and 256 bits, respectively. These hash functions are based on the Merkle–Damgård construction and can be used for digital signatures, message authentication, and random number generators. SHA-2 not only performs better than SHA-1 but also provides better security due to improved collision resistance. SHA-224 and SHA-256 are designed for 32-bit registers, while SHA-384 and SHA-512 are designed for 64-bit registers. SHA-512/224 and SHA-512/256 are truncated versions of SHA-512, also using a 64-bit digest, with output sizes equal to the 32-bit register variants (224-bit and 256-bit digests and better performance on 64-bit CPUs). Table 4.14 gives a comparison of different SHA-2 hash algorithms.

In Python 3, taking SHA-256 as an example, the following code can be used to calculate:

```
import hashlib
hashval = hashlib.sha256(b"datasecurity").hexdigest()
print(hashval)
```

Among the attacks on the SHA-2 hash algorithm, some academic papers have pointed out that the number of rounds of SHA-256 and SHA-512 can be reduced by (pseudo) collision attacks

Table 4.14. Comparison of SHA-2 hash functions.

Function	Output size/bit	Block size/bit	Output size/bit
SHA-224	<264	512	224
SHA-256	<264	512	256
SHA-384	<2128	1024	384
SHA-512	<2128	1024	512
SHA-512/224	<2128	1024	224
SHA-512/256	<2128	1024	256

(Lamberger and Mendel, 2011) and preimage attacks (Khovratovich *et al.*, 2012). Of course, these attacks only reduce the number of rounds and do not constitute a complete attack on the algorithm in the technical sense.

4.7.4 *Keccak and SHA-3*

Keccak, a family of hash functions, was designed by Guido Bertoni, Joan Daemen, Michaël Peeters, and Gilles Van Assche and won the 2012 NIST Secure Hash Algorithm Competition (Bertoni *et al.*, 2009). Since then, Keccak has been standardized as the forms of the SHA3-224, SHA3-256, SHA3-384, and SHA3-512 hash functions.

Although SHA-3 may sound similar to SHA-2, the design of the two is quite different. The hardware implementation of SHA-3 is very efficient, but the software implementation of SHA-3 is less efficient than SHA-2.

The SHA-3 hash function was introduced in Python version 3.6 and can be used as follows:

```
import hashlib
hashval = hashlib.sha3_256(b"datasecurity").hexdigest()
print(hashval)
```

4.7.5 *Password storage*

A very important and easily abused scenario for cryptographic hash functions is password storage. In most cases, users log in to a system with their account name and password. The system usually needs to store the password somewhere in order to verify the correctness of the password entered by the user.

If the password is stored directly in a file or database in plain text, it will cause fatal problems. Even ciphertext storage has similar attack methods. For example, when comparing password strings, the attacker can guess the length of the password based on the time difference. This method is called a "timing attack". Of course, the most direct way for an attacker to obtain passwords is to illegally access the password database.

In the security world, this method is known as "dragging" or "database enumeration". Typically, the user's account name and password are stored together, and many users will reuse the same password on different sites for convenience, leading to account leakage and illegal access. More seriously, if a mobile phone number or email address is used as an account name, or the password database also contains such information, the attacker may try to hijack other services registered with the mobile phone number or email address. This approach is commonly referred to in security circles as "credential stuffing".

A simple workaround is to hash the password with a cryptographically secure hash function. Since hash functions are easy to compute, as long as the user provides a password, the hash value can be quickly computed and compared to the hash value stored in the database. Although the attackers stole the user database, they could only see the hashes, not the actual passwords. According to the preimage resistance of the hash function mentioned in this chapter, it is difficult for the attacker to obtain the corresponding plaintext by using the hash value, so it will be difficult for the attacker to obtain the original password. But is this really true?

In reality, people use very few passwords. Research shows that 17% of users use 123456 as a password for their accounts, and 91% of users use passwords that are in the top 1,000 most common passwords.

Analysis of clear text passwords leaked in numerous data breaches has shown that the vast majority of passwords follow the less secure conventions described in the following:

(1) Use personal date of birth or other meaningful personal information, such as "810326".
(2) Use names of idols, pets, hobbies or other related information, such as "messi10", "jordan", "baseball", and "mustang".
(3) Use numerical sequences or combinations with specific meanings, such as "123456" and "246810".
(4) Use certain forms of words, phrases, such as "password", "batman", and "iloveyou".
(5) Use a specific keyboard layout, such as "qwerty", "qazwsx", and "123qwe".
(6) Use a combination or variation of the above forms, such as "passw0rd" and "loveyou4ever".

As mentioned many times in this chapter, hash functions guarantee the same output value for two identical input values. This valuable feature exacerbates the problem.

Attackers can list all common passwords and their combinations, and even include all possible English words and their combinations, and calculate their hash values to build a huge corresponding table. Such a table is called a "rainbow table". The name of this table probably comes from the colorful colored cards. The output of a hash function is a bit like random assignment. When the output of the hash function is expressed in hexadecimal format, there are some color-like representations, such as 0xFFC90E, which is both the output of the hash function and the representation of the golden color, and the outputs of different hash functions correspond to different colors.

With the popularization of distributed computing power and storage capacity, rainbow tables of billions of plaintext passwords and their hash values can be easily found on the Internet, and the storage size can even reach tens of TBs (terabyte, $1TB = 1024GB = 2^40B$). Obviously, it is not secure enough to simply hash the password and then store it.

Typically, people use salt to defend against the rainbow table attack on password storage. Adding a random "salt" before or after a password string before hashing it can produce completely different hashes. The salt can be stored in clear text next to the password hash.

Let the hash function be H, and the hash with salt value can be expressed as

$$Hashvalue = H(Password||Salt)$$

When a user authenticates with a password, simply combine the salt with the password, hash it, and then compare it to the stored hash.

If you choose a large enough cryptographic random number as the salt value (such as 32 bytes), you can achieve better protection against precomputed attacks (such as "rainbow table" attacks). To successfully perform a rainbow table attack, the attacker must compute a separate table for each salt value. Computing and storing tables with different salt values is very difficult because individual tables are usually very large. Computing a single table takes a lot of time, so computing 2^256 different tables is out of the question.

It's worth noting that many systems use the same salt value for all users. While this prevents rainbow table attacks to some extent, it is still possible to attack all passwords at once if the attacker knows the salt value. All the attacker needs to do is compute a rainbow table for that salt value and compare the result to the hashed password in the database to accomplish the goal of the attack.

4.7.6 *Hash tree*

A hash tree is a tree in which each node is identified by a hash value. The hash value includes the contents of the node and the hash values of its ancestors. A root node with no ancestors hashes only its own contents. A practical hash tree might be a binary tree, or perhaps only the leaf nodes carry their own data, and the parent nodes carry only derived data. A Merkle tree is a kind of hash tree. Such hash trees or their variants are used by many systems, especially distributed systems. Examples include distributed version control systems (such as Git), digital currencies (such as Bitcoin), distributed peer-to-peer networks (such as BitTorrent), and distributed databases (such as Cassandra).

4.8 Message Authentication Code

A Message Authentication Code (MAC) is a small piece of information that can be used to verify the authenticity and integrity of a message. These pieces are often referred to as "tags". The MAC algorithm takes a message of arbitrary length and a key of fixed length and generates a tag. The MAC algorithm also includes an associated authentication algorithm, which takes a message, key, and tag and judges whether the message has been tampered with based on the tag. A simple verification algorithm is to recompute the tag based on the message and key, and check that the calculated tag matches the input tag. However, this algorithm is not always sufficient; many secure MAC algorithms are random and produce a different tag each time they are computed.

Note that the term "message" is used here instead of "plaintext" or "ciphertext". This ambiguity is intentional. In the real world, MAC is used as a method of implementing authenticated encryption, so messages are always ciphertext. However, there is nothing wrong with applying MACs to plaintext messages. In fact, the modes of partially secure authenticated encryption schemes explicitly allow authenticating (but not encrypting) messages to be sent along with the authenticating ciphertext.

4.8.1 *Secure MAC*

The security requirements for MAC functions are similar to those for cryptographic hash functions.

"Security" is relative to a potential adversary or threat. Then it is important what kind of adversary the "security" can deal with.

In the case of a digital signature or a secure MAC, the goal of an adversary can be abstracted as: generate a valid signature/MAC (denoted as σ) for the message m.

Assuming that the adversary can obtain the private key of the signature or the key of the MAC function, the attacker can perform a complete attack, i.e., the security of the digital signature is completely compromised.

Assuming that the adversary does not obtain the private key of the signature or the key of the MAC function, the following attack methods can be defined according to the adversary's capabilities:

Universal forgery: This refers to the fact that an adversary can create a valid signature σ for any given message m. Adversaries capable of universal forgery can sign messages of their choice (e.g., selective forgery), randomly selected messages, or even arbitrarily provided messages.

Selective forgery: This refers to the adversary's ability to create a valid signature pair (m, σ) for a limited set of messages. m, which may be related to the interesting mathematical properties of the signature algorithm. However, the essence of selective forgery is that the scope of a particular selective forgery attack has been determined. Adversaries who can successfully perform selective forgery attacks also have the ability to perform existential forgery attacks.

Existential forgery: An adversary is able to create at least one message/signature pair (m, σ) where m is not signed by a legitimate signer, but σ is valid. The adversary is free to choose any m, and as long as the σ is valid, the attacker successfully constructs an existential forgery. Therefore, existential forgery is easier than selective forgery because the attacker can choose the plaintext m that is most likely to forge the signature due to the weakness of the algorithm.

Weak existential forgery: An adversary with a weak existential forgery is not as capable as an adversary with the ability to construct an existential forgery. Weak existential forgery means that an adversary is able to create at least one valid message/signature pair (m', σ') on the premise of obtaining a certain number of message/signature pairs (m, σ).

A secure MAC function must resist existential forgery under chosen-message attacks. That is, suppose the adversary can arbitrarily choose m as the input to the MAC algorithm to obtain a valid signature σ. For m', which was not queried, the corresponding σ' cannot be computed.

4.8.2 *Keys in MAC algorithm*

Common practices for verifying the integrity of a message involve calculating some sort of digest of the message using a checksum (such as CRC32 or Adler32) or even a cryptographically secure hash (such as the SHA family).

Assuming a software package is released on the Internet, the general form of distribution includes a compressed package of source code and may also include binary packages for some popular operating systems. Then you put a (cryptographically secure!) hash next to it so that anyone who downloads it can verify the hash and confirm that they have downloaded what they think they have.

Of course, this solution has relatively large security risks. Computing these hashes is something anyone can do. In fact, the publisher of the package and even the intended downloader can verify the integrity of the package this way. The problem is that an attacker who modifies the downloaded file can recompute the hash of the modified downloaded file and store the value. After the user downloads the modified file, its hash value is calculated and compared to the modified hash value to determine that the download was successful. Regardless of whether the attacker attacks the storage of the software package or the transfer process, this scheme cannot completely defend against the attack.

Some possible improvements include applying the signature algorithm directly to the binary, or signing the message; as long as the hash function used to sign the message is secure, double preimage attacks can be defended. The important difference is that creating a signature (using a public key signature algorithm) is not something an attacker can do, only someone who has the corresponding private key.

4.8.3 *HMAC introduction*

Hash-based Message Authentication Code (HMAC) is a standard for generating MACs that take a cryptographic hash function as a parameter. The main purpose of HMAC is to verify data integrity and message authenticity at the same time. The HMAC algorithm was invented by Mihir Bellare, Ran Canetti, and Hugo Krawczyk in their 1996 paper and then standardized in RFC 2104 (Krawczyk *et al.*, 1997). Prior to this, many protocols attempted to implement their own message authentication using hash functions, but these attempts were not as successful as expected. The paper proposes a provably secure MAC that requires nothing more than a key and a hash function.

The hash function used in HMAC calculations can be any cryptographically secure hash function, such as SHA-2 and SHA-3. The naming format of HMAC is "HMAC-" plus the name of the hash function used. For example, if SHA3–256 is used, the HMAC will be named HMAC-SHA3-256.

A good feature of HMAC is that it has a fairly strong proof of security. As long as the underlying hash function is a pseudorandom function, HMAC itself is a pseudorandom function. The security of HMAC does not even depend entirely on the collision resistance of the underlying hash function. The proof was introduced after the invention of HMAC and is consistent with real-world observations: even though MD5 and SHA-0 (first published in 1993 and not yet a formal standard, the predecessor of SHA-1) have serious collision attacks, HMACs constructed from these hash functions still appear to be secure.

The characteristic of HMAC is that the message passes through the hash function twice and is combined with the key before each pass.

Assume that the cryptographically secure hash function that HMAC relies on is H, the key is K, and the input plaintext is $plaintext$. The data block size of the hash function H is B bytes (for most hash functions, $B = 64$). L represents the length of the output bytes of the hash function ($L = 16$ for MD5, $L = 20$ for SHA-1, $L = 32$ for SHA-256).

If the length of key K is greater than B, then H is first performed once on the key and the resulting byte array K' of length L is used as the key for the HMAC function.

HMAC defines two additional constant byte arrays:

$$ipad = 0x3636...(0x36Btimes)$$
$$opad = 0x5C5C...(0x5CBtimes)$$

According to the RFC 2104 standard, the HMAC function can be expressed as

$$HMAC(K, plaintext) = H((K' \oplus opad)||H((K' \oplus ipad)||plaintext))$$

$$K' = \begin{cases} H(K) & \text{if } K > Blocksize \\ K & \text{if } K \leqslant Blocksize \end{cases} \tag{4.11}$$

The "||" symbol represents a connection.

For the HMAC calculation of the Python 3 version, you can refer to the following code example:

```
import hashlib, hmac, binascii
mac = hmac.new(b'secret', b'datasecurity',
      hashlib.sha256) .hexdigest()
print(mac)
```

The above code will give the output "b7 86 a3 04 9b 78 62 95 33 73 23 98 0a 08 73 02 14 06 79 ec 41 15 7d 89 85 c2 44 99 31 c2 5c 32".

4.8.4 *Authentication encryption mode*

So far in this book, encryption and authentication have been discussed separately, and their different roles have been introduced. Encryption and authentication are also considered fundamentally different steps in most secure communications protocols.

Alternatively, authentication can be made an essential part of how the encryption algorithm works. Unauthenticated encryption presents many security risks and is not a security best practice. It makes sense to use new models that guarantee not only the confidentiality of arbitrary data flows, but also their integrity and authenticity.

In addition to protecting the integrity and confidentiality of messages, authenticated encryption provides security against "chosen ciphertext" attacks. In chosen ciphertext attacks, the attacker attempts to gain deep insight into the cryptosystem (such as information about the decryption key) by submitting carefully chosen ciphertexts to some "decryption oracle" and analyzing the decryption results. An authenticated encryption scheme can detect incorrectly constructed ciphertext and refuse to decrypt it. This, in turn, prevents an attacker from requesting decryption of any ciphertext unless it was generated correctly using the encryption algorithm, thereby avoiding possible disclosure of the plaintext. Properly implemented, authenticated encryption removes the usefulness of a decryption oracle by preventing an attacker from obtaining any useful information that he does not already have.

Authenticated Encryption with Associated Data (AEAD) is an authenticated encryption mode (Rogaway, 2002). The scheme assumes that the message consists of the following two parts:

(1) The actual content itself.
(2) **Metadata:** Data about the content.

In many cases, the metadata should be in clear text, but the content itself should be encrypted. The entire message should be authenticated: an attacker should not be able to tamper with the metadata and still have the resulting message be considered valid.

Figure 4.31 shows a typical use case for AEAD. AEAD is a variant of AE that allows recipients to check the integrity of encrypted and unencrypted information in messages. AEAD binds associated data (AD) with ciphertext and the context in which it should appear so that attempts to "cut and paste" valid ciphertext into other contexts are detected and rejected.

The following scenario provides a visual illustration. Email systems began in the 1960s, before the concept of "Internet security" existed. From a modern security perspective, email systems have many security and privacy issues:

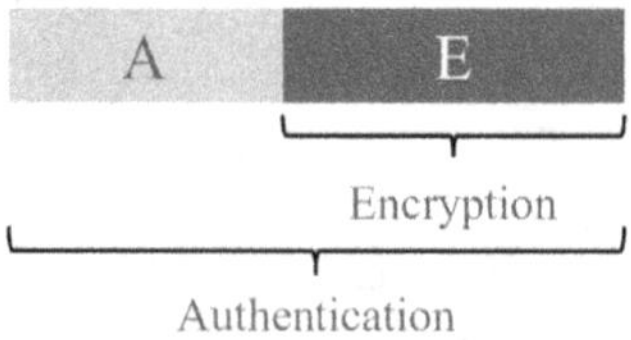

Fig. 4.31. Typical application scenarios of AEAD.

(1) Email content is not encrypted by default.
(2) Email can be transmitted between multiple servers, and it is relatively easy to intercept and steal email.
(3) The recipient and sender of the email are both in clear text, which is not easy to keep secret.

If you are designing a cryptographic system for a next-generation email system, the metadata of the message can include the intended recipient. Of course, the email content itself must be encrypted and authenticated so that only the recipient can read it. However, the metadata needs to be in plain text: the email server delivering the message needs to know which recipients to send the message to.

Additional scenarios, such as network packets or frames, require visibility of the header, confidentiality of the payload, and integrity and authenticity of the above.

4.9 Digital Signature

Hash functions can be used to ensure the integrity and authenticity of messages, but in real-world communication scenarios, it is also necessary to confirm the authenticity and integrity of each message associated with each user to ensure user non-repudiation. This is the main use case for digital signatures.

A digital signature is a digest of a message, encrypted with the sender's private key, which is appended to the message to authenticate the sender in the same way that a handwritten signature appended to a written document is used to authenticate the signer (typically the sender). Digital signatures are used to confirm the identity of the sender and the integrity of the document, providing non-repudiation of the sender.

We can create a digital signature using a combination of public key encryption and a one-way secure hash function according to the following steps:

In the first step, the sender of the message uses the message digest function to generate a message authentication code (MAC).

In the second step, the MAC is encrypted using a public-private key encryption algorithm, and the encrypted MAC is attached to the message as a digital signature.

The message is then sent to the recipient. Once the message is received by the recipient, the digital signature is decrypted using the sender's public key to verify that the message is indeed from the intended sender. The detailed steps are as follows:

(1) The receiver separates the received message into two parts: the original document and the digital signature.
(2) The receiver uses the sender's public key to decrypt the digital signature to obtain the original MAC.
(3) The receiver uses the original document and enters it into a hash function to generate a new MAC.
(4) The new MAC is compared to the sender's MAC for a match.

If these numbers match, the message has not been tampered with, data integrity is ensured, and the authenticity of the sender can be verified.

The workflow of digital signature verification is shown in Fig. 4.32.

Since the digital signature is extracted from the email as a digest and then encrypted, the signature is tied to the content of the email.

Since digital signatures are used to authenticate messages and identify the sender of those messages, they can be used in a variety of areas where double-checking is required. Anything that can be digitized can be digitally signed. This means that digital signatures can be used with any type of message, encrypted or not, to establish the authenticity of the sender and ensure the integrity of the message

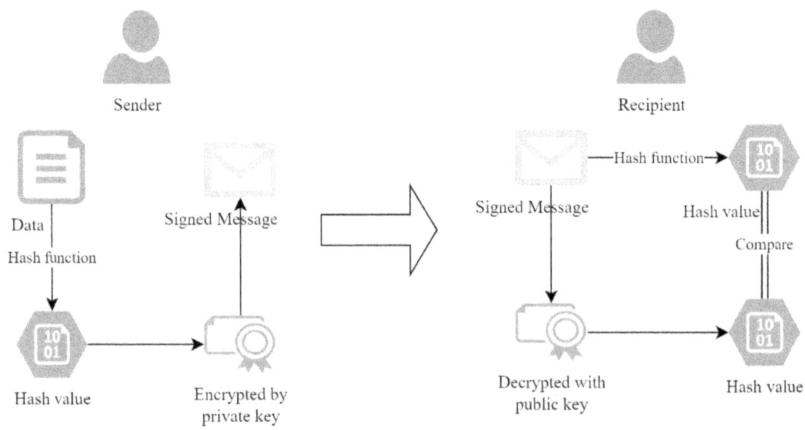

Fig. 4.32. Workflow of digital signature verification.

during transmission. However, digital signatures cannot be used to ensure the confidentiality of message content.

The ElGamal public key encryption algorithm (ElGamal, 1985) and the Digital Signature Standard (DSS) proposed by NIST in the FIPS 186-4 standard (FIPS Publication, 2000) are the most widely used digital signature algorithms today. DSS encrypts faster than RSA.

Despite the popularity of digital signatures, they are not the only way to verify the authenticity of senders and the integrity of messages. Since the digital signature algorithm is very complex and the public and private key certificates are relatively expensive, the online community uses other less secure but very simple verification methods, including cyclic redundancy checks (CRC).

In CRC, a fixed number of check bits, known as a checksum, are appended to the message being transmitted. The data receivers check the check bits for errors by evaluating the attached checksum, which is obtained by finding the remainder of the polynomial division of the transmitted content. If an error is detected, a negative acknowledgement is sent requesting the data to be retransmitted.

4.10 Security Protocols Based on Cryptography

Many security mechanisms and controls in modern computing and communications are based on cryptographic security protocols.

4.10.1 *Identification and login protocol*

In network-based communication and interaction, one of the most fundamental problems is the identity or login problem. The problem can be abstracted as participant A having to prove his identity to participant B in order to obtain the right to use or access certain resources owned by participant B.

Give some intuitive examples. According to the conventions of cryptography, Party A's name is Alice.

If Alice needs to enter a building, she may need to enter the access code to prove her identity. A very similar scenario is the login interface of smartphones and laptops, where users also need to enter their passwords. The attack scenario is mainly password leakage or the attacker impersonating Alice's identity.

When Alice needs to open her car door, she usually uses her car key, which has a contactless password. The attack scenario is mainly that someone eavesdrops on the communication channel and observes the interaction data between the car key and the car. The main security requirement is that an attacker cannot open the car door without getting the car key.

Alice takes her bank card to an ATM to withdraw money. The attack scenario has two parts: from the bank's perspective, there are risks such as forgery of Alice's identity and bank card; from Alice's perspective, there is the risk of forgery of the ATM. Counterfeit ATMs can be used to steal a card's real information and impersonate the card in subsequent transactions.

The most complex problem is that Alice wants to log into her online banking. As a remote login method, the browser must first negotiate a secure channel with the bank's website, and then Alice must prove her identity to the bank. A traditional way is to use a bank password. There are many possible attack scenarios. Malicious websites can impersonate bank Web sites and trick Alice into entering her bank password to prove her identity. This method is called a "phishing attack". Phishing attacks are a type of active attack. On the banking side, the attack scenarios are more complex. There are fake users who steal user passwords, there are also attacks that try to brute-force crack user passwords, and they even include denial-of-service (DoS) attacks on website servers. This section focuses on identification and login problems that rely on identification protocols, one of the basic protocols that cryptography can provide.

The identification protocol consists of two parties: the prover and the verifier. A prover can be a human user or a service requester; a verifier can be a computer, a server, or a service provider. In the example above, when Alice goes to the ATM to withdraw money, Alice is the prover and the ATM is the verifier.

The prover provides some kind of credentials (such as Secret Key, SK) to the verifier, and the verifier verifies that the credentials are valid. The Verification Key (VK) is then returned to the prover to complete the identification process.

According to how the verifier and prover interact, attack scenarios can be divided into three types:

(1) **Direct attack scenario:** This refers to the building access control scenario. At this point, there is physical interaction between the verifier and the prover. If the attacker cannot

eavesdrop, he must impersonate the prover himself. There are many security methods at this point, and in general, a simple password authentication protocol can satisfy this scenario.

(2) **Eavesdropping attack scenario:** See the car key remote unlocking scenario. At this time, there is relatively close interaction between the verifier and the prover, but the communication channel may be monitored. Simple password authentication protocols are not secure enough in this scenario. However, a more complex protocol based on one-time authentication passwords may be sufficient for this scenario.

(3) **Active attack scenario:** See the online banking login scenario. In this case, the verifier and the prover have a remote, non-direct interaction. Both the verifier and the prover can be impersonated. A fake verifier can obtain the proofs provided by the prover and then impersonate the prover to the real verifier. This type of attack is called an "active attack". In this scenario, a more complex interaction protocol between the verifier and the prover is required, such as an interaction protocol based on a challenge-response mechanism.

Of course, Alice's online banking credentials can be obtained not only by spoofing bank websites but also by some other attack methods. For example, malware on a local computer could also attempt to overwrite or impersonate an online banking login screen. Identification protocols can be classified in several ways.

Based on whether VK can be made public, identification protocols fall into two categories: private VK agreement and public VK agreement. Most identification protocols require verifiers not to disclose the VK. However, it is obvious that the way of exposing the VK is more secure. Since security is not compromised against supposedly malicious verifiers.

Based on whether VK and SK are variable, identification protocols are divided into stateful protocols and stateless protocols. Each time a stateful protocol is executed, VK and SK can be updated. Obviously, stateful protocols are more secure. However, stateful protocols require some kind of synchronization mechanism between prover and verifier, so their implementation and use are more complicated.

Based on the authentication object, the identity authentication protocol can be divided into two categories: one-way authentication protocol and two-way authentication protocol. Two-way authentication (mutual identification) means that not only resource user A must prove its identity to resource owner B but also resource owner B must prove its identity to resource user A.

In short, the identity authentication protocol can prevent attackers from impersonating resource user A. This impersonation usually occurs without resource user A's knowledge. Possible attackers have the ability to monitor or even tamper with or block the communication channel, or have the ability to impersonate resource owner B.

Simply relying on the identity authentication protocol is not enough to establish a secure session in remote interaction scenarios. At this point, man-in-the-middle (MITM) attacks must also be considered. In the man-in-the-middle attack scenario, the attacker controls the communication channel between Alice and the Verifier and forwards all authentication messages between the two, thereby monitoring the conversations between both parties without their knowledge. To overcome man-in-the-middle attacks, identity authentication protocols can be used in conjunction with session key exchange protocols.

4.10.2 *Authenticated key exchange*

Suppose Alice and Bob wish to establish a secure channel over an insecure network for secure communication, which begs the question: how can Alice and Bob mutually identify each other and establish such a shared session key? The protocol used for this purpose is called the Authenticated Key Exchange (AKE) protocol.

Simply put, the AKE protocol should allow two users to establish a shared session key. At the end of a successful run of this protocol, User A should have a clear idea of which user (User B) he is communicating with, i.e., with which user a shared session key has been established. A secure AKE protocol should ensure that User A's session key is a newly generated random key known only to User B.

If there is a public Trusted Third Party (TTP), things are simpler. Each user in the system must perform some kind of registration protocol with the TTP; after performing the registration protocol, the user has established its own long-term key. In some protocols, the TTP can be offline, and the operation of the AKE protocol does not depend on communication with the TTP. The role of TTP in these protocols is that of a Certificate Authority (CA). There is also an online TTP-based AKE protocol, each run of the AKE protocol must contact the TTP, and the TTP shares secret information with the user. Obviously, the offline TTP-based AKE protocol is more secure and convenient in general scenarios. A notable difference is that the online TTP protocol can be built using symmetric key primitives instead of relying on public key encryption algorithms and tools. In addition, the key revocation of the online TTP protocol is simpler.

The AKE protocol must support multiple runs by the user. Each run invokes an instance of that user. Although there is only one long-term key for a given user, each run of the AKE protocol should generate a new session key. In this way, even if an attacker can steal a particular session key, the security of other sessions is not significantly compromised. In addition, some methods of implementing secure channels rely on the freshness of the session key to maintain its security over multiple sessions. For example, suppose a stream cipher is used to maintain the confidentiality of data sent over a secure channel between a user and his bank. If two different streams are encrypted with the same key, an attacker can perform a two-time pad attack to obtain information about the encrypted data. Session key freshness ensures that keys used in different sessions are effectively independent of each other.

Authentication key exchange is based on two different types of protocols:

(1) Authentication mechanism, such as MAC or digital signature.
(2) Key encapsulation, usually by some kind of Diffie-Hellman key agreement.

A simple example of AKE is the modern TLS handshake mechanism, which first uses digital signatures (X.509 certificates signed by a Certificate Authority) to sign an ephemeral Elliptic Curve Diffie–Hellman (ECDH) public key, which is then used in the derivation of a shared key to encrypt and authenticate network traffic. Note that the above explanation omits parts of the complex implementation mechanisms, such as not describing the underlying mechanisms of mutual-auth TLS or protocol negotiation.

The AKE mechanism used by some instant messaging protocols is more complex. For example, the X3DH and Double Ratchet protocols used by the Signal end-to-end encrypted communications software (Cohn-Gordon *et al.*, 2020).

Currently, the IETF is also working hard to promote the Messaging Layer Security (MLS) standard (Alwen *et al.*, 2020), which uses a binary tree of ECDH handshakes to manage state and optimize group operations (TreeKEM).

Standardization of the Password-Authenticated Key Exchange (PAKE) protocol is also underway. Cryptographers in the IETF Crypto Forum Research Group (CFRG) are working to standardize a family of cryptographically authenticated key exchange protocols.

There are two flavors of PAKE: balanced (mutual authentication) and boosted (prover on one side and verifier on the other). Balanced PAKE is good for encrypted tunnels where the user controls two endpoints (such as a Wi-Fi network), while enhanced PAKE is good for eliminating the risk of password theft in client/server applications if the server is hacked. The CFRG selected a balanced Composable PAKE (CPace) (Abdalla *et al.*, 2021) and an Enhanced PAKE (OPAQUE) (Jarecki *et al.*, 2018), both of which are currently at the draft standard or call for comments stage and have not yet officially become published standards.

4.10.3 *Zero-knowledge proof*

Recalling what was discussed in Section 1.1, "Scope of the Data Security Field", the primary purpose of data security is to protect data that is collected, stored, created, received, or transmitted. The security of the data transmission channel is very important, and cryptography provides the foundation for building a secure transmission channel.

Modern cryptography is based on key secrecy. In the private key encryption system, the two communicating parties must contact each other before communication and agree on a common private key. In a public key cryptographic system, each party has a pair of public and private keys. The public key can be made public, while the private key must be kept strictly confidential. The public key encryption scheme can eliminate the need for the communicating parties to negotiate a common key in advance. This scheme relies on the fact that users have limited computing resources, whether for legal or illegal purposes, i.e., public key encryption systems rely on the problem of computational difficulty. The security of zero-knowledge proofs is based on the fact that if you want to compute the private key from the public key, you have to solve computationally hard problems. A sender can encrypt a message using a public key, and a receiver can decrypt a message using a private key. Only the owner of the private key can decrypt the message, but anyone who knows the public key can send the message.

Modern cryptography is based on computational complexity, and most security algorithms and protocols based on cryptography have their own limitations. If there is a means to reduce the computational difficulty, it will effectively crack the security algorithm and security protocol. For example, if a quantum computer can really quickly compute the corresponding private key from the RSA public key, the entire public key encryption system will be at great risk.

Thus, the idea of proving knowledge of some assertion without revealing any information about the assertion itself is very attractive. Zero-knowledge protocols are designed for this scenario.

Several terms commonly used in zero-knowledge protocols include assertion, knowledge, and proof. An assertion is a statement that states whether a condition is true or false. Proof means confirming in some way whether an assertion is correct. And knowledge in this context refers to the information contained in the assertion itself. For example, an assertion in elementary geometry is "the sum of the interior angles of a triangle is 180 degrees". Its proof is the basic content of elementary geometry and will not be explained in this book. Goldbach's conjecture is also an assertion or assumption in the field of mathematics. It can be described as "any even number greater than 2 can be factorized into the sum of two prime numbers". In mathematical terms, this assertion has not yet been proven. In other disciplines, too, large amounts of evidence (or statistical features) are often accumulated to prove or disprove certain assertions.

The zero-knowledge protocol, as an encryption protocol, does not reveal the secret itself during the operation, the secret will not be transmitted to the other party, but the user can still prove to the other party that he knows the secret. Zero-knowledge protocols can be a good way to prove the identity of communicating parties, or to prove possession of a particular key during the key exchange step.

Zero-knowledge protocols typically have three to four different roles: Prover (P), Verifier (V), Eavesdropper (Eve), and Malicious User (M). In a zero-knowledge protocol, the prover tries to prove knowledge of the secret to the verifier without revealing the secret itself. The verifier can ask questions to determine whether the prover actually knows the secret, but even if the verifier does not follow the rules of the protocol, it cannot discover any information about the secret. An eavesdropper is a third party who listens to a conversation. If the zero-knowledge protocol is secure, an eavesdropper cannot learn anything about the secret and cannot convince others that he knows the secret. A malicious user can also be expected to be able to send, modify, or destroy messages. A good protocol should protect against malicious users. The protocol must take into account that both P and V may have malicious intentions. P may try to trick V into receiving a false message, and V may try to obtain information and misuse it. A good protocol must be structured in the following way:

(1) If P does not know the secret information, he cannot pretend to have such knowledge. A decent zero-knowledge protocol guarantees that P will not deceive V with close to 100% probability through multiple rounds of Q&A.
(2) V can conclude that P knows the secret, but V cannot obtain any further information. V also cannot convince others that he knows the secret. In particular, V cannot gain anything from the interactions of the protocol without directly asking P questions. This is the meaning of "zero knowledge".

In the following, a vivid but less precise example is used to describe the zero-knowledge protocol.

The assertion made by P is "I can remember the suits and numbers of 1000 playing cards". Obviously, V can choose to shuffle the order of the 1000 playing cards and let P speak out one by one to verify the assertion made by P. But doing so will take a long time and is uneconomical as a verification protocol.

So, what is the appropriate verification protocol? A possible verification protocol can consist of several rounds, each of which can be divided into two rounds as follows:

(1) P closes his eyes. V is allowed to randomly replace any one of the 1,000 playing cards with other playing cards in his hand. Of course, V can also do nothing.
(2) P opens his eyes. P observes these 1,000 playing cards and then says whether there are any replaced playing cards. If so, he states which playing card was replaced, along with the number and suit of the original playing card.

After the first round, even if P is correct, V will be skeptical. Maybe P is just lucky? If the same is true after the second round, then V's level of belief will increase. Allowing the above rounds to be repeated several times, V will be able to verify that the assertion made by P is true with almost 100% probability. Of course, during this process, P did not let V know how he remembered the playing cards (this was a secret) nor did he even have to reveal the contents of each playing card.

The concept of "Zero Knowledge" was proposed in 1985 by MIT researchers Shafi Goldwasser, Silvio Micali, and Charles Rackoff in their paper "Knowledge Complexity of Interactive Proof Systems" (Goldwasser *et al.*, 1989). In this theoretical system, a prover (first party) exchanges messages with a verifier (second party) to convince the verifier of the correctness of some mathematical statements.

Before Goldwasser *et al.* introduced the concept of "zero knowledge", most of the work in this area focused on proving the reliability of interactive proof systems. That is, the zero knowledge concept considers situations where a malicious prover tries to "trick" a verifier into believing a false statement. Goldwasser *et al.* analyze this problem from a different perspective. They suggest that we should not only worry about provers but also about what happens when verifiers are not trusted.

The specific problem scenario is information leakage. That is, in addition to proving that the facts are correct, how much additional information will the verifier know during the proof process?

Goldwasser *et al.* proposed in the paper three key properties that any zero-knowledge proof must satisfy:

(1) **Integrity:** If the prover is honest, then he will eventually convince the verifier.

(2) **Soundness:** The prover can only convince the verifier if the statement is correct.
(3) **Zero-knowledge:** The verifier cannot learn any information except that the statement is true.

The development of "zero-knowledge" proofs has taken relatively large twists and turns. It has long existed only as a theory and has not been applied and popularized in engineering in the expected scenarios. In 2007, another paper by Goldwasser *et al.* presented a delegated computation validator that runs in polynomial time, which was a major leap forward. Although still at the theoretical level, the work has the potential to become a reality. In 2010, another seminal paper by Gennaro, Gentry, and Parno *et al.* proposed the concept of non-interactive verifiable computation (Gennaro *et al.*, 2010).

Since then, research and innovation in this direction have continued to advance. Eventually, the paper "Pinocchio: Verifiable Computing with Almost Practicality" (Parno *et al.*, 2016) was published. Pinocchio is "a built system for efficiently verifying general computations while relying only on cryptographic assumptions". The paper defines and details all the elements of building a specific Zero-Knowledge Succinct Non-interactive Arguments of Knowledge (zkSNARK), which it considers "nearly practical". The paper "Succinct Non-Interactive Zero-Knowledge for Von Neumann Architectures" (Ben-Sasson *et al.*, 2014), first published by Ben Sassen *et al.* at the end of 2013, shows that the Pinocchio zero-knowledge proof system is indeed feasible. The 2012 paper "From extractable collision resistance to succinct non-interactive proofs of knowledge and back" (Bitansky *et al.*, 2012) was the first to define an acronym for zero-knowledge succinct non-interactive proofs of knowledge zk-SNARKs. The first widespread application of zkSNARKs was in the Zerocash blockchain protocol, where zero-knowledge cryptography provides the core computational functionality of the protocol by mathematically proving that a party possesses certain information without revealing what that information is. An article on the Forbes website (Bambysheva, 2021) states that in 2021, "82 cryptocurrencies with a combined value of $8.85 billion had their transactions encrypted with zero-knowledge proofs or similar privacy techniques".

4.10.4 *Secure multi-party computation*

Secure Multi-Party Computation (SMPC), also known as Secure Computing, Multi-Party Computing (MPC), or Privacy Preserving Computing, is a subfield of cryptography whose goal is to provide methods for parties to jointly compute the result of a function on inputs provided by the parties, while keeping those inputs private. SMPC is not a typical cryptographic problem. Modern cryptography ensures the security and integrity of the communication, and the attacker is generally not a participant in the communication (e.g., the eavesdropper is not a sender or receiver of the communication). The main goal of secure multiparty computation is to protect the privacy of the participants in the presence of malicious participants.

The foundations of secure multiparty computation began in the late 1970s, focusing on how to encrypt and compute in online gaming/computing scenarios where a trusted third party was not required. The method of traditional cryptography is to hide content, but the method of this new type of computation and protocol is to hide some of the information in the data, use data from multiple sources to perform computations simultaneously, and generate the output correctly.

Secure multi-party computation originated from the millionaire problem proposed by Academician Andrew Chi-Chih Yao in his 1982 paper (Yao, 1982). The problem can be vividly described as two millionaires who want to know who is richer on both sides, but neither wants to reveal his real wealth to the other party, and of course neither wants to reveal it to a third party. For this scenario, it is necessary to find a secure computational method or protocol that can compare and return results. The problem is abstracted to a Boolean comparison problem, i.e., whether $X1 > X2$ is true. The first solution proposed in the paper was exponential in time and space, and several performance improvements followed. Oded Goldreich generalizes the problem to multiparty comparison and computation (Goldreich, 1998).

The problem of secure multiparty computation can be abstracted as how multiple participants who do not trust each other can, in the absence of a Trusted Third Party (TTP), compute together using each participant's private information without revealing that information. If there is a TTP, it is straightforward. For example, in the average wage calculation scenario, each participant tells the TTP his or her monthly wage, and the TTP calculates the average wage.

In essence, secure multiparty computation is part of the problem of distributed multiparty protocols. The protocol describes how multiparty participants interact with information in a distributed network. And most of the target scenarios of secure multiparty computation are easily solved in the presence of TTP. Therefore, the goal of secure multiparty computation is how to "equate" a TTP to get the same result in the absence of a TTP.

The reason it is called "secure" multiparty computation is that the protocol must define the abstract functions and security goals required by each scenario. For the goal, all possible attack scenarios of malicious parties must be considered, and security proofs with strict mathematical dimensions are required.

Secure multiparty computation can be summarized by the following mathematical model: There are n mutually distrustful participants $P1, P2, \ldots, Pn$, each participant P_i secretly inputs x_i, and they must jointly execute a known function:

$$f : (x_1, x_2, \ldots, x_n) \to (y_1, y_2, \ldots, y_n)$$

Among them, y_i is the corresponding output obtained by P_i. During the calculation of the function f, any participant P_i is required to except y_i, neither can get any input information from other participants $P_j (j \neq i)$.

In most cases, $y_1 = y_2 = \cdots = y_n$, so the function can be expressed simply as

$$f : (x_1, x_2, \ldots, x_n) \to y$$

more abstractly described as

$$f : (\{0, 1\}^*)^k \to (\{0, 1\}^*)^k$$

That is, for k mutually distrustful participants, the computational function f receives k bit strings from different participants as input and generates k bit strings as output.

Secure multi-party computing focuses on protecting the privacy of participants and is closely related to outsourced computing, homomorphic encryption, and other scenarios, but the business goals and implementation mechanisms are not the same.

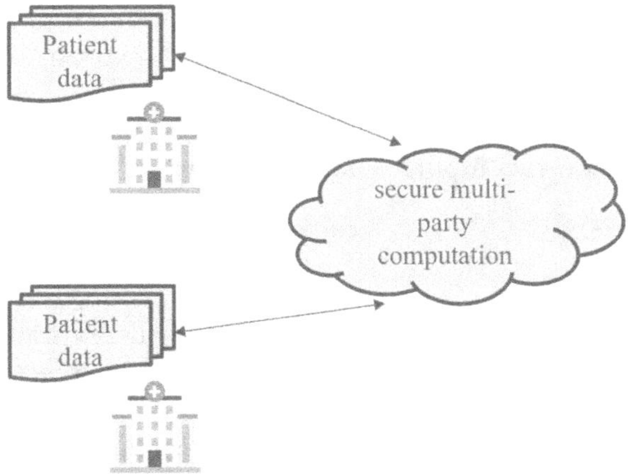

Fig. 4.33. Schematic diagram of a secure multi-party computing scenario for medical data.

Secure multi-party computing technology can better solve the usage requirements of medical data privacy scenarios. Based on medical data research, secure multi-party computing can achieve research results without violating patient privacy, as shown in Fig. 4.33. For example, in genetic disease research based on genetic data, secure multi-party computing can still obtain overall research results even without collecting the original data of the genome.

With the popularization of cloud computing and big data, building privacy-preserving secure computing is not only a theoretical research topic in academia, but also has received more and more practical applications in the industrial field. Secure multi-party computation is one of the hot technologies in the industrial field.

4.10.5 *Homomorphic encryption*

Homomorphic encryption is a form of encryption that allows one to perform a certain form of algebraic operation on ciphertext, and the result is still encrypted, and the decryption result is the same as the result of performing the same operation on plaintext. In other words, this technology allows people to perform operations such as search and comparison on encrypted data and obtain correct results without having to decrypt the data during the entire process. Its significance is that it really fundamentally solves the confidentiality problem when entrusting data and its operations to a third party. It is a data mining technology used for privacy protection, such as for various cloud computing applications.

An encryption function with homomorphic properties means that the operation on two inputs: a and b satisfy

$$Dec(Enc(a) \odot Enc(b)) = a \oplus b$$

Enc is the encryption operation, *Dec* is the decryption operation, $\odot$ and $\oplus$ correspond to the operations on the plaintext domain and the ciphertext domain, respectively. When $\oplus$ represents addition, the encryption is called additive homomorphic encryption; when $\odot$ represents multiplication, the encryption is called multiplicative homomorphic encryption.

Fully Homomorphic Encryption (FHE) refers to an encryption function that satisfies the properties of additive homomorphic encryption and multiplicative homomorphic encryption and can perform any number of additions and multiplications. Expressed in a mathematical formula, that is,

$$Dec(f(Enc(m_1), Enc(m_2), \ldots, Enc(m_k))) = f(m_1, m_2, \ldots, m_k)$$

or

$$f(Enc(m_1), Enc(m_2), \ldots, Enc(m_k)) = Enc(f(m_1, m_2, \ldots, m_k))$$

If f is an arbitrary function, it is called **Fully Homomorphic Encryption**.

Homomorphic encryption has always been an important topic in the field of cryptography, and in the past only a few methods have been found to partially implement this operation. In September 2009, Craig Gentry proposed a feasible method of "fully homomorphic encryption" (Gentry, 2009) from a mathematical point of view, i.e., any operation that can be performed on the encrypted data in plaintext without decryption, which made a decisive breakthrough in this technology. Therefore, on this basis, people can study more complete practical technologies, which is of great importance for the protection of information security.

The advantages of homomorphic encryption technology are as follows:

(1) Multiple ciphertexts can be computed directly without the high computational cost of decrypting each ciphertext.
(2) It allows the ciphertext to be computed by the keyless party, which reduces the cost of communication and shifts the computational task, thus balancing the computational cost for all parties.
(3) It can be used to allow the decryptor to know only the final result, but not the plaintext corresponding to each ciphertext, thus improving the security of the information.

Due to the advantages of homomorphic encryption technology in terms of computational complexity, communication complexity, and security, more and more research efforts are devoted to exploring its theory and applications.

If there is a real breakthrough in this technology, service providers that store confidential data about other parties could be entrusted by users to fully analyze the data without having to interact with them frequently or see any private data. Homomorphic encryption allows organizations to store sensitive information in cloud services, avoiding the high administrative costs and risk of leaks from internal servers, while ensuring that the information can be used and searched; users can also use search engines to make queries and get results without worrying that the search engine will leak the records of their query.

In order to improve the efficiency of fully homomorphic encryption, the cryptography community continues to promote its research and exploration, which will bring fully homomorphic encryption closer and closer to practicality. But for now, fully homomorphic encryption is still not technically feasible.

4.11 Cryptography and Key Management in the Quantum Era

The world is on the verge of a quantum revolution. Quantum computers can provide unprecedented and disruptive processing capabilities. In the field of cryptography and key management, based on the peculiarity of quantum theory and quantum computers, there are both theoretically perfect security solutions and unprecedented challenges.

The application of quantum theory to the field of cryptography is called "quantum cryptography". Quantum cryptography focuses on theoretical and practical issues in quantum key distribution scenarios. Unlike traditional symmetric and asymmetric cryptographic techniques, quantum cryptography is more like physics than mathematics. It relies more on physical concepts as key aspects of its security model. This type of cryptography uses photons (particles/-light waves) and their intrinsic properties to develop "unbreakable" cryptographic systems. It works by sending photons over an optical link, relying on the quantum inability to measure the system without perturbing it. This typical property of the state guarantees a high level of security that is theoretically impossible to eavesdrop on.

Another related research direction is to study the negative impact of the development of quantum computing on existing cryptography and key management technologies and thereby determine cryptographic algorithms and key management technologies that are still secure in the quantum computing era. This line of research is commonly referred to as "post-quantum cryptography" or "quantum-safe cryptography".

The security performance of the Internet, most computers, and communication systems relies on secure and effective cryptographic algorithms, especially identity authentication, integrity verification, transmission encryption, and other algorithms of the communicating parties. For example, Diffie–Hellman key exchange algorithm, RSA digital signature algorithm, and AES symmetric encryption algorithm, which uses a shared key for encryption. In recent years, asymmetric algorithms such as Elliptic Curve Cryptography have also been increasingly used. If a breakthrough in quantum computing is achieved in the coming decades, the security of the above asymmetric algorithms is likely to be significantly compromised.

4.11.1 *Quantum key distribution*

As mentioned earlier, quantum cryptography mainly studies quantum key distribution (QKD). In theory, this is a reliable way to exchange keys without authentication.

QKD is like a public key exchange system but exploits the unique properties of individual photons. The QKD systems currently being explored are typically deployed on standard fiber optic cables, but instead of using them to send data signals, they are used to send individual photons.

Since these individual photons are entangled with the photons remaining in the sender's system, any interception of them will cause the wave function collapse (a term in quantum physics), allowing the sender to realize that its communication may be intercepted or monitored.

It sounds like something out of science fiction, but in fact such systems are already in practical use. China is a world leader in QKD research and practice. Academician Pan Jianwei of the University of Science and Technology of China and his team led the construction of a dedicated optical cable for quantum communication between Beijing and Shanghai. In the fall of 2019, Quantum Xchange in the United States announced the construction of the first commercial quantum distribution network in the United States, connecting financial firms in New York City with data centers in New Jersey. The network can exchange quantum keys over existing fiber-optic networks.

To date, QKD applications have not become mainstream. The cost of individual transmitters and receivers can exceed \$100,000, not including the potential cost of a dedicated fiber-optic network. On the other hand, there are more obvious limitations to QKD systems: because QKD relies on the exchange of entangled single photons, it cannot be practically used as a communication system. Instead, it is limited to the exchange of cryptographic keys.

At the same time, the vast majority of threats do not intercept encryption keys on the transmission link and decrypt the ciphertext directly. In addition to the key exchange scenario, the encryption, integrity protection, or authentication process performed on the device is a major vulnerability of QKD. The quantum key distribution system cannot provide effective security protection for these scenarios.

The UK government's "Quantum Security Technologies" (NCSC, 2020) white paper discourages the use of QKD, stating that it appears to introduce new potential attack vectors, its dependence on hardware is not cost-effective, and the limited range of QKD makes it unsuitable for future challenges where post-quantum cryptography is a better choice. Before more business scenarios require quantum networks, QKD may still be a niche solution.

4.11.2 *Asymmetric cryptography in the quantum era*

In essence, the security of the RSA and Diffie–Hellman algorithms is based on computational difficulty. For example, RSA is based on the large integer factorization problem and the computationally hard problem of discrete logarithms. The large integer factorization problem means that the computer can quickly compute the product of two very large prime numbers, but it is difficult to factorize a very large composite number into two prime factors.

In 1994, the American mathematician Peter Shor invented the Shor algorithm (Shor, 1994), which can use the ability of an ideal quantum computer to solve the factorization of large integers. Shor's algorithm can complete prime factorization in approximately polynomial time (approximately $log(N)$, where N is the length of the input composite number). If a sufficiently large quantum computer can be built to run Shor's algorithm, large integers can be factorized very efficiently, and the related algorithms of RSA and Diffie–Hellman can be broken completely. This will have huge implications for Internet security.

Matteo Mariantoni predicted in 2014 that within 15 years, a quantum computer costing $1 million could break all existing cryptographic algorithms (Mariantoni, 2014). Fortunately, as of the end of January 2024, there is no confirmation of when practical quantum computers will become a reality. Therefore, the cryptographic algorithm based on the public key will not face the risk of being cracked immediately.

4.11.3 *Symmetric cryptography in the quantum era*

An ideal quantum computer would only be able to halve the effective length of a symmetric key. Cryptographic algorithms based on symmetric keys of sufficient length, such as AES-256, remain secure in the theoretical quantum age and need not be considered for replacement in the short term.

Grover's algorithm applied to AES-128 requires about 2^64 iterations, which cannot be effectively parallelized. Since quantum computers are very slow (in terms of operations per second), very expensive, and difficult to transfer quantum states from a malfunctioning quantum computer, even quantum computer clusters are hardly a real threat to symmetric algorithms. Both AES-128 and SHA-256 are quantum resistant according to the evaluation criteria of the NIST PQC (Post Quantum Cryptography) standardization project (Moody and Chen, 2019).

4.11.4 *Post-quantum cryptography algorithms and their standardization*

Even if the impact of quantum computers on traditional cryptography is somewhat exaggerated, research is still going on in academia and industry to design and develop algorithms that are secure in the post-quantum era. The goal is to make the algorithms run on classical computers rather than quantum computers, in order to withstand attacks from powerful quantum computers. The National Institute of Standards and Technology (NIST) is at the forefront of this research.

NIST is currently standardizing algorithms for stateless quantum-resistant signatures, public-key encryption, and key establishment, with first draft publications expected in 2024–2026. Thereafter, new standardized algorithms can be added to security protocols, such as X.509, IKEv2, TLS, and used in various industries.

The IETF Cryptography Forum research group has completed standards for two stateful hash-based signature algorithms, XMSS (Huelsing *et al.*, 2018) and LMS (McGrew *et al.*, 2019), which are expected to be standardized by NIST. EXtended Merkle Signature Scheme (XMSS) and Leighton–Micali Signature (LMS) are currently the only post-quantum cryptographic algorithms that can be used in production systems for scenarios such as signature verification when updating device firmware.

4.11.5 *The rise of quantum supremacy*

Quantum supremacy or quantum supremacy was proposed by Preskil in 2012 (Preskil, 2012), but its earliest idea can be traced back to the early 1980s, when Yuri Manin published "Computable and Uncomputable" (Manin, 1980), painting a vision for the future of computing. Scientists such as Richard Feynman and Paul Benioff have further outlined the promise and potential of quantum computing. In theory, tasks that take a classical computer months or years to perform using binary bits could be done exponentially faster by quantum computers using qubits.

In quantum computing, the goal of quantum supremacy is to demonstrate that programmable quantum devices can solve problems that classical computers cannot, at least not in a feasible time. In contrast, a weaker quantum computing advantage or quantum advantage (Feynman *et al.*, 2018) is the proof that a quantum device can solve a problem faster than a classical computer. Conceptually, quantum supremacy involves both the engineering task of building a powerful quantum computer and the computational complexity theory task of discovering the problems that such a quantum computer can solve. In recent years, some scientists have disputed the term "quantum supremacy" because they believe that the word "supremacy" may be offensive to certain groups of people. These scientists published an open letter in the journal Nature suggesting the use of "quantum superiority" as an alternative to "quantum supremacy" (Palacios-Berraquero *et al.*, 2019). This point of view has been recognized by more and more people. In the scenarios presented in this book, "quantum supremacy" and "quantum superiority" refer to the same thing.

In both academia and industry, it is widely believed that it will take many years to build a quantum computer that can solve any practical problem more cheaply than a classical computer. In October 2019, Google published a paper "Quantum Supremacy Using a Programmable Superconducting Processor" (Arute *et al.*, 2019) in the journal Nature, claiming that "to our knowledge, this experiment marks the first computation that can be performed only on a quantum processor".

The "quantum supremacy" computer consists of 54 physical quantum bits (qubit/quantum bit). The problem Google has set for its quantum supremacy computer is to determine the randomness of numbers generated by a random number generator. According to Google, "...obtaining a million samples on the quantum processor takes 200 seconds, whereas classical sampling of the same fidelity would take 10,000 years on a million cores, and verifying fidelity would take millions of years".

The scientific community generally welcomed the paper. In an article in Nature (Ball, 2016), Professor Daniel Lidar, a quantum computing expert at the University of Southern California, commented: "It's terrific work in many respects, and is filled with valuable lessons for the quantum computing community". *The New York Times* even quoted scientists and compared Google's research breakthrough to the Wright Brothers' first airplane flight in 1903 (Porter, 2019).

IBM disputed Google's claims. In a paper (Pednault, 2019), IBM said the same task could be done in 2.5 days on a classical system, not the 10,000 years claimed by Google. IBM said Google "failed to fully account for plentiful disk storage" when estimating how long its traditional supercomputers would take to perform the calculation.

Back to post-quantum cryptography.

In a "quantum supremacy" computer, the quantum bits corresponding to quantum error correction are only a fraction of a single logical quantum bit, which is a far cry from a quantum computer capable of cracking any real-world cryptographic algorithm, which would require thousands of logical quantum bits and hundreds of billions of quantum gates. Scaling up the number of quantum bits will not be easy, and Kevin Hartnett suggests in his paper (Hartnett, 2019) that the number of quantum bits in a quantum computer will grow exponentially, similar to Moore's law. It is likely that we will see undisputed "quantum supremacy" in the coming years.

4.11.6 *Prospects of post-quantum cryptography*

Quantum computers that break modern cryptography on a large scale are extremely unlikely within the next decade, according to several reports from academia and industry. It is also widely believed that quantum computers do not pose a major threat to symmetric algorithms. Standardization organizations such as IETF, 3GPP, and various industries are calmly awaiting the results of the NIST PQC standardization.

Quantum computers are likely to be very disruptive to certain industries but will probably not pose a real threat to asymmetric cryptography for decades and probably never to symmetric cryptography.

Organizations that need to protect information over a long period of time should start thinking about post-quantum cryptography. For many cryptography usage scenarios, the currently deployed elliptic curve cryptography and RSA algorithms can still be used safely for an extended period of time.

4.12 Cryptography Technology Practice

Most of the modern cryptography practices utilize a combination of symmetric encryption and public/private key encryption and make a lot of delicate designs in key management to comprehensively ensure the security of data. In the following section, typical cases are selected for further elaboration.

4.12.1 *Apple iOS data security*

Take iOS (Apple Inc., 2019) for the iPhone, an operating system designed by Apple Inc., as an example.

iOS employs many encryption and data protection features to keep user data secure. Its security design follows the defense-in-depth architecture principles described in Section 3.4.1 and implements security defense layer by layer through hardware and firmware security, operating system security, and application security to ultimately protect user data security.

In the hardware security layer, iOS uses high-strength chip-level encryption and supports hardware-level device keys and root certificates. The Application Processor and Secure Enclave store the device's Unique ID (UID) and Device Group ID (GID). These IDs are determined at manufacturing to be used as AES 256-bit keys and do not support plaintext reading by any software or direct hardware debug interface.

In the OS security layer, through tight integration with hardware, every process from secure boot to secure operation to system update ensures the security and reliability of system components through signature and encryption.

In the application security layer, iOS uses features such as application signing, verification, and sandboxing to ensure that the

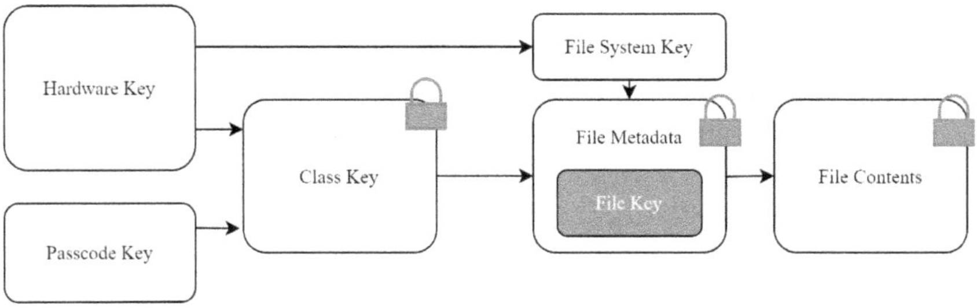

Fig. 4.34. iOS three-tier key architecture.

functions of the application are not tampered with and that the application can only access its own data or data obtained through permission requests. Data security is ensured by implementing perfect data isolation and access control mechanisms.

In the data security layer, iOS sets up different data classifications, and designs different encryption and access control mechanisms according to the usage scenarios of each type of data. For example, the encryption mechanism and usage scenarios of answering calls and protecting photos are obviously different.

The encryption mechanism of iOS relies on the encryption engine and device key provided by the hardware security layer. Each iOS device is equipped with a dedicated AES-256 encryption engine built into the DMA path between flash memory and main system memory, which enables efficient file encryption.

Data protection for file storage is realized by constructing and managing a multi-layer key architecture (iOS three-layer key architecture is shown in Fig. 4.34), and is based on the hardware encryption technology of each iOS device. iOS abstracts the access rights of files into the access rights of file keys, combined with file classification, to achieve file encryption protection. Whether a file can be accessed depends on whether the key of this type is unlocked and the file key is correct. The APFS file system even supports setting different keys for different parts of each file.

Each time a file is created, iOS creates a new 256-bit key (a file-unique key, or simply "file key", which is different for each file) and provides it to the hardware AES encryption engine. The AES encryption engine uses this key to encrypt files written to flash memory in AES-XTS mode. Older iOS systems may use AES CBC mode.

The initialization vector is calculated using file block offsets and encrypted using the SHA-1 hash of the file's unique key.

Depending on the classification of each file, the file's unique key is encapsulated using the appropriate class key. The specific encapsulation method follows the RFC 3394 standard. The encapsulated file's unique key is stored in the file's metadata.

When a file is opened, the file system key is used to decrypt the file metadata to reveal the encapsulated file-unique key and the file's type tag. The file-unique key is unsealed with the corresponding class key and passed to the hardware AES encryption engine, which decrypts the file as it is read from flash memory. All wrapper file key processing takes place in the secure enclave, which is inaccessible to the application processor.

The metadata for all files in the file system is encrypted using a file system key that is created when iOS is first installed or when the user wipes the device. The file system key is also encrypted with an "erasable key" stored in flash memory. This key does not protect the confidentiality of the data, but can be quickly erased if necessary. For example, if the user uses the "Erase All Content and Settings" option, all files in the file system will be rendered inaccessible by erasing the key. For scenarios such as mobile office, lost or stolen mobile phone, iOS supports a remote wipe command issued by an MDM solution or iCloud to wipe the key.

4.12.2 *AWS key management service*

AWS (Amazon Web Services) cloud service is a public cloud service provided by Amazon Inc. AWS Key Management Service (AWS KMS) provides encryption keys and operations that are secured by FIPS 140–2 certified hardware security modules (HSMs) for cloud extension. The key management capabilities of AWS KMS can be used by multiple AWS cloud services to protect data in applications.

AWS KMS is shown in Fig. 4.35. AWS KMS, as an encryption service provider, provides a KMS interface to generate and manage encryption keys that are used to protect data. The traditional key management service provided by AWS KMS integrates with AWS services to provide consistent, complete, and centralized protection of customer keys across AWS cloud services.

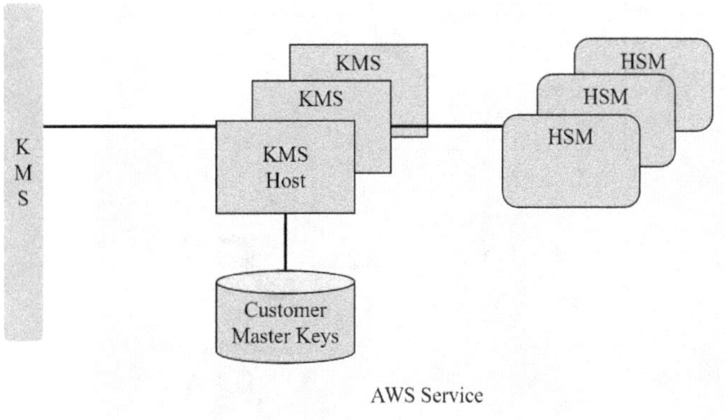

Fig. 4.35. AWS KMS.

AWS KMS's key management and encryption and decryption operations rely on a cluster of hardware security modules (HSMs) that are FIPS 140-2 certified. An HSM is a hardware encryption device based on a hardware security chip (described in Section 3.6.1) that provides dedicated encryption functions to meet the security and scalability requirements of AWS KMS. Customers of AWS cloud services can establish their own HSM-based encryption key hierarchy based on the Customer Master Key (CMK). These keys are only available on the HSM for the necessary period required to process the key request. Customers can create multiple CMKs, each represented by their key ID. By creating a policy attached to a key, customers can define access control policies, i.e., who can manage and/or use the CMK.

The Web API provided by AWS KMS supports secure communication protocols; all requests to AWS KMS must be made over Transport Layer Security (TLS), and only TLS with perfect forward secrecy cipher suites is allowed. This prevents attacks such as eavesdropping, theft, and forgery during key transmission. AWS KMS authenticates and authorizes each access request using the same identity and authentication mechanisms as other AWS API operations.

When used, the CMK can be used to protect symmetric keys or key pairs, and form a layered architecture similar to a two-layer key protection.

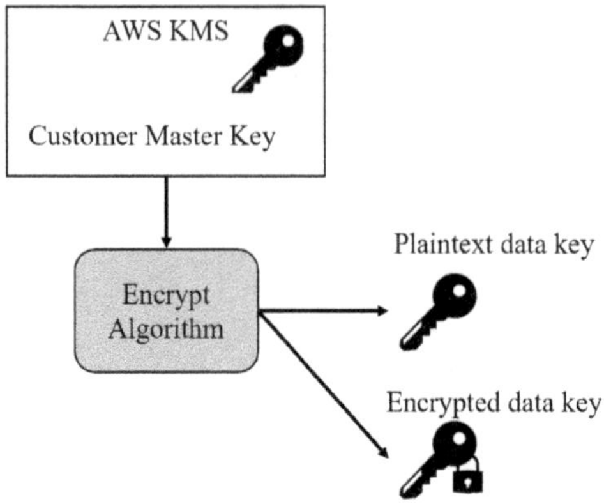

Fig. 4.36. AWS KMS data key generation process.

When data is encrypted with a symmetric key, the symmetric key is referred to as a data key. Data keys can be generated and protected by a given CMK.

A call to the key generation API returns the plaintext data key and the ciphertext data key. After the plaintext data key is used to encrypt data, the plaintext data key must be removed from memory as soon as possible. Callers can securely store ciphertext data keys and encrypted data to decrypt the data as needed. The AWS KMS data key generation process is shown in Fig. 4.36.

In addition, AWS KMS also supports the generation of public and private key pairs and the encryption of the private keys. This public/private key pair can be used for encryption or signature mechanisms.

From a security perspective, AWS KMS guarantees that no one, including service providers, can retrieve the plaintext of the master key from the service. To comply with various cryptography and data security laws and regulations in different countries and regions, the AWS KMS master key is not transmitted outside the AWS region in which it is created and can only be used within that AWS region.

With respect to availability, to ensure that keys are not lost and that data can always be retrieved, AWS KMS stores multiple encrypted copies of keys in systems designed to be 99.9999999999% uptime. AWS KMS is designed to use a redundant architecture that spans multiple Availability Zones in each region to provide highly available services. Since most AWS services rely on AWS KMS to encrypt and decrypt customer data, AWS KMS is designed to provide the availability required to support AWS service level agreements.

AWS KMS supports automatic key rotation, which automatically rotates the master key once a year without the need to re-encrypt encrypted data. Once the master key is rotated, AWS KMS automatically stores the old version of the key material for easy decryption of previously encrypted data.

Chapter 5

Privacy Protection and Data Security Compliance

This chapter introduces the concepts, risks, and challenges of privacy protection. From the perspective of the risks and challenges, it presents the privacy principles, the privacy solution requirements, and the related privacy technologies, and finally, it presents the scenario-based privacy measures.

5.1 The Concept and Regulation of Privacy

The concept of privacy has a long history. In the Western world, the concept of privacy can be traced back to ancient Greek times, when Aristotle proposed two distinct spheres: the public sphere (polis), associated with political life, and the private sphere (oikos or oikia), associated with domestic life.

As far as the definition of privacy itself is concerned, privacy refers to things or events, emotional activities, etc. that have occurred that are ethical and legitimate but cannot or are unwilling to be revealed. Personal privacy is a secret that citizens do not want to be made public or known by others (people outside a certain range) in their personal lives, and this secret has nothing to do with other people and social interests. Information considered "private" may be inherent, specific, or sensitive to a particular individual.

The concept of "personal privacy" in the modern sense was first published by Samuel D. Warren and Louis Brandeis in "On the Right of Privacy" in the "Harvard Law Review" in the United States in 1890 (Warren and Brandeis, 1890) first presented in the article. The two proponents of the concept of privacy (see Fig. 5.1) are initially described this concept as "the right of the individual to be let alone".

The article proposes that the right to be alone without interruption includes the right of individuals to choose to live in seclusion, to be free from censorship or surveillance in a private setting such as their own home. Although the definition of privacy by these two scholars is relatively vague and cannot be regarded as a legal concept in the strict sense, they have pioneered research in this field.

There are many descriptions of privacy and privacy rights from different dimensions and perspectives. For example, a description of privacy-based restricted access rights. Sissela Bok proposed in 1982 (Bok, 1982) that privacy is "the condition of being protected from unwanted access by others — either physical access, personal information, or attention". Closely related to the right to restrict access to privacy is the right to control private information. In other words, individuals or groups have the right to decide for themselves how to share information about themselves with others, and they must also be responsible for their own decisions. For example, personal information that is voluntarily disclosed to the media may no longer be

Samuel D. Warren Louis Brandeis

Fig. 5.1. Proposer of the concept of privacy.

protected by the "right to privacy". In addition, there are studies and discussions in the academic and legal circles about the interaction between privacy and personality, autonomy, self-identity, and personal growth.

As technology advances, so do the ways in which privacy is protected and violated. With some technologies, such as printing or the Internet, the increased ability to share information can create new ways to invade privacy. In 1890, Samuel D. Warren and Louis Brandeis published their essay "The Right to Privacy" in the context of advances in printing technology that made newspapers an important vehicle for the widespread dissemination of information. The article primarily discusses the impact of the proliferation of print media in the form of text and photographs on people's privacy. Subsequently, the rise and popularity of cable television and other forms of communication have been accompanied by concerns about whether they violate users' privacy. Cable television operators can obtain information such as the time users watch, the channels and programs users like, and thus infer users' personal interests, work and rest patterns, and much more.

The development of the Internet has created greater challenges to privacy. Personal activities on the Internet, such as blogs, microblogs, and friends' photo updates, can be stored, correlated, and analyzed for long periods of time. The integration of online/offline services such as payment, QR code, navigation, and AR/VR complicates privacy protection.

With the development of the Internet, the concept of information privacy (or data privacy) has been proposed. Information privacy focuses on the stages of collecting, processing, and disseminating personal information. The main content includes meeting the public's expectation of privacy through technical and management means and complying with relevant laws, regulations, and legal requirements. Data privacy, sometimes referred to as personal data protection, is one of the main topics discussed in this chapter.

5.1.1 *Privacy challenges*

With the development of computer and Internet technology, the protection of personal privacy is facing more and more challenges, especially the large-scale collection and processing of personal information, as well as the user's right to know and control private information. The confidentiality of e-mail and the privacy of online browsing and searching were among the first to come under public scrutiny.

Intel co-founder and former CEO Andy Grove expressed his views on Internet privacy in a May 2000 interview: "Privacy is one of the biggest problems in this new electronic age. At the heart of the internet culture, is a force that wants to find out everything about you".

In fact, there are many prominent contradictions in privacy protection in the Internet age, which constitute a multi-dimensional privacy protection challenges and contradictions:

(1) There is a contradiction between the user's claim to be concerned about privacy and his actual behavior.
(2) The contradiction between personal privacy and convenience and functional requirements.
(3) The contradiction between privacy sovereignty and effective use.
(4) The contradiction between strong privacy oversight and the difficulty of organizational compliance.

One is the user dimension: On the one hand, users claim to be concerned about privacy, but their actual behavior is inconsistent. According to one survey, 84% of Americans don't know how to encrypt email, 41% of children set their status on social networks to be visible to all, and 56% of millennials are willing to share their location to receive coupons from merchants. On the other hand, individual privacy awareness is very high. For example, 42% of Americans (56% of Brits) will not give out their credit card number over the phone, nearly 100% of people say they are concerned about privacy, especially online privacy, and only 8% of Americans believe that large Internet companies will protect their privacy.

Another user behavior is to actively share their personal information. According to a study, by the fourth quarter of 2020, Facebook, the largest social networking site, currently has nearly 2.8 billion accounts, and more than 5 billion pieces of content are uploaded every day. These data can be used to directly observe relevant behavioral records of users, such as browsing records, search query records or Facebook profiles, published photos and articles, and to infer derived information about users, such as sexual orientation, political orientation, and religion, race, drug use, intelligence, and personality.

The second is the personal dimension: the contradiction between personal privacy and convenience and functional requirements. The further integration of online/offline offers many conveniences for life and work. Behind these conveniences lies a risk to privacy. For example, scanning QR codes to order and pay at restaurants and grocery stores will generate big data trend insights on private data such as consumer behavior and even portraits of individual consumer preferences. Connecting to free Wi-Fi provided by merchants, using map navigation, etc., while obtaining related functions and services, etc. will also provide some personal information. How to ensure that their private information is known and controllable while experiencing the convenience of the Internet life era is a major challenge for consumers.

Third, there are complex issues such as the ownership of private data of individuals and others, organizational interests, and public interests. Data is valuable because it can be used, and digital privacy is valuable because it can be used. When individuals participate in social life and communicate with others, it means that they may voluntarily disclose some personal information. The main challenge of privacy is how to protect personal data and personal information while collecting and using private data.

Fourth, under the current international political and economic trends, legislators and administrative agencies in various countries and regions generally adopt a strict regulatory attitude toward privacy protection. How to conduct Internet business and data services without touching the red line of supervision, and how to use and protect privacy in accordance with technology and business dimensions is an unavoidable problem for every organization.

5.1.2 *Privacy regulatory trends*

Countries and regions around the world have a long history of legislation on privacy protection, and the connotation of privacy and privacy protection is a process of continuous evolution.

Some countries grant citizens the right to privacy in their constitutions. Representative examples include the following: The Brazilian Constitution stipulates that "the privacy, private life, honor and image of the people shall not be violated" and the South African Constitution stipulates that "everyone has the right to privacy". The Constitution of the People's Republic of China does not directly propose the abstract concept of privacy or the right to privacy but clarifies it as rights such as personal freedom, personal dignity, freedom of communication, and confidentiality of communication. The Constitution of the United States of America also does not explicitly propose and define the right to privacy but covers some principles of privacy protection through the First, Third, Fourth, and Fifth Amendments.

Many countries also have separate privacy protection laws. For example, Argentina promulgated the "Personal Data Protection Act" in 2000, Australia promulgated the "Privacy Act" in 1988 (Doyle and Bagaric, 2005), and Canada promulgated the "Personal Information Protection and Electronic Documents Act" in 2000 (referred to as PIPEDA or PIPED Act), Japan's "Personal Information Protection Law" of 2003 (Sakai and Oshima, 2020), and so on.

In the legislation of various countries and regions, there are many concepts describing privacy and personal information, and different terms such as "personal information", "personal data", and "privacy" are mixed, and their meanings are similar. For example, the Japanese law uses "Personal Information", and the EU GDPR uses "Personal Data", the Hong Kong Special Administrative Region of China uses the "Personal Data (Privacy)", while the United States, Australia, and other countries directly use the word "Privacy".

In addition to national privacy laws, there are privacy agreements of various international organizations and national associations. Article 12 of the United Nations Universal Declaration of Human Rights stipulates: "No one shall be arbitrarily interfered with in his private life, family, home or correspondence, and his honor and reputation shall not be attacked...". The Organisation for Economic Co-operation and Development (OECD) published the Privacy Guidelines Organization for Economic Co-operation & Development (1980) in 1980 and updated them in 2013. In 2004, the "Privacy Framework" of the Asia-Pacific Economic Cooperation adopted the core values and basic concepts of the OECD "Privacy Guidelines" and served as the privacy protection agreement of the member countries of the organization.

Ever since human society entered the electronic age, privacy professionals have begun to consider how technological changes will lead to changes in the connotation of privacy and the protection of privacy. To this day, the protection of information privacy in the Internet age is one of the focal points of legislation in various countries.

Internationally, the legal requirements for privacy regulation have gradually become more stringent. Among them, the construction of relevant legal systems in European and American countries has shown a significant growth rate in recent years. Take the United States as an example. In 1998, the U.S. Federal Trade Commission (FTC) considered the risks to children's privacy on the Internet and formulated the Children's Online Privacy Protection Act (COPPA). The law aims to give parents the right to control their children's use of information online, while limiting the ability to collect information from children and requiring warnings about potentially harmful information or content. In 2000, the U.S. federal government enacted the Children's Internet Protection Act (CIPA), which requires federally funded schools and libraries to implement technological safeguards to filter or prevent children from accessing Internet images that are harmful to them. In the area of public health, the U.S. federal government enacted and implemented the Health Insurance Portability and Accountability Act (HIPAA) in 1996. The Act clearly states that Protected Health Information (PHI) is a protected personal privacy. The most famous post-2000 privacy law is the Fair and Accurate Credit Transactions Act (FACTA), which was promulgated and implemented by the U.S. federal government in 2003. Other laws at the federal and state levels in the United States also involve the protection of personal information, such as the Fair Credit Reporting Act, the Gramm-Leach-Bliley Act (GLBA), and State Breach Notification Laws.

Different countries have different definitions of privacy and different regulatory requirements for privacy protection, which makes it difficult for multinational companies to operate and for industry organizations to share information. The European Union has formulated a unified privacy protection law, among which the more important ones are the "Data Protection Act" promulgated in 1995 and the "General Data Protection Regulation" (GDPR) promulgated in 2016, which came into force in May 2018. In the field of global data protection legislation, the most influential is the GDPR promulgated by the European Union. GDPR is known as the most stringent personal data protection regulation in history.

A multilingual banner issued by the European Commission, "For a Union That Strives For More" (see Fig. 5.2), hung in the EU headquarters building at the end of 2019, represents the commitment of the EU's leading institutions to advance security and privacy protection in the digital world.

GDPR is an EU law regulation on the protection of personal data and privacy in the European Union (EU) and the European Economic Area (EEA). The main purpose of the regulation is to give individuals control over their personal data, and to simplify the regulatory environment for international business activities by harmonizing regulations within the EU. The GDPR defines the concepts of personal data controllers (organizations that collect data from EU residents), processors (organizations that process data on behalf of data controllers, such as cloud service providers), and data subjects (individuals) and requires that personal data controllers and

Fig. 5.2. Banners on the EU headquarters building.

processors take into account the legal principle of privacy protection and provide safeguards of a technical or organizational dimension (such as the use of pseudonymization or anonymization, where appropriate) when designing and establishing business processes for the processing of personal data.

GDPR gives individuals as data subjects eight rights:

- The right to be informed.
- The right to rectification.
- The right of access.
- The right to erasure (right to be forgotten).
- The right to restrict the processing of your data.
- The right to data portability.
- The right to object.
- Rights regarding automated profiling and decision-making.

The GDPR has been called "the most stringent law in history" because of its features such as extraterritorial jurisdiction, mandatory requirements, and high fines. The GDPR requires that all organizations that process personal data within the geographic scope of the European Economic Area, regardless of whether they are located in the European Economic Area, are subject to the law. The law requires government agencies and businesses that process personal data to hire a data protection officer (DPO). If a data breach affects user privacy, companies must report it to national data protection authorities within 72 hours. Under certain circumstances, organizations that violate the GDPR can be fined up to 20 million euros or, if the GDPR is violated by a company, the company can be fined up to 4% of its annual global turnover for the previous financial year or 20 million euros, whichever is higher.

As an earlier and more stringent privacy law, the GDPR has far-reaching implications around the world. It has become the model for legislation in many countries outside the EU, including Japan, South Korea, Brazil, Chile, Argentina, and Kenya. The California Consumer Privacy Act (CCPA) (CCPA, 2018), which was passed in California on June 28, 2018, also has many similarities to GDPR.

Across privacy legislation and law enforcement practices in various countries, privacy oversight is becoming increasingly stringent. On January 21, 2019, CNIL, the French data protection authority, announced that Google was fined 50 million euros for violating GDPR (Netter, 2019).

5.2 OECD Eight Principles of Privacy Protection

As early as the 1970s, with the popularity of computers, the collection and processing of personal information was gradually transferred to computers. How to protect electronically obtained personal information began to be considered by the legal profession at that time.

In 1973, the Department of Health, Education, and Welfare (HEW) issued a report entitled "Records, Computers, and the Rights of Citizens" (Secretary's Advisory Committee on Automated Personal Data Systems, 1973) (the "HEW Report"). The HEW Report was the first comprehensive study of the privacy risks posed by the increasing use by organizations of electronic information technologies to replace traditional paper-based systems for creating, storing, and retrieving information. "Safeguards for personal privacy based on our concept of mutuality in record-keeping would require adherence by record-keeping organizations to certain fundamental principles of fair information practice", the report states. The report then sets out five fundamental principles for the processing of personal data. This is one of the founding principles of Fair Information Practices.

Fair Information Practices, also known as Fair Information Privacy Practices (FIPPs), are a set of principles and best practices that describe how information-based societies approach the storage, management, and flow of information to maintain fairness, privacy, and security in a rapidly evolving global technology environment.

In 1980, the International Organization for Economic Cooperation and Development (often referred to as the OECD) issued its Guidelines on the Protection of Privacy and Transborder Flows of Personal Data (Organization for Economic Co-operation & Development, 1980), which expanded on the five principles of the HEW report to set out eight principles of privacy protection. The guidelines are considered to be a minimum standard for the protection of privacy and individual liberties.

The eight principles of privacy protection proposed by the OECD are the most widely accepted privacy principles. The OECD Fair

Information Practices are also the basis for privacy laws and related policies in many other countries (e.g., Sweden, Australia, Belgium).

In 1995, a variant of these principles became the basis for the EU Data Protection Directive, which evolved into the subsequent EU GDPR. OECD member countries, including the United States, have also agreed to the FIPP through a consensus and formal ratification process, and they form the basis of many modern international privacy agreements and national laws.

In addition to the preamble, the OECD "Guidelines on the Protection of Privacy and Transborder Flows of Personal Data" contain five chapters and 22 paragraphs:

- **Chapter 1 (Paragraphs 1–6):** General Provisions, Definitions of Specific Terms and Scope of Application, states that personal information means any information about an identified or identifiable natural person.
- **Chapter II (Paragraphs 7–14):** This sets out eight principles for the processing of personal information.
- **Chapter III (Paragraphs 15–18):** Basic Principles of International Application, Free Movement and Legal Restrictions, stipulates that Member States shall take all appropriate measures to ensure the freedom of the international movement of personal information and the conditions under which such free movement may be restricted.
- **Chapter IV (Paragraph 19):** This provides that in order to ensure the implementation of the principles of Chapter II within each country, Member States shall ensure the privacy of personal information and individual freedoms by enacting domestic laws and establishing institutions.
- **Chapter V (Paragraphs 20–22):** International Cooperation deals with international cooperation.

The preamble of the OECD Guidelines states "...that automatic processing and transborder flows of personal data create new forms of relationships among countries and require the development of compatible rules and practices; ...that transborder flows of personal data contribute to economic and social development; ...that domestic legislation concerning privacy protection and transborder flows of personal data may hinder such transborder flows; ...". Therefore, the purpose of the Guidelines is to promote the continued flow of data and thereby the continued growth of the global economy, on the basis of adequate protection of personal information.

Based on the above purposes, the OECD Guidelines put forward the following basic principles:

(1) **Collection limitation principle:** The collection of personal information should be limited and such information should be obtained by lawful and fair means and, where appropriate, with the knowledge or consent of the individual.
(2) **Data quality principle:** Personal data should be relevant to the purposes for which it is used and, to the extent necessary for those purposes, accurate, complete, and up to date.
(3) **Purpose specification principle:** The purpose for which personal data is collected should be specified before the data is collected, and subsequent use should be limited to the fulfillment of that purpose or other purposes that are not incompatible with that purpose and that are specified each time the purpose changes.
(4) **Use limitation principle:** Personal data should not be disclosed, made available or otherwise used for purposes other than those specified, unless:

    (a) With the consent of the data subject.
    (b) Authorized by law.

(5) **Security safeguards principle:** Personal data should be protected by reasonable security safeguards to prevent the risk of data loss or unauthorized access, destruction, use, modification or disclosure.
(6) **Openness principle:** There should be an overarching policy of openness covering the development, practice and policy dimensions of personal data. It should be possible to ascertain at any time the existence and nature of personal data, the principal purposes of use, and the identity and usual place of residence of the controller.
(7) **Individual participation principle:** An individual should have the following rights:

    (a) To obtain from the data controller or otherwise ascertain whether the data controller holds data relating to him or her.

(b) The data controller should provide personal data relating to an individual within a reasonable time and at a reasonable cost (if any). The data should be provided in an appropriate manner and in an easily understandable form.
(c) If the request under (a) and (b) is refused, to be informed of the reasons and to have the right to challenge the refusal.
(d) To challenge data relating to him or her and, if the challenge is successful, to request that such data be deleted, corrected, supplemented or amended.

(8) **Accountability principle:** Data Controllers are responsible for compliance with the measures taken to implement the above principles.

The above principles are not independent; some are interrelated and even partially overlapping. They should be understood in their entirety rather than interpreted individually.

5.2.1 *Collection limitation principle*

Collection Limitation Principle contains two aspects:

The first aspect is the limitation on the collection of personal data itself.

Much information is particularly sensitive and its collection should be limited or prohibited because of the way it is processed, its nature, the context in which it is used, or for other reasons.

The OECD believes that it is largely impossible to precisely define sensitive information. However, the Guidelines it has developed include a general statement that the collection of personal information should be limited to indicate its position that the collection of information should be limited.

The Guidance's Explanatory Memorandum states that collection limitations should relate to the following:

(1) Data quality aspects (i.e., that it should be possible to derive information of a sufficiently high quality from the data collected and that the data should be collected within an appropriate information framework).
(2) Limitations related to the purpose of data processing (i.e., that only certain categories of data should be collected and,

where appropriate, that data collection should be limited to the minimum necessary to fulfill the specified purpose).
(3) "Earmarking" of particularly sensitive data in accordance with the traditions and attitudes of each Member State.
(4) Restrictions on the data collection activities of certain data controllers.
(5) Civil rights concerns.

The second aspect is the requirements for data collection methods. The knowledge and consent of the data subject during the data collection process is a basic rule, with knowledge being the minimum requirement. The use of covert recording devices (e.g., pin-hole cameras) or tricking individuals into providing data is clearly unlawful. Consent is not required only in certain scenarios due to practical difficulties or infeasibility, and sometimes even knowledge is not required. A typical example is the collection of information in criminal investigations.

From a legislative perspective, the two aspects contained in the principle of limitation of collection focus on limiting the collection of data relating to the personality and interests of natural persons from the two dimensions of purpose and means.

5.2.2 *Data quality principle*

The Data Quality Principle requires that "personal data should be relevant to the purposes for which they are to be used, and, to the extent necessary for those purposes, should be accurate, complete and kept up-to-date".

Innovations in information and communication technologies are having a major impact on business operations, government regulation, and personal activities. The amount of personal data collected, used and stored is enormous and growing. Personal data plays an increasingly important role in business, society, and everyday life. New technologies and the responsible use of data have generated and will continue to generate enormous social and economic benefits.

The abundance and persistence of personal data also increase the risks to privacy. Personal data is increasingly being used in ways that may not have been anticipated at the time the data was collected.

With regard to the data quality principle, the OECD states that data should be linked to the purpose for which it is used. For example, data containing ratings and opinions may be misinterpreted if used for unrelated purposes. Accuracy, completeness, and timeliness of data are elements of the data quality principle and must be linked to the purpose for which the data is used.

5.2.3 *Purpose specification principle*

The Purpose Specification Principle means that the collection, processing, and use of personal data should be based on specific and clear purposes, and that the collection, processing, or use of personal data for purposes other than those intended is prohibited.

The Purpose Specification principle is closely related to the two principles before and after it (the data quality principle and the use limitation principle). The premise of this principle is that the purpose of data collection should be lawful and not violate the principles of good faith and public policy in civil law.

In real life, when personal data is collected, the purposes need to be specified in advance. For example, government agencies may collect personal information as necessary to perform their functions. When personal information is collected for commercial purposes, the purpose must be clearly stated and the user's knowledge and consent must be obtained. If the purpose of the data collection changes, a new request and authorization is required.

According to the description in the Explanatory Memorandum of the Guidelines developed by the OECD, the purpose specification principle also includes "when data no longer serve a purpose, and if it is practicable, it may be necessary to have them destroyed (erased) or given an anonymous form". The rationale is that "control over data may be lost when data are no longer of interest; this may lead to risks of theft, unauthorised copying or the like".

5.2.4 *Use limitation principle*

The use limitation principle means that data users should use personal data within the scope of the purpose of data collection. The use limitation principle is closely related to the purpose specification principle, and there is even a view that the two can be combined.

It should be noted that the transfer of personal data, especially outside the organization, may result in a violation of this principle.

It is necessary to clarify the need for the transfer of personal data and to use technical, business and other means to ensure that the recipient's use of personal data does not exceed the scope of the original purpose for which the data was collected.

5.2.5 *Security safeguards principle*

The Security Safeguards Principle refers to the limitations on the use and disclosure of personal information that should be protected and enhanced by security controls. Possible safeguards include physical controls (such as physical access control and dedicated data centers) and organizational controls (such as authorization levels for access to personal data and confidentiality obligations of data processors). Computer systems also include information controls (encryption, anomalous activity and threat monitoring, and disaster recovery).

In modern computer networks, the cost of copying and manipulating data is extremely low. Data transmission over various types of networks is also very fast. Data security is therefore extremely important.

There is some overlap in the terms used in the security principle, such as "access" and "disclosure" of data. "Loss" of data includes accidental deletion of data, destruction of media (and thus destruction of data), and theft of media. "Alteration" means unauthorized data entry, while "use" means unauthorized copying.

5.2.6 *Openness principle*

The Openness Principle defines that "there should be a general policy of openness about developments, practices and policies with respect to personal data. Means should be readily available of establishing the existence and nature of personal data, and the main purposes of their use, as well as the identity and usual residence of the data controller".

The Openness Principle can be seen as a prerequisite for the Individual Participation Principle. Individuals are only likely to participate only if they are informed about the collection, storage or use of their personal data by the organization.

The Openness Principle requires that individuals should be able to obtain from the controller a description of the controller's activities in collecting, processing and storing their personal data in an "easily accessible" manner.

5.2.7 *Individual participation principle*

The Individual Participation Principle means that the data subject (individual) can confirm with the data controller whether he or she is the owner of the data concerning him or her. The individual participation principle, including the right of access and rectification, is considered one of the most important principles of privacy protection.

Data controllers should allow data subjects (individuals) to access, modify, and add to data about them.

5.2.8 *Accountability principle*

The Individual Participation Principle means that a data subject (individual) can confirm to the data controller whether or not they have data about themselves. The principle of individual participation, which includes the right to inquire and the right to challenge, is considered one of the most important privacy principles.

The data controller shall allow the data subject (individual) to view, amend, and supplement the data concerning himself/herself.

5.3 Privacy Protection Technologies

Privacy protection involves the following foundational technologies, as well as the combined use or associated use of these technologies. Some technologies are also known as Privacy Enhancing Technologies (PETs):

(1) **Data minimization:** Only collect data that is clearly related to the purpose, and do not collect or immediately delete irrelevant data.
(2) **Pseudonymization:** Delete all elements that can directly identify individuals through hashing and polymorphic pseudonyms.
(3) **Anonymization:** Various methods are used for data anonymization and data aggregation.
(4) **Encryption:** Encryption of the collected data, including the encryption of data transmission and storage.

(5) **Access control:** Physical and logical means, as well as authentication and authorization, are used to ensure that only necessary personnel have access to relevant data and that its operation is fully controlled.
(6) **Default data protection:** All settings are privacy-friendly by default, and the user interface is transparent and rights management is clear.
(7) **Data retention period and deletion:** Strictly adhere to the data retention period, and immediately delete the data that has reached the purpose of use, or mark the data as deleted and unavailable.
(8) **Privacy notice and user communication:** Maintain smooth communication with users, ensure transparency, and publish privacy policies in a user-friendly manner using simple and understandable language.

5.3.1 *Overview of anonymization and pseudonymization technologies*

As mentioned in the previous chapters, with the vigorous development of the Internet and the gradual maturity of big data technology, data has become the most valuable asset of various organizations and the basis of new business models. Especially with the popularization of the mobile Internet, web browsing, and application tracking, user profiling and precise advertising marketing provide users with more and more rich content, but also bring privacy concerns to users.

Personal data that is not sufficiently anonymized and made public is a frequent privacy risk that organizations encounter. According to a report by ZDNet in August 2019 (Asha, 2019), Public Transport Victoria (PTV) has been found in breach of the Privacy and Data Protection Act 2014 (PDP Act) by the Office of the Victorian Information Commissioner (OVIC) for releasing data that exposed the travel history of 15,184,336 myki cards. Some telecom operators have also made headlines for releasing insufficiently anonymized datasets.

The various types of data held by organizations involve a wide range of personal information. The contradiction between the two dimensions of data protection and effective utilization has gradually become prominent.

By using various privacy technologies to desensitize and de-identify users' personal information, companies or organizations can perform data-driven insights and analysis based on this data to achieve business goals.

Several typical business scenarios are described as follows:

(1) **Data as a Service (DaaS):** Cloud service providers can access anonymous user portrait data for advertising purposes; telecom operators can access anonymous location data for urban planning purposes.
(2) **Telemetry scenario:** Realize user behavior statistics similar to iPhone and Chrome browser usage statistics features.
(3) **IoT scenarios:** Manufacturers can obtain anonymous car usage patterns, or energy suppliers can provide smart meter analysis based on anonymous usage data.
(4) **Healthcare scenarios:** Hospitals can provide anonymous patient data to researchers and insurance companies.
(5) **Archiving scenarios:** Insurance companies can store anonymized historical data so that it can be retained even after the legal retention period has expired.

Anonymization and pseudonymization technologies attempt to reduce violations of user privacy while utilizing data. In order to meet the requirements of privacy protection, it can also maximize the value of data.

In the above scenarios, anonymization is mainly used to protect personal privacy. Anonymization also has other application scenarios, such as analyzing confidential business data. Companies in similar industries or groups can benchmark their performance against each other without revealing detailed financial or operational data.

It should be noted that anonymization is not intended to replace data access control and system security configuration. Its primary function is to enable the collection and analysis of data that cannot be collected because of laws and regulations. Anonymization complements, rather than replaces, other privacy technologies such as masking, authentication, and encryption, which can work independently or synergistically in different scenarios.

There is no specific definition of anonymization in the GDPR, but Recital 26 defines the requirement as "the principles of data protection should therefore not apply to anonymous information, namely information which does not relate to an identified or identifiable natural person or to personal data rendered anonymous in such a manner that the data subject is not or no longer identifiable. This Regulation does not therefore concern the processing of such anonymous information, including for statistical or research purposes". The anonymized data is exempted from strict personal data protection requirements, so anonymization has a very high standard in legal semantics: even the data controller should not be able to re-identify the data subject in an appropriately anonymized dataset. Another term in the legal system that corresponds to anonymization, pseudonymization, the risk that a data subject can be re-identified is relatively high. For example, simply replacing the names in a payroll record with "Alice" and "Bob" as a simple "pseudonymization" is still easily identifiable.

As defined by the European Union Agency for Cybersecurity (ENISA) (Stalla-Bourdillon and Knight, 2016), "Pseudonymisation is the processing of personal data in such a manner that the personal data can no longer be attributed to a specific data subject without the use of additional information, provided that such additional information is kept separately and is subject to technical and organisational measures to ensure that the personal data are not attributed to an identified or identifiable natural person".

Directly associated with pseudonymization is the definition of pseudonyms. A pseudonym, also known as a cryptonym, is information that is associated with personally identifiable information or any other kind of personal data such as location data. Pseudonyms may have varying degrees of linkability from original identifiers. When evaluating the strength of a pseudonym, the different levels of linkability of different types of pseudonyms are taken into account, and the design of a pseudonym system may require a certain level of linkability (such as a log file system or a reputation system, obviously a reputation system requires a higher level of linkability).

Therefore, it can be seen that the biggest difference between pseudonymization and anonymization is whether a specific individual can be identified under the premise of linked data reference. From a legal point of view, pseudonymized data, or directly referred to as a pseudonym, is still personal data, and the relevant laws on personal data and privacy protection still apply.

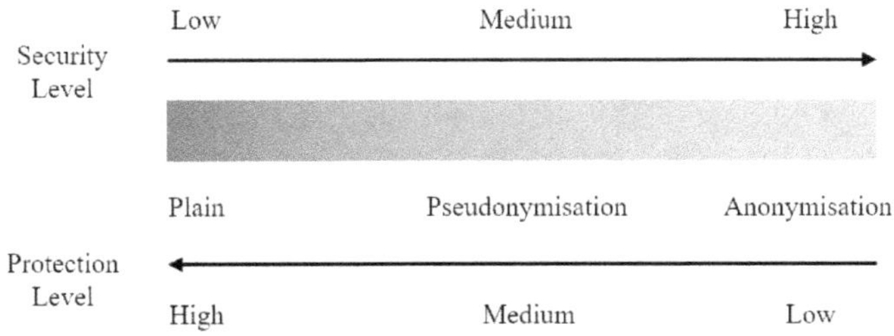

Fig. 5.3. Comparison between pseudonymization and anonymization.

However, pseudonymization can reduce privacy risks, help organizations better meet their privacy obligations, reduce data distortion, and preserve the value and meaning of information.

The comparison between pseudonymization and anonymization is shown in Fig. 5.3, which can vividly illustrate the difference and connection between plaintext, pseudonymization, and anonymization.

From the perspective of security of data storage, use, and transmission, the security of personal data is lowest for plaintext, followed by pseudonymized data, and highest for anonymized data. In terms of data protection requirements, the protection requirements are highest for cleartext personal data, moderate for pseudonymized data, and low for anonymized data, which is no longer considered personal data in some jurisdictions and requires almost no additional protection. Therefore, a combination of technical, administrative, and other safeguards is needed to protect personal information according to the different formats of personal information, and the appropriate laws, regulations, and legal requirements must be taken into account.

The anonymization techniques listed by the EU WP29 Working Party in Opinion 05/2014 on Anonymization Techniques include randomization techniques and generalization techniques (see Fig. 5.4). Randomization techniques include noise addition, permutation, and differential privacy; generalization techniques include aggregation and k-anonymity, l-diversity, and t-closeness.

Similarly, in the Opinion, the WP29 Working Party listed commonly used pseudonymization techniques (see Fig. 5.5), including the following:

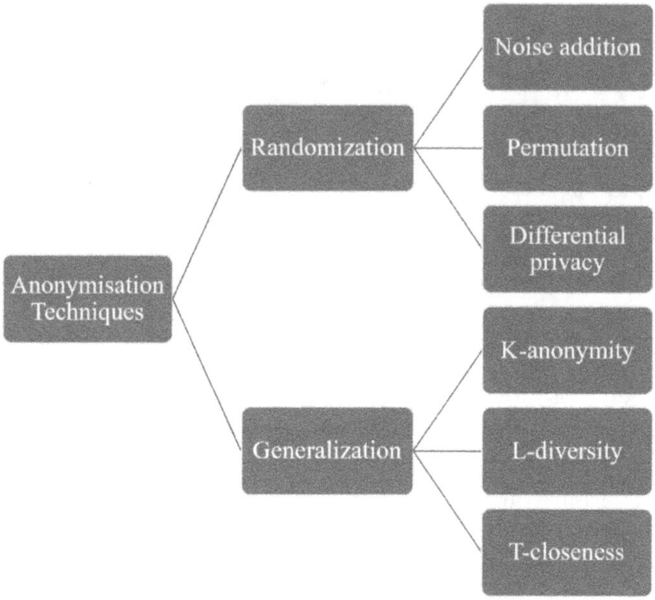

Fig. 5.4. Anonymization technologies.

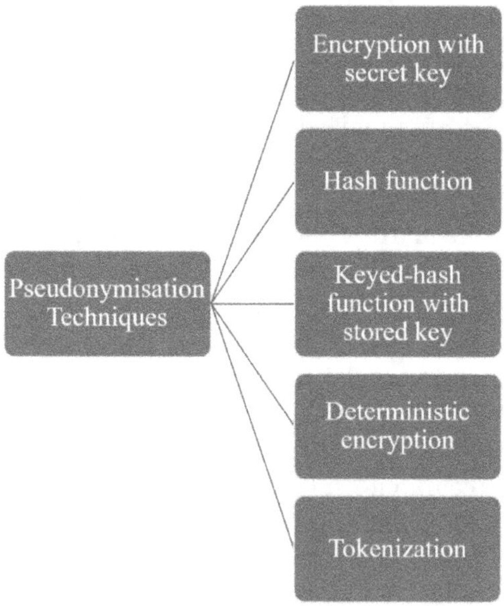

Fig. 5.5. Pseudonymization technologies.

(1) Encryption with secret key.
(2) Hash function without key (Hash Function).
(3) Keyed-hash function with stored key.
(4) Deterministic encryption or hash functions that discard keys.
(5) Tokenization.

5.3.2 *Randomization techniques*

Randomization techniques include noise addition, permutation, differential privacy, and other techniques.

Noise addition technique, refers to the technique of adding noise: a certain amount of noise is added to the original data to make the data less accurate. For example, the weight table of hospital patients, after adding noise, the error of the displayed weight is ± 5 kg. In this way, the statistical analysis for the dimension of weight still has a certain feasibility, even for each patient, and also helps the doctor to judge the appropriate dosage of medication, and to a certain extent protects the patient's privacy.

Permutation technology refers to the rearrangement of the value of an attribute, and its different values can still be associated with different data main body. This technique is useful when the exact distribution of attributes in a dataset needs to be preserved.

The permutation technique can be considered as a special form of noise addition techniques. In classical noise addition techniques, the values of the attributes are replaced by random values. Generating consistent noise is difficult, and modifying the original values slightly is difficult to ensure privacy. As an alternative, replacement techniques can change the values in a dataset by swapping data from one record to another. This swapping ensures that the range and distribution of the values remain the same, but there is a change in the correlation between the values and the individuals.

It is worth noting that if two or more attributes have a logical relationship or statistical correlation and are permuted independently, this logical relationship will be broken. Therefore, it is important to permutate all relevant attributes, otherwise an attacker can identify the shifted attributes and reverse the permutation.

For example, consider a subset of attributes in a medical record. In most cases, there is a close logical relationship between the

Table 5.1. Cases of inappropriate permutation.

Age	Gender	Position	Monthly salary/$
25	Female	Clerk	5,000
30	Male	Engineer	1,500
51	Male	General Manager	8,000
38	Male	Department Manager	2,000
28	Female	Assistant engineer	13,000

attributes of "Cause/Symptom/Department of Hospitalization", so it is possible to detect or even reverse the substitution of only one of the values.

Similar to noise addition techniques, permutation techniques do not provide sufficient anonymization guarantees on their own. When used, they are accompanied by the replacement of other attributes associated with the replaced attribute or the removal of associated attributes in the dataset to reduce the risk of identifying an individual through correlation analysis.

Table 5.1 shows an example of an inappropriate permutation. In the average salary statistics scenario, the actual salaries of the employees are permuted. However, it is still easy to infer the true salary of each employee from the logically related attributes ("position").

Another slightly different permutation technique is to use one set of attributes for another, often combined with noise techniques. For example, for the height attribute, "green" is used for heights between 160 and 170 cm, and "blue" is used for heights between 170 and 180 cm.

Differential privacy is also a randomization technique, but its implementation is slightly different. Rather than attempting to change the value of the attribute, differential privacy technology adds a certain amount of random noise before generating the query view of the dataset. In this way, an anonymized view of the dataset can be generated while retaining a copy of the original dataset. With some differential privacy techniques, it is even possible to analyze how much noise needs to be added and in what form to achieve the necessary privacy guarantees. In this case, attention must be paid to the possibility of identifying individuals through frequent queries and correlations across different result sets. Differential privacy technology does not alter the original data, data controllers can identify individuals through the results of differential privacy queries, and such results must also be treated as personal data.

One of the benefits of differential privacy techniques is that by responding to specific queries, records are made available to authorized third parties rather than being published. To facilitate auditing, data controllers can maintain lists of all queries and results to ensure that unauthorized data cannot be accessed by third parties. Queries can also be subjected to anonymization techniques, including noise addition and permutation, to further protect privacy. How to better design interactive query-response mechanisms to answer questions fairly accurately (with less noise) while preserving privacy is still under research in academia.

To limit inference and linkability attacks, it is necessary to track the requests made by query entities and to observe the information obtained by the subjects. Therefore, differential privacy datasets should not be used on public search engines that do not provide traceability of query entities.

5.3.3 *Differential privacy*

For a long time, differential privacy appeared as an academic concept in various academic papers. It was only after Craig Federighi, Apple's vice president of engineering, announced during the 2016 Worldwide Developers Conference (WWDC) keynote that Apple was using the technique to protect the privacy of iOS users that the technical term gradually entered the public domain. In fact, prior to that, in 2014, a Chrome browser feature implemented by Google to collect behavioral statistics also used the Randomized Aggregatable Privacy-Preserving Ordinal Response project (RAPPOR project) (Erlingsson *et al.*, 2014).

In a randomized response, random noise is added before it is submitted to the collector. For example, if the actual statistic is 0, the browser will replace the 0 with a randomly chosen 0 or 1 with some probability. Each user can deny his or her response because it may be a random value. Overall, however, the superimposed signal distribution should still be approximate or be equivalent to the original signal distribution, and the random noise in the statistics can be ignored by organizations that collect statistics (such as Google or Apple Inc.).

From the perspective of privacy protection, the subject of privacy is an individual, and only those involving a specific user are called privacy leaks. Publishing information about a group of users (aggregated information) is not considered a data breach.

The following is a simple example to illustrate the concept of differential privacy.

Table 5.2 shows a salary table of a company used to analyze the average salary.

As we all know, everyone's salary is personal information. How can we make salary statistics available to those who really need it, while avoiding leakage? One of the intuitive ideas is to hide the name and replace it with a pseudonym, see Table 5.3.

But this scenario could still be risky. Based on the date of birth and city, it may be possible to infer who each person is, leading to the risk of privacy violations.

To analyze the average salary scenario, adding a different random value to each person's salary can maximize anonymity, see Table 5.4.

Table 5.2. Sample Payroll statements.

Name	Date of birth	Place of work	Monthly salary/$
Abby Norris	1991–02	New York	13,000
Tyne Spencer	1987–11	Chicago	15,000
Craig Hale	1980–06	New York	12,000
Gloria Ortega	1993–05	Hoston	9,000
Simon Hunter	1985–07	San Diego	11,000

Table 5.3. Payroll statements.

Name	Date of birth	Place of work	Monthly salary/$
Alice	1991–02	New York	13,000
Bob	1987–11	Chicago	15,000
Chris	1980–06	New York	12,000
David	1993–05	Hoston	9,000
Eva	1985–07	San Diego	11,000

Table 5.4. Noise addition on monthly salary.

Name	Date of birth	Place of work	Monthly salary/$
Alice	1991–02	New York	13,000 + X1 = 10,529
Bob	1987–11	Chicago	15.000 + X2 = 10,184
Chris	1980–06	New York	12,000 + X3 = 18,816
David	1993–05	Houston	9,000 + X4 = 3,032
Eva	1985–07	San Diego	11,000 + X5 = 17,439

In the selection of X_i, it is only necessary to ensure that

$$\sum_{i=1}^{N} X_i = 0$$

Then the average value of the final salary must be accurate.

Differential Privacy (Dwork, 2008) is a mathematical framework and set of techniques designed to protect the privacy of individuals while allowing the analysis of aggregate data. It ensures that statistical analyses performed on datasets cannot be used to infer information about any specific individual within it.

Differential Privacy works by adding "noise" to the data in such a way that an adversary cannot distinguish between the original dataset and a slightly perturbed version of it. This noise is carefully crafted so as not to significantly affect the statistical properties of the data or the results of analysis. A decent differential privacy implementation can resist attacks such as de-anonymization and re-identification of individuals in the dataset. An example of differential privacy is shown in Fig. 5.6.

How to analyze and count data in datasets without compromising personal privacy has been a topic of discussion for a long time. Swedish statistician Tore Dalenius proposed a strict definition of data privacy in the 1970s (Dalenius, 1977): After using sensitive datasets, attackers should not be able to obtain information they did not know before.

In 2006, computer scientist Cynthia Dwork proved that this definition is unattainable (Dwork *et al.*, 2006). In other words, any way to access sensitive data violates the strict definition of data privacy. The problem she found was that certain types of background information can lead to new conclusions about a person.

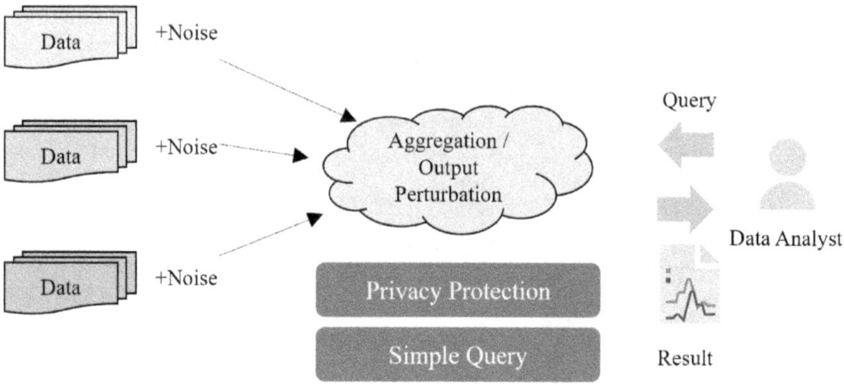

Fig. 5.6. Example of differential privacy.

This background information may not even be present in the target dataset. A prominent example is the "anonymized" Netflix rental data published on the Kaggle website. According to the 2008 paper published by Narayanan and Shmatikov (2008), at least two participants were found to be identifiable by probabilistic data association.

Based on this, Cynthia Dwork *et al.* provided a weakened version of the mathematical definition that can be used for privacy protection, namely differential privacy (Dwork, 2008). Informally, differential privacy guarantees the following for each individual who contributes data to an analysis: the output of a differentially private analysis will be roughly the same, whether you contribute your data or not.

Qualitatively, noise-added statistics or other outputs can be said to be "differentially private" if the inclusion or exclusion of a source in the dataset does not change the probability of the outcome by more than a specified factor. This factor is related to the parameter ϵ (Greek letter epsilon). The parameter ϵ quantifies how sensitive the aggregated result is to an individual's data, also known as the "privacy loss" or "privacy budget". If the factor is small, it means that the output is highly "privacy protected", i.e., it is difficult to identify the original private data on this basis; if the factor is large, it means that the reconstruction dataset and the identification dataset have a greater impact on the results individuals, relatively easily.

Consider two databases that differ by only one record (assuming the record belongs to Alice), denoted by D_1 and D_2, respectively. Perform some analysis on the database, denoted by function M, and the output results are Res1 and Res2, respectively. Differential privacy means that the two resulting outputs should be indistinguishable. Anyone looking at the output has no way of knowing whether Alice's data was used, or what Alice's data was.

Expressed mathematically as

$$\frac{Pr(M(D_1) = Res1)}{Pr(M(D_2) = Res2)} \leq e^{\varepsilon}$$

Among them, $Pr()$ represents the probability distribution. e is the natural logarithm.

Differential privacy technology has the following advantages:

(1) It is resistant to privacy attacks based on background information, and can effectively prevent linkage attacks that may occur during data de-identification.
(2) **Composability:** The privacy loss of performing two differential privacy analyses on the same data can be determined by simply summing the individual privacy losses of the two analyses. Composability means that meaningful privacy guarantees can be provided even when the results of multiple analyses of the same data are published. Some de-identification technologies are not composable so that datasets published using these technologies result in catastrophic privacy losses.
(3) **Security of post-processing:** The results of differential processing can be further processed or transformed, and differential privacy can still be maintained.

Differential privacy provides a rigorous mathematical definition and proof of privacy risk, and since its introduction, it has gained considerable support in the fields of privacy computation and statistical analysis, and has also been widely applied in related fields such as data mining and machine learning.

5.3.4 *Generalization techniques*

Generalization is another family of anonymization techniques that is distinct from randomization techniques. These techniques involve generalizing or diluting the attributes of subjects by changing the scale or order of magnitude (e.g., a street rather than a detailed home address, a week rather than a day). While generalization can be effective in preventing singling out, it does not allow for effective anonymization in all cases; in particular, it requires specific and sophisticated quantitative approaches to prevent linkage and inference.

k-anonymity, l-diversity, and t-closeness are statistical approaches that quantify the degree of identifiability within a tabular dataset, especially when variables within the dataset are combined. They are complementary approaches: a dataset can be k-anonymous, l-diversity, and t-closeness, where k, l, and t are all numbers.

Privacy models like k-anonymity, l-diversity and t-closeness distinguish between three types of variables in a dataset:

(1) **Identifiers (key attributes):** These are direct identifiers, such as full names, student numbers, and email addresses. These variables should in principle not be collected at all or should be removed from the dataset if they are not necessary for the research project.
(2) **Quasi-identifiers:** These are indirect identifiers that can lead to identification when combined with other quasi-identifiers in the dataset or external information. These are often demographic variables, such as age, gender, and location, but can also be something else entirely such as physical characteristics and timestamps. In general, quasi-identifiers are usually variables that are likely to be known to someone in the outside world.
(3) **Sensitive attributes:** These are variables of interest which should be protected, and which cannot be changed, because they are the main outcome variables. For example, it might be "medical condition" in a health record, or "income" in a financial record.

It is important to correctly categorize the variables in the dataset as one of these variable types if you want to apply k-anonymity, l-diversity, and t-closeness because they determine how the dataset will be de-identified.

The concept of k-anonymity was introduced in 2002 by Dr. Latanya Sweeney in her paper "k-anonymity: A Privacy-Preserving Model Type" (Sweeney, 2002). k-anonymity ensures that

Table 5.5. A sample dataset of patients maintained by a hospital.

Name	Gender	Age	Weight/kg	Diseases
Philip Massy	Male	52	75	Diabetes
Kyla Webb	Female	36	57	Pneumonia
Cary Meyer	Male	65	85	Diabetes
Jasmine Fuller	Female	77	60	Influenza
Celia Hill	Female	46	55	Hepatitis

each individual in a dataset cannot be distinguished from at least $k - 1$ other individuals with respect to the quasi-identifiers in the dataset. To achieve this, the attribute values are generalized to such an extent that each individual shares the same value. For example, by reducing the granularity of a location from a city to a country, a larger number of individuals are included. Individual dates of birth can be generalized into a range of dates, or grouped by month or year. Other numeric attributes (e.g., salaries, weight, and height) can be generalized by interval values (e.g., annual salary \$30,000–\$40,000). These methods can be used when the correlation of point values of attributes can create quasi-identifiers.

k-anonymity can be illustrated with a vivid example. Consider a sample dataset with some sensitive attributes maintained by a hospital (see Table 5.5).

Generally, only the attending physician, nurse, and patient can view a patient's information, so it should be protected by strong encryption, access control, and authorization mechanisms.

At the same time, patient data is an extremely valuable source of information that can help physicians gain insight into the causes of disease or find effective treatments. Medical researchers can use the data to analyze questions such as whether there is a correlation between body weight and diabetes. Legal and regulatory requirements to protect sensitive data and patient privacy may prevent such analysis because of the risk of exposing personal information.

How can this dataset be used for research without violating privacy?

First, the real name of each patient is not the concern of the medical researcher and should be completely anonymized. Anonymization of patient names is shown in Table 5.6.

Table 5.6. Anonymization of patient names.

Name	Gender	Age	Weight/kg	Diseases
Patient A	Male	52	75	Diabetes
Patient B	Female	36	57	Pneumonia
Patient C	Male	65	85	Diabetes
Patient D	Female	77	60	Influenza
Patient E	Female	46	55	Hepatitis

Of course, this is not enough. Suppose a medical researcher knows an overweight and retired (age > 60) colleague, and knows that his colleague is in the dataset, he can directly identify his colleague from the above dataset, and know what disease he suffers from (privacy!).

An immediate measure would be to remove "quasi-identifiers" that could lead to the identification of a specific person. Specifically for this record, the weight field should be deleted. However, the dataset obtained in this way will not be able to support the study of the relationship between body weight and diabetes.

How to strike a balance between business needs and privacy protection? Anonymization is a viable option.

A popular approach is not to delete data that can identify an individual, but to generalize it in a structured way. The goal is to divide the dataset into groups that can still provide valid statistical insights into the data being analyzed (in this case, the Weight and Diseases fields) but cannot provide sensitive information about individuals.

In Table 5.7, the quasi-identifiers "gender", "age", and "weight" have been generalized into two groups, each of which is guaranteed to have two or more members. In practical use cases, the groups will of course be larger, a good empirical number is 10, which is common in questionnaire scenarios. K (in this case the value is 2) in k-anonymity represents the size of the smallest group.

Under k-anonymity, even researchers who know an overweight and retired (age > 60) colleague, and who know that he is in this dataset, have no way of guessing which data his colleague corresponds to.

Formally, k-anonymity can be defined as each *equivalence class* (i.e., a set of records that are indistinguishable from each other with respect to certain "identifying" attributes) containing at least k records.

Table 5.7. Example of k-anonymity.

Name	Gender	Age	Weight/kg	Diseases
Patient A	*	60	70	Diabetes
Patient B	*	40	55	Pneumonia
Patient C	*	60	70	Diabetes
Patient D	*	60	70	Influenza
Patient E	*	40	55	Hepatitis

k-anonymity is not a sufficient defense against several types of privacy attacks, such as homogeneity attacks and background knowledge attacks.

l-diversity (Machanavajjhala *et al.*, 2007) is an extension of k-anonymity to ensure that every attribute in every equivalence class has at least l different values, thereby providing protection against deterministic inference attacks.

A fundamental goal to be achieved is to limit the occurrence of equivalence classes with poor attribute variability so that an attacker with background knowledge about a particular subject will always be left with significant uncertainty.

l-diversity can be used to protect data from inference attacks when attribute values are well distributed. However, it must be emphasized that this technique does not prevent information leakage if the attributes within a partition are unevenly distributed or belong to a small range of values or semantic meanings. Finally, l-diversity is subject to probabilistic inference attacks.

Still using the above scenario as an example. For Table 5.8, if l-diversity is imperfect and the attribute "disease" is unevenly distributed, the attacker learns from other sources that a man aged about 60 years and weighing about 70 kg is in this dataset, there will be a high probability of inferring his actual disease.

t-closeness (Li *et al.*, 2006) is a refinement of l-diversity that aims to create equivalent classes that resemble the original distribution of attributes in the table. This technique is useful when it is important to keep the data as close to the original as possible; to that end, an additional constraint is placed on the equivalence class, namely that not only should there be at least 1 different values within each equivalence class but also each value be represented as many times as necessary to mirror the initial distribution of each attribute.

Table 5.8. Example of l-diversity.

Name	Gender	Age	Weight/kg	Diseases
Patient A	*	60	70	Diabetes
Patient B	*	40	55	Pneumonia
Patient C	*	60	70	Diabetes
Patient D	*	60	70	Diabetes
Patient E	*	60	70	Diabetes
Patient F	*	60	70	Diabetes
Patient G	*	60	70	Influenza
Patient H	*	40	55	Hepatitis

The cases mentioned above are only some figurative examples. When using them in practice, it is important to keep in mind that when handling personal data, you must carefully analyze the applicable laws and regulations, implement mechanisms to protect personal data, and communicate appropriately with roles, such as legal and privacy officers.

5.3.5 *Encryption technologies*

Encryption technology is the key technology for pseudonymization, including keyed encryption, deterministic encryption, etc.

In a keyed encryption scenario, the owner of the key can easily recover the plaintext data from the pseudonymized data. However, as long as modern encryption methods are used correctly, non-owners of the keys cannot decrypt the data.

Deterministic encryption technology generally refers to using a particular encryption algorithm, randomly generating keys, encrypting data, and then discarding the keys. This technique can be compared to choosing a random number to pseudonymize each attribute in a data table and then deleting the original table. This solution reduces the risk of correlation between different datasets. If an appropriate encryption algorithm is used, it is computationally difficult for an attacker to perform decryption or replay operations because the attacker does not know the key.

5.3.6 *Hash technologies*

Hashing is a pseudonymization technique. See Section 4.7 "Hash Algorithm" for an explanation. A hash function accepts an input

Table 5.9. Example of rainbow table.

Country	MD5
Germany	d8b00929dec65d422303256336ada04f
Russia	5feb168ca8fb495dcc89b1208cdeb919
...	...

of arbitrary size and returns an output of fixed size that is very difficult to reverse. The problem is that if the range of the input value is known, the corresponding input value can be obtained by reverse lookup using the rainbow table attack method described in Section 4.7.5 "Password Storage".

For example, if the range of input values is the names of all countries, the attacker can construct the names of all countries in advance. The corresponding relationship between the name and its hash value and then reverse query based on the hash value. See Table 5.9 for examples of rainbow representations.

To mitigate such attacks, a salted-hash function can be used, i.e., a random value, called a "salt", is added to the hashed input data. This method can reduce the possibility of inferring the input value, but there is still the possibility of obtaining the original attribute value through the rainbow table attack.

Keyed hash functions provide slightly better security. This type of function takes a secret key as an additional input. It differs from a salted hash function in that the salt is usually made public. Assuming that the key is known, the owner of the data can recompute it many times to obtain pseudonymized data. It is much more difficult for an attacker to perform either a rainbow table attack or a replay attack without knowing the key.

5.3.7 *Tokenization technology*

Tokenization is the process of replacing sensitive data elements with non-sensitive equivalents (called "tokens") that have no external or exploitable meaning or value when applied to data security. Tokens are references (identifiers) to sensitive data that can be mapped back to sensitive data through tokenization. The method used to map raw data to tokens must ensure that the tokens cannot be reversed without tokenization technology. Tokenization technology provides permissions and interfaces for data processing applications to request tokens or de-tokenize tokens for sensitive data. Tokenization technologies must be secured and validated using security best practices for protecting sensitive data, secure storage, auditing, authentication, and authorization.

Tokenization is commonly used in the financial sector to replace credit card numbers with values that are less useful to attackers. It comes from one-way encryption technology, or assigns a serial number through an indexing function, or randomly generates a number unrelated to the original data to replace the original data itself.

Security and the ability to mitigate risk depend on the proper implementation of tokenization technology in processing systems that process or store sensitive data. Use tokenization technology to tokenize data to create tokens or detokenize to obtain sensitive data under strict security controls. A good tokenization method should be unable to reverse-engineer tokens into real-time data through direct attack, cryptanalysis, side-channel analysis, token mapping table theft, or brute force techniques.

The purpose of replacing real-time data with tokens in the system is to minimize the exposure of sensitive data to applications, storage, people, and processes, and to reduce the risk of inadvertent disclosure and unauthorized access to sensitive data. Applications can use tokens instead of real-time data to operate and also support the decryption of tokens under certain business-related and strictly audited conditions. Tokenization technology can be operated in a secure, isolated area of a data center or provided by a security service provider.

Tokenization can be used to protect sensitive data such as bank accounts, financial statements, medical records, criminal records, driver's licenses, loan applications, stock transactions, and other types of personally identifiable information (PII). The most common scenario, however, is credit card processing, where the PCI Council (Scoping Special Interest Group, 2011) defines tokenization as "a process by which the primary account number (PAN) is replaced with a surrogate value called a token. De-tokenization is the reverse process of redeeming a token for its associated PAN value. The security of an individual token relies primarily on the infeasibility of determining the original PAN when only the surrogate value is known. The appropriateness of choosing tokenization as an alternative to privacy technologies such as encryption depends on various regulatory requirements, compliance, and acceptance by individual audit or rating organizations.

While tokenization and encryption are similar in some respects, they differ in some important ways. When implemented properly, both tokenization and encryption can effectively protect data, and computer security systems can use both. Both are approaches to data security that perform essentially the same function, but they are processed differently and have different effects on the data they protect.

One difference between the two is that tokenization is a non-mathematical method of replacing sensitive data with a non-sensitive substitute without changing the type or length of the data. This is an important distinction from encryption because changes in the length and type of data can render the information unreadable in intermediate systems such as databases. Tokenized data can still be processed by legacy systems, making tokenization more flexible than traditional encryption.

Another difference is that tokenization requires fewer computing resources to process. With tokenization, certain data can be made fully or partially visible for processing and analysis while hiding sensitive information. As a result, data can be processed more quickly and with less strain on system resources. This can be an important benefit in systems that rely on high performance.

There are some limitations to the development of tokenization technology, and there are limitations to its use in both the compliance and technical dimensions.

The first limitation is that using tokenization to securely process database fields containing sensitive data in areas such as finance still presents potential privacy and legal risks. In other words, the tokenized data is still not anonymized or pseudonymized data in the legal sense, and regulatory requirements for personal data must still be followed. In addition, if decryption is required on necessary occasions, the mapping relationship from token to plaintext must be stored in some form of mapping table, and this mapping table itself is a valuable attack target and will introduce implementation-dimensional security risks.

The second limitation is that the level of security of the tokenization solution needs to be measured through independent verification. Due to the lack of standards, independent verification is essential to confirm the strength of the token when used for regulatory compliance. The PCI Council recommends independent review and verification of any claims regarding security and compliance (Scoping Special Interest Group and Tokenization Taskforce, 2011): "Merchants considering the use of tokenization should perform a thorough evaluation and risk analysis to identify and document the unique characteristics of their particular implementation, including all interactions with payment card data and the particular tokenization systems and processes".

The method of generating tokens may also have limitations from a security perspective. Considering attacks on random number generators (which are a common choice for generating tokens and token maps), careful consideration must be given to ensure that proven and certified methods and designs are used. Aspects such as speed, entropy, seed, and bias of RNGs must be carefully analyzed and their security measured.

With the increasing popularity of tokenization technology, new tokenization technologies such as vaultless tokenization and stateless tokenization have emerged to eliminate the risks and complexities of traditional token generation methods. complexity and scaling to accommodate emerging big data and high performance transaction processing scenarios in the financial services and banking industries. Stateless tokenization uses a random mapping of live data elements to replace values without the need for a database, while maintaining the isolation properties of tokenization.

5.3.8 *Summary of privacy protection technologies*

Each privacy technology has its advantages and disadvantages. Appropriate privacy technologies must be evaluated based on the dataset and its risks. In many cases, anonymized datasets may still pose risks to individuals. Even if an individual's records can no longer be accurately retrieved, the individual's information can still be collected using other available sources of information (public or private). Incomplete anonymization and large-scale publication of "anonymized" datasets can have a variety of direct or indirect negative effects on organizations and individuals.

Thus, while anonymization techniques can provide privacy assurances, this requires that their application scenarios be designed appropriately. For example, the context and goals of the anonymization process must be clearly defined to enable targeted anonymization operations.

Some anonymization techniques have inherent limitations. These limitations must be carefully considered before an anonymization process using a particular technique is adopted by the data controller. It is also important to consider the purpose of the anonymization, such as protecting an individual's privacy when releasing a dataset, or only allowing information to be retrieved from the dataset on an item-by-item basis.

None of the techniques described in this chapter is certain to satisfy the three basic requirements for effective anonymization: the inability to identify a single individual, the lack of linkage between records associated with an individual, and the inability to identify an individual by inference. However, because certain risks may be addressed in whole or in part by a particular technique, careful engineering is required when designing the application of individual techniques to a given situation and when using them in combination to increase the robustness of the results.

Table 5.10 shows a comparison of the privacy-preserving technologies under consideration with respect to three basic requirements.

Which anonymization or pseudonymization technology is actually used must be determined according to specific scenarios. If any of the baseline requirements cannot be met, an appropriate risk assessment is required. If national law requires the competent authority to assess or approve an anonymous procedure, this assessment shall be made available to the competent authority.

Table 5.10. Strengths and weaknesses of privacy-preserving techniques.

Technologies	Is Singling out still a risk?	Is Linkability still a risk?	Is Inference still a risk?
Pseudonymisation	Yes	Yes	Yes
Noise addition	Yes	May not	May not
Substitution	Yes	Yes	May not
Aggregation or k-anonymity	No	Yes	Yes
l-diversity	No	Yes	May not
Differential privacy	May not	May not	May not
Hashing or Tokenization	Yes	Yes	May not

5.4 General Data Security Compliance Requirements

Compliance is an issue that cannot be avoided or bypassed when building a data security system, and its importance continues to grow. Whether it is data security and privacy protection for consumer devices, data security and information security for enterprises and institutions, or new forms such as business migration to the cloud, there are many compliance requirements. Especially in recent years, various countries and regions have frequently introduced various laws, regulations, and regulatory requirements, and the enforcement of regulations has become more and more stringent, and the penalties for violations have become more and more severe. Increasingly stringent regulatory requirements pose increasing challenges to the design, implementation, development, and operation of a compliance framework for information security systems and the operation and delivery of related products and services, especially Internet-related services.

Therefore, when designing and implementing various products and services, identifying regulated data classifications and protecting them in accordance with appropriate regulatory requirements is the first step in meeting compliance requirements, based on the descriptions in Section 1.2 "Principles and Implementation of Data Classification". On this basis, organizations need to identify the relevant laws and regulations involved in a targeted manner according to the characteristics of the business domains in which they operate and follow the compliance lifecycle of compliance source identification, compliance policy and standards, compliance plan formulation and implementation, compliance outcome evaluation and certification, and compliance reporting, so as to build compliance capabilities and business. The compliance lifecycle is shown in Fig. 5.7.

This section focuses on the interpretation of compliance and regulatory requirements related to data security.

According to the interpretation of "Data Security Requirements" in Section 1.1.3, "regulations" are divided into four levels: laws of the country and region, regulations of the region where the business is located, international and national standards, and industry standards.

There are different regulatory requirements for different types of data. For example, for personal data regulation, the EU has the GDPR, Canada has the "Privacy Act", and China has the "Personal

Fig. 5.7. Compliance lifecycle system.

Information Protection Law" (Calzada, 2022); for financial data regulation, there is the GLBA Act and the PCI DSS industry standard (Payment Card Industry, 2010), among others; for health data regulation, the United States has the HIPAA Act; for children's personal data, countries such as China and the United States also have independent laws or regulations. In addition, there are also complex regulatory requirements for systems that carry data protection (cloud service systems, IT systems, information systems, computing equipment, etc.). For example, for cloud services, there are various international standards and industry standard requirements such as CIS benchmark, CSA-STAR certification, and ISO 27001 standard (International Organization for Standardization, 2022).

This section interprets the compliance requirements from three aspects: national and regional legal compliance requirements, international and national standard compliance requirements, and industry standard compliance requirements, so as to guide the analysis and design of corresponding compliance requirements and ensure compliance with laws and regulations.

5.4.1 *National and regional legal compliance requirements*

5.4.1.1 *EU GDPR*

The European Union's General Data Protection Regulation (GDPR), passed by the European Parliament in April 2016, is a law on privacy and personal data protection in European Union law that applies in the European Union (EU) and the European Economic Area (EEA). The main purpose of GDPR is to give individuals control over their personal data, unify regulations within the EU, and simplify the regulatory environment for international business. The GDPR, which replaces the outdated 1995 Data Protection Directive (95/46/EC), contains rules and requirements for the processing of personal data of individuals (known as data subjects) in the European Economic Area and applies to any business that processes personal data within the European Economic Area. The GDPR also addresses requirements for the transfer of personal data outside of the EU and EEA.

Article 4 of the GDPR defines the relevant terms:

(1) **Personal data and data subject:** Any information relating to an identified or identifiable natural person (data subject). For example, name, ID number, home address, and network identification.
(2) **Processing:** Any operation or set of operations which is performed on personal data or on sets of personal data.
(3) **Data controller:** The natural or legal person, public authority, agency or other body which, alone or jointly with others, determines the purposes and means of the processing of personal data.
(4) **Data processor:** A natural or legal person, public authority, agency or other body which processes personal data on behalf of the controller.

The GDPR defines the types of personal data described in the following (referred to as "special categories of personal data") and imposes additional processing restrictions:

(1) **Genetic data:** Personal data relating to inherited or acquired genetic characteristics of a natural person, obtained from the analysis of a biological sample taken from that natural person, in particular chromosomes, deoxyribonucleic acid (DNA) or ribonucleic acid (RNA), or from the analysis of any other element making it possible to obtain equivalent information.
(2) **Biometric data:** Personal data resulting from specific technical processing relating to the physical, physiological or behavioral

characteristics of a natural person, which allow or confirm the unique identification of that natural person, such as facial images or fingerprints.

(3) **Health data:** Personal data relating to the physical or mental health of a natural person, including the provision of health care services, which reveals information about his or her state of health.

Article 9 of the GDPR requires that for the above-mentioned types of data, the processing should be prohibited by default, unless the specified exemption conditions are met.

With reference to the "OECD Guidelines on the Protection of Privacy and Transborder Flows of Personal Data", Article 5 of the GDPR defines the principles of "lawfulness, fairness, and transparency" in the processing of personal data and further defines and explains the principles of purpose specification, collection limitation, data quality, and accountability.

In addition to specifying the requirements for the various organizations that control and process personal data, Article 3 of the GDPR defines the rights of data subjects with respect to personal data:

(1) **Right to know:** Organizations must inform users what is being collected, what it will be used for, how long it will be kept, whether it will be shared, and with whom.
(2) **Access rights:** The organization must provide a means for individuals to contact the organization to request a copy of the information the organization holds about them.
(3) **Right to correction:** The individual must be able to verify that the information the organization holds about him or her is accurate, and if it is found to be inaccurate, the individual has the right to request that the organization update it.
(4) **Right to erasure:** Sometimes referred to as the "right to be forgotten". This is not an absolute right, but individuals may request that personal data held by the organization about them be removed.
(5) **Right to restrict processing:** This is also not an absolute right, but individuals can declare that they refuse to consent to the processing of their data.

(6) **Right to data portability:** Data subjects have the right to extract personal data held by an organization for use by other organizations in other places. For example, downloading profile information on Facebook for use on another social media platform.
(7) **Right to object:** Individuals have the ability to ask an organization to stop using their information in a way they object to. For example, making a marketing call or sending marketing materials by mail.
(8) **Rights related to automated decision making and profiling:** With the tremendous growth of profiling, individuals can object to or appeal automated decision making, such as user profiling and targeted online advertising.

Organizations, especially data controllers, need to be able to demonstrate compliance with the GDPR, so the following measures are very important:

(1) Designate a team within the organization that is responsible for data protection.
(2) Keep detailed documentation of the data collected, including how it is used, where it is stored, which employee is responsible for it, etc.
(3) Provide cybersecurity and privacy awareness training to the organization's employees and business partners.
(4) Implement organizational and technical security measures.
(5) Enter into a data processing contract with the data processor.
(6) Appoint a data protection officer, if the type of organization as defined by the GDPR is met.

Articles 24, 25, 28, and 32 of the GDPR require organizations to process personal data securely by implementing "appropriate technical and organizational measures". In the context of rapid technological development, the GDPR does not specify which technologies can be used. However, Article 32 of the GDPR lists the following requirements for technical and organizational measures:

(1) the pseudonymization and encryption of personal data;
(2) the ability to ensure the ongoing confidentiality, integrity, availability and resilience of processing systems and services;

(3) the ability to restore availability of and access to personal data in a timely manner in the event of a physical or technical incident;
(4) a process for regularly testing, assessing and evaluating the effectiveness of technical and organizational measures to ensure the security of the processing.

Technical measures are needed to cover comprehensive personal data protection requirements. Whether it's using two-factor authentication for administrator accounts or using end-to-end encryption when transferring personal data to cloud services, these are some basic technical measures. The technical frameworks presented in this book, such as the ISO/IEC 27001 series of standards, the NIST Cybersecurity Framework, and the CIS Key Control Measures Set, can be used as a reference.

Organizational measures are things like employee training, adding a data privacy policy to the employee handbook, or following the principle of minimum necessity by limiting the number of employees and positions that have access to personal data.

In addition, in the event of a data breach, organizations must notify affected individuals within 72 hours or face penalties. This notification requirement can be waived if technical safeguards, such as encryption, are used to render the data useless to attackers.

Article 25 of the GDPR calls for "data protection by design" and "data protection by default". This means that data protection principles must be considered in the design of any new product or campaign. For example, if a company is launching a new app, it must first consider what personal data the app might collect from users, then how to minimize the amount of data and how to protect it with the latest technology.

It's worth noting that GDPR regulations are becoming increasingly stringent. In January 2019, the French regulator CNIL decided to impose a huge fine of 50 million euros on Google (Netter, 2019) because Google did not explain clearly enough to Android users how to process their personal data. CNIL found that Google did not legally obtain user consent when serving personalized ads. According to the GDPR, when user consent is used as a basis for processing personal data, it must be the user's informed, explicit consent and not coerced. In this case, the French court found that Google had not provided sufficiently clear information, including the design of its default checkboxes, to lawfully obtain the user's consent. French courts have confirmed that this does not comply with the requirements of the GDPR.

CNIL's fine against Google is the largest to date against a tech giant under Europe's General Data Protection Regulation. The French court found that the size of the fine was proportionate, given the seriousness and ongoing nature of the breach.

Google subsequently appealed, but in June 2020, France's highest administrative court (the "Conseil d'Etat") ruled to reject the appeal and uphold the original judgment.

5.4.1.2 *Canadian privacy protection laws*

Canada has a number of privacy laws designed to protect the privacy of individuals and give them access to the information that has been collected about them. The Office of the Privacy Commissioner of Canada (OPCC) oversees compliance with these laws.

The Privacy Act regulates how Canadian federal government organizations, including federal employees, collect, use, and disclose personal information. One that has received much attention is the Personal Information Protection and Electronic Documents Act (PIPEDA). It applies to the operations of commercial or for-profit businesses, as well as federally regulated businesses, such as banks, airlines, and telecommunications companies.

PIPEDA is based on the Canadian Standards Association's Model Code for the Protection of Personal Information, which includes 10 principles.

The 10 Information Fairness Principles are as follows:

1. **Accountability:** An organization is responsible for the personal information under its control and must designate one or more individuals to be responsible for the organization's compliance with the following principles.
2. **Identification of purposes:** The purposes for which personal information is collected shall be identified by the organization at or before the time the information is collected.
3. **Consent:** The knowledge and consent of the individual is required for the collection, use, or disclosure of personal information, except when inappropriate.

4. **Limitation of collection:** The collection of personal information shall be limited to that which is necessary for the purposes identified by the organization. Information shall be collected by fair and lawful means.
5. **Limiting use, disclosure, and retention:** Personal information shall not be used or disclosed for purposes other than those for which it was collected, except with the consent of the individual or as required by law. Personal information will be retained only as long as necessary to fulfill those purposes.
6. **Accuracy:** Personal information will be as accurate, complete, and current as necessary for the purposes for which it is to be used.
7. **Safeguards:** Personal information will be protected by safeguards appropriate to the sensitivity of the information.
8. **Openness:** An organization shall make readily available to individuals' specific information about its policies and practices relating to the management of personal information.
9. **Individual access:** Upon request, an individual shall be informed of the existence, use, and disclosure of his or her personal information and shall be given access to that information. An individual must be able to challenge the accuracy and completeness of the information and have it corrected, if necessary.

PIPEDA grants individuals the following rights:

- Know why an organization collects, uses, or discloses their personal information.
- Expect an organization to collect, use, or disclose their personal information reasonably and appropriately and not use the information for any purpose other than that to which they have consented.
- Know who in the organization is responsible for protecting their personal information.
- Expect an organization to protect their personal information by taking appropriate security measures.
- Expect the personal information an organization holds about them to be accurate, complete, and up-to-date.
- Obtain access to their personal information and ask for corrections if necessary, and complain about how an organization handles their personal information if they feel their privacy rights have not been respected.

Similar to GDPR, PIPEDA requires organizations to do the following:

- Obtain consent when they collect, use, or disclose their personal information.
- Supply an individual with a product or a service even if they refuse consent for the collection, use, or disclosure of your personal information unless that information is essential to the transaction.
- Collect information by fair and lawful means.
- Have personal information policies that are clear, understandable, and readily available.

Canada is a federal country, and some provinces have also enacted their own privacy legislation. In general, PIPEDA applies to business activities in all provinces and territories, except for activities in provinces that have their own privacy legislation that has been declared "substantially similar" to the federal legislation. For example, the private sector privacy legislation in British Columbia, Alberta, and Quebec is considered to be substantially similar to PIPEDA and therefore the provincial laws supersede the federal law.

5.4.1.3 *U.S. privacy protection laws*

For companies that need to do business across borders, especially those from mainland China, it should be noted that many countries are federalized, and different states, municipalities, or autonomous regions have their own laws and regulations. Among them, the California Consumer Privacy Act (CCPA) in the United States has a relatively wide impact, which can be used as a reference for compliance.

The CCPA is the first state-level privacy law in the U.S. federal system. It provides various privacy rights to consumers in the state of California.

A business is subject to the CCPA if its annual revenue exceeds $25,000,000, or it derives more than 50% of its annual revenue from sales to individual California consumers, or it acquires or discloses the personal information of more than 50,000 California consumers annually.

The CCPA went into effect on July 1, 2020. The law provides California residents with rights similar to those set forth in the

European Union's GDPR, such as disclosure requirements, data subject requests (DSRs), "opt-out" of certain data transfers, and "opt-in" requirements for minors. In addition, there are some access, erasure, and portability requirements.

To comply with the requirements of the CCPA, similar to the requirements of the GDPR, you can start with the following five dimensions:

(1) **Discovery:** Determine what personal data the organization currently holds and where it resides.
(2) **Mapping:** Identify how the organization currently shares personal data with third parties. Clarify whether those third parties are subject to the CCPA's opt-out requirements.
(3) **Management:** Manage how personal information is used and accessed.
(4) **Protection:** Establish security controls to prevent, detect, and respond to exploits and data breaches.
(5) **Documentation:** Document the breach response plan and ensure that the organization's contracts with relevant third parties can be classified as exceptions to the opt-out requirement.

5.4.2 *International and national standards compliance requirements*

There are a number of international and national standards for data security and privacy. Some standards are country-specific entry thresholds known as "mandatory certification standards". When providing services to these countries and regions, companies should not only meet the requirements of local laws and regulations but also meet the requirements of compulsory certification standards. This is also known as "standards compliance". Other standards are recommended and can be used as a reference. Some standards also provide certification functions to certify that an organization or products or services provided by the organization meet a certain level of security.

Authoritative international standards are formulated primarily by two standards organizations: ISO and IEC. The International Organization for Standardization (ISO) is an international standards body composed of representatives from various national and regional standards organizations. Founded in 1946 and headquartered in

Geneva, Switzerland, it operates in 165 countries and regions. The International Electrotechnical Commission (IEC) is an international standards organization with 89 member countries that prepares and publishes International Standards for all electrical, electronic and related technologies (collectively, "electrotechnology").

ISO/IEC JTC 1/SC 27 is responsible for international standards in the area of information security. SC 27 is a security technology subcommittee of JTC 1, a joint ISO/IEC committee established in 1990 to develop international standards, technical reports, and technical specifications. SC 27's standardization activities include common concepts, management system requirements, technologies, and related guidelines for information security, cybersecurity, and privacy.

SC 27 has five working groups and two management groups. See Table 5.11 for its organizational structure and work content.

As a horizontal business area, data security and privacy are related to the work and international standards of the above-mentioned working groups.

The ISO/IEC 27000 series of standards published by SC 27 covers best practice recommendations for information security management, i.e., managing information security risks and ensuring the security of information assets through the design and implementation of security controls. These international standards provide a high-level framework for information security that encompasses the legal, physical, and technical controls involved in an organization's information risk management process.

ISO/IEC 27001, published by the ISO and IEC Joint Subcommittee (SC 27) of the International Organization for Standardization (ISO) and the International Electrotechnical Commission (IEC), is one of the ISO/IEC 27000 series of information

Table 5.11. SC 27 working groups.

SC 27		Scope
Working Group	WG 1	Information security management system (ISMS) standards
Working Group	WG 2	Cryptography and security mechanisms
Working Group	WG 3	Security evaluation, testing and specification
Working Group	WG 4	Security controls and services
Working Group	WG 5	Identity management and privacy technologies

security standards (International Organization for Standardization, 2022). It was first published in 2005 and revised in 2013. ISO/IEC 27001 describes the requirements for establishing, implementing, maintaining, and continually improving an information security management system (ISMS). Organizations that meet the requirements of the standard have the option to be certified by an accredited certification body following successful completion of an audit.

ISO/IEC 27001 requires management to perform the following tasks:

(1) Systematically examine the organization's information security risks, taking into account threats, vulnerabilities, and impacts.
(2) Design and implement a coherent and comprehensive set of information security controls and other forms of risk management (such as risk avoidance or risk transfer) to address those risks that are deemed unacceptable.
(3) Adopt an overall management program to ensure that information security controls continue to meet the organization's information security needs.

The security management standard is primarily ISO 27005: "Information Security Risk Management" (International Organization for Standardization, 2018). The implementation guide standards are mainly ISO 27110: Cybersecurity Framework Development Guidelines". The main standard for designing security controls is ISO 27002: "Practical Guidelines for Information Security Controls" (International Organization for Standardization, 2022).

The top-level privacy framework standard is ISO/IEC 29100:2011 "Privacy Framework" (International Organization for Standardization, 2011), which specifies a common privacy terminology, defines the actors and their roles in processing personally identifiable information (PII), describes privacy safeguarding considerations, and provides references to known privacy principles for information technology.

Other important criteria related to privacy can be grouped into four different levels, as shown in Fig. 5.8.

In addition to the series of standards, ISO has also published a number of supporting documents, which are not described in detail in this book.

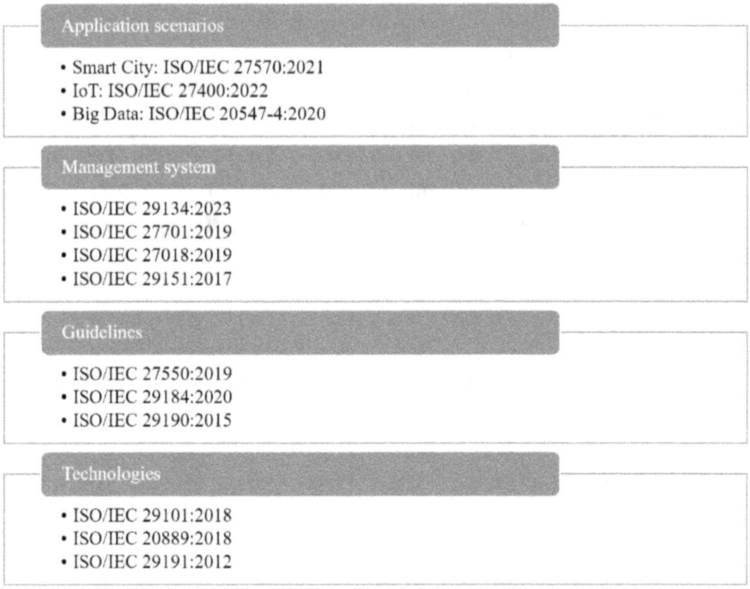

Fig. 5.8. Other important criteria related to privacy.

In the management series of standards, ISO 27005 provides security risk management, ISO 29134 provides implementation guidance for conducting privacy impact assessments (PIAs), and ISO 29151 provides a collection of control measures for personally identifiable information (PII).

Other international standards related to the protection of personal information include:

* ISO/IEC 17788/89;
* ISO/IEC 19086;
* ISO/IEC 19941;
* ISO/IEC 19944.

For different scenarios of personal data collection and processing, different international standards focus on different objects and fields. The ISO international standards related to personal data are shown in Fig. 5.9.

As mentioned above, standards such as ISO 9001, ISO 27001, ISO 27017, and ISO 27701 also cover information security, data security, and privacy.

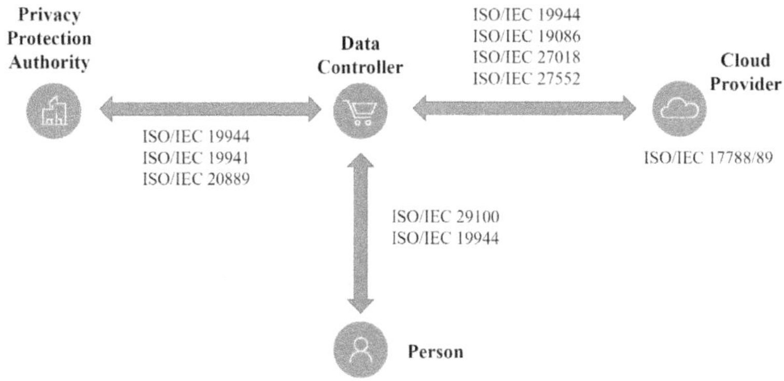

Fig. 5.9. ISO international standards related to personal data.

5.4.3 *Industry standard compliance requirements*

This section describes two industry standards, CIS Benchmark and CSA STAR, which apply to security configuration and cloud security, respectively. This section also describes Common Criteria (CC) certification and FIPS cryptographic module certification.

5.4.3.1 *CIS benchmark*

The Center for Internet Security (CIS) is a non-profit organization whose mission is to "develops, validates, and promotes best practices and tools to help protect IT systems and organizations from cyber threats". It draws on the expertise of cybersecurity and IT professionals from government, business, and academia from around the world. CIS develops a variety of standards and best practices, including CIS benchmarks, controls, and hardened images.

CIS benchmarks (Center for Internet Security, 2023) are configuration baselines and best practices for securely configuring various software systems, such as server operating systems (e.g., CentOS, Amazon Linux, and Debian Linux), cloud platforms (e.g., Docker and Kubernetes), and database systems (e.g., MongoDB). Each benchmark references one or more CIS controls, which are designed to help organizations improve their cyber defenses.

The CIS Control Set V7.1 defines 20 security control measures, see Table 5.12.

CIS controls can be mapped to many established regulatory and standards frameworks, including HIPAA, NIST Cybersecurity

Table 5.12. Security control measures defined by CIS Controls V7.1.

No.	Catalog	Controls
1	Basic	Inventory and control of hardware assets
2		Inventory and control of software assets
3		Continuous vulnerability management
4		Controlled use of administrative privileges
5		Secure configuration for hardware and software on mobile devices, laptops, workstations, and servers
6		Maintenance, monitoring and analysis of audit logs
7	Foundation	Email and web browser protections
8		Malware defenses
9		Limitation and control of network ports, protocols, and services
10		Data recovery capabilities
11	Foundation	Secure configuration for network devices, such as firewalls, routers, and switches
12		Boundary defense
13		Data protection
14		Controlled access based on the need to know
15		Wireless access control
16		Account monitoring and control
17	Organization	Implement a security awareness and training program
18		Application software security
19		Incident response and management
20		Penetration tests and red team exercises

Framework (CSF) and NIST SP 800–53, ISO 27000 series of standards, PCI DSS standards, and more.

For example, Clause 1.2 of the CIS Amazon Web Services Foundations is "1.2 Ensure multi-factor authentication (MFA) is enabled for all IAM users that have a console password". The CIS control to which this clause refers is CIS Control 4: "Controlled Use of Administrative Privileges".

The CIS benchmarks provide two levels of security settings:

Level 1 recommends essential baseline security requirements that can be configured on any system and should cause little or no service interruption or reduced functionality.

Level 2 recommends security settings for environments that require greater security, which may result in some reduced functionality.

CIS Hardened Images are securely configured virtual machine images based on CIS benchmarks that have been hardened to either a Level 1 or Level 2 CIS benchmark profile. Hardening is a process that helps protect against unauthorized access, denial of service, and other cyber threats by limiting potential vulnerabilities that make systems susceptible to cyber attacks.

Most cloud service providers support the integration and automated deployment of CIS hardened images. For example, the Microsoft Azure cloud service supports Microsoft Windows Server Edition, Red Hat Linux, Ubuntu Linux and other similar Linux flavors. Amazon Web Services (AWS), Google Cloud Platform (GCP) also provide similar support.

5.4.3.2 *CSA STAR*

The Cloud Security Alliance (CSA), an international non-profit organization founded in 2008, released the Cloud Control Matrix (CCM) in 2010 as an implementation guide for cloud computing security (covering IaaS, PaaS, and SaaS). The CCM covers fundamental security principles across 16 domains to help cloud customers assess the overall security risk of cloud services. The most notable feature of the CCM is that each of its cloud security controls aligns with internationally recognized information security standards such as ISO 27001, PCI DSS, ISACA, COBIT, and NIST.

CSA officially released the CSA STAR (Security, Trust & Assurance Registry) certification (Cloud Security Alliance, 2013) on September 25, 2013. The CSA STAR certification uses the security management and control requirements of the ISO 27001 management system standard, combined with CSA's own proposed CSA CCM, to demonstrate the security process of cloud computing through assessment.

CSA STAR certification includes three levels of assurance that align with the control objectives in the CSA CCM:

Level 1: STAR Self-Assessment.

Level 2: STAR Attestation, STAR Certification and C-STAR Evaluation (based on an independent third-party review).

Level 3: STAR Continuous Monitoring (CSA is still developing procedures and requirements; this level has not yet been implemented).

CSA STAR certification uses the SOC 2 Type 2 audit mechanism, based on the CCM standard, to conduct rigorous independent audits of cloud providers' security posture. Independent auditors who evaluate cloud providers' STAR-certified offerings must be certified public accountants (CPAs) and hold the CSA CCSK Certificate of Cloud Security Knowledge (CCSK).

5.4.3.3 *CC certification*

The Common Criteria for Information Technology Security Evaluation, abbreviated Common Criteria or CC, is an international standard (ISO/IEC 15408) used for computer security certification.

Common Criteria evaluations are performed on computer security products and systems. The object of the evaluation is called the Target of Evaluation (ToE). The process of evaluation is to verify that a pre-defined target conforms to the statement based on its security properties.

The first related concept is the Protection Profile (PP). The PP identifies the security requirements for specific types of security devices (e.g., smart cards or network firewalls used to provide digital signatures). Product vendors select one or more PPs, depending on the type of product, against which their products are evaluated. In this way, customers can select PP-certified products for the type of equipment they intend to purchase. In this case, the PP can be used as a template for the product ST (Security Target, as defined in the following), or the ST author should at least ensure that all the requirements in the relevant PP appear in the ST documentation of the target.

A security target (ST) is a document which identifies the security attributes of a ToE. In general, the generation of an ST conforms to a particular PP. Of course, an ST may claim to conform to several PPs. The ST defines the security requirements and expectations of the system, including protection against threats, attacks and vulnerabilities. They are often expressed in the form of a hierarchy, such as EAL4+, which indicates the system's ability to protect confidential information and data integrity. The ToE is evaluated against the Security Functional Requirements (SFR) described in the ST. Vendors may customize the evaluations to accurately reflect the intended functionality of their products. For example, the security functional requirements of a network firewall and a database management system are very different.

The security feature requirements (SFRs) specify the individual security features that the product may provide. Common Criteria provides a standard catalog of such features. For example, an SFR may describe how to authenticate a user playing a particular role.

In addition to evaluating the security features of the product, the quality assurance process of the product can also be evaluated to increase confidence in the security features. This is where the Security Assurance Requirement (SAR) comes in. The SAR describes the measures taken during product development and evaluation to ensure compliance with claimed security features. For example, an assessment might require that all source code be maintained in a change management system, or that full functional testing be performed.

5.4.3.4 *FIPS cryptographic module certification*

The Federal Information Processing Standards (FIPS) are public standards developed by the U.S. federal government to govern the use of computer systems by non-military government agencies and government contractors.

FIPS 140–2 is a security standard for hardware, software, and firmware solutions jointly sponsored by the United States and Canada. During the procurement process, the U.S. government requires that all solutions using encryption technology be FIPS 140–2 validated to ensure that end users receive a high level of security and reliability.

Many national government procurement and tender requirements around the world clearly state that products with cryptographic modules must meet the requirements of FIPS 140–2, and some commercial products have also passed FIPS 140–2 certification. For example, the kernel module (Apple CoreCrypto Kernel Module v9.0) of Apple's iOS 12 operating system has received CMVP (Cryptographic Module Validation Program) certification number 3438 (Apple Inc., 2019).

FIPS 140–2 defines four levels of security (see Table 5.13).

Table 5.13. FIPS 140-2 security levels.

Level	Description
Security Level 1	Cryptographic module
Security Level 2	Cryptographic module, Role-based or identity-based operator authentication, Referenced PPs evaluated at EAL2+, locks or tamper evidence.
Security Level 3	Complete resistance to physical attacks, identity-based operator authentication, Referenced PPs plus trusted path evaluated at EAL3 plus security policy modeling.
Security Level 4	Tamper detection and response envelope. EFP or EFT. Referenced PPs plus trusted path evaluated at EAL4.

5.5 Industry Data and Privacy Compliance Requirements

Unlike the data security oversight requirements described in Section 5.4, "General Data Security Compliance Requirements", which apply to all industries, some laws, regulations, and industry standards define separate regulatory requirements for data security compliance based on different industries and domains.

When designing and implementing various products and services, based on the description in Section 1.2 "Principles and Implementation of Data Classification", it is necessary to identify the regulated data classification and protect it according to the appropriate regulatory requirements to meet the compliance requirements.

5.5.1 *Financial data supervision requirements*

Among the many types of data, financial data is a very sensitive type of data, involving various requirements such as personal privacy, financial supervision, payment supervision, and cross-border data. In terms of data value and sensitivity to breach, financial data is clearly at the top of the list.

5.5.1.1 *SOX Act (USA)*

SOX is short for the Sarbanes-Oxley Act, and its full name is the Public Company Accounting Reform and Investor Protection Act. The SOX Act is a U.S. federal law that includes reforms in accounting oversight, corporate governance, securities market oversight, and other areas. Provisions of the Act address the responsibilities of public company boards of directors, increase criminal penalties for certain misconduct, and require the Securities and Exchange Commission to develop rules to define how public companies will comply with the Act.

Closely related to public company financials are the following three key provisions: Section 302, Section 404, and Section 802.

Section 302 of the SOX Act of 2002 requires a company's principal executive officer to personally certify in writing that the company's financial statements "fairly present, in all material respects, the financial condition and results of operations of the issuer as of and for the periods presented therein. Corporate officers who sign financial statements they know to be false are subject to criminal penalties.

Section 404 of the SOX Act of 2002 requires management and auditors to establish a system of internal controls and reporting procedures to ensure the adequacy of those controls. This requires management to define and implement clear policies, procedures, and rules to ensure the integrity of financial and accounting information, ensure accountability, and prevent fraud.

Section 802 of the SOX Act of 2002 contains three rules that affect recordkeeping. The first rule addresses the destruction and falsification of records. The second rule strictly regulates the retention period of stored records. The third rule outlines the specific business records that companies are required to retain, including records of electronic communications.

In addition to the financial aspects of an organization, such as auditing, accuracy, and controls, the SOX Act of 2002 outlines requirements for information technology (IT) departments regarding electronic records. The Act does not prescribe a set of business practices but rather what business records must be retained and for how long. The standards set forth in SOX 2002 do not dictate how companies should store their records, only that it is the responsibility of the company's IT department.

The effectiveness of internal control and the accuracy, integrity, and efficiency of financial information depend on the support of IT systems. As part of preventive internal control activities, best practices such as authorization and segregation of duties, and network and information security measures can help prevent inadvertent errors or intentional tampering.

In the implementation dimension, IT controls are a subset of an organization's internal controls whose primary objective is to ensure the confidentiality, integrity, and availability of data. IT controls also include the overall management of an organization's IT function.

Information technology controls fall into two categories: general IT controls and IT application controls. General IT controls include controls over the IT environment, computer operations, program and data access, program development, and program changes that help ensure the reliability of data produced by IT systems and support the system's ability to function as intended and reliably output the result. IT application control refers to transaction processing control, which focuses on the "input, processing, output" process of transaction processing to ensure the integrity and accurate processing of data from input to output.

The "SOX Act" has made information technology control more important in U.S. listed companies. The COBIT framework is a widely used framework published by ISACA that defines the objectives and recommended assessment methods for various IT general controls and IT application controls. It is generally the responsibility of a company's IT department to ensure that effective information technology controls are designed and implemented.

In summary, it is necessary for public companies to design and implement appropriate internal controls (including IT controls) based on the requirements of SOX when handling important data related to the company's operations, especially financial data.

5.5.1.2 *GLBA Act (USA)*

The Gramm-Leach-Bliley Act (GLBA), also known as the "Financial Modernization Act" of 1999, is a federal law of the United States. The Act requires financial institutions (companies that provide consumers with financial products or services, such as loans, financial or investment advice, or insurance) to explain their information sharing practices to customers and to protect sensitive information.

To comply with GLBA standards, a financial institution must disclose to customers how the institution shares customers' sensitive information and provide customers with specific protections for their private information in accordance with a written information security plan prepared by the institution. If customers do not want their personal information shared with third parties, financial institutions must inform customers of their right to opt out.

GLBA creates the Consumer Financial Information Privacy Rule and issues additional privacy and security requirements to facilitate GLBA implementation. GLBA is enforced by the Federal Trade Commission, the Federal Banking Agency and other federal regulators, and state insurance regulators.

Compliance with GLBA allows financial institutions to reduce penalties or reputational damage for unauthorized disclosure or loss of customer information. GLBA also requires that certain privacy and security safeguards be provided to customers, some of which include the following:

(1) Private information must be protected from unauthorized access.
(2) Customers must be notified of the sharing of private information between financial institutions and third parties, and customers must be able to opt out of the sharing of private information.
(3) track user activity, including attempts to access protected records.

GLBA compliance helps build and strengthen consumer confidence and trust by protecting consumers and customers. Customers can be assured that the financial institution will keep their information secure; security fosters customer loyalty, which in turn enhances the financial institution's reputation and enables the financial institution to obtain more business and other benefits.

5.5.1.3 *PCI DSS industry standard (US)*

The Payment Card Industry (PCI) Data Security Standard (DSS) is a global information security standard designed to protect credit card data and transactions from fraud. The PCI DSS industry standard applies to organizations that process credit card transactions for the five major credit card brands [Visa, MasterCard, American Express, Discover, and Japan Credit Bureau (JCB)]. PCI DSS specifies the requirements for storing, processing, or transmitting payment and cardholder data, specifically data such as the primary account number (PAN), cardholder name, and credit card expiration date.

The PCI Data Security Standard specifies 12 compliance requirements, which are grouped into 6 logically related groups or 6 "Control Objectives". The six groups are as follows:

(1) Establish and maintain secure networks and systems.
(2) Protect cardholder data.
(3) Maintain a Vulnerability Management Plan.
(4) Implement strong access control measures.
(5) Regularly monitor and test the network.
(6) Maintain an information security policy.

The first set of control objectives are secure network requirements, which include firewall requirements for fixed and wireless networks to prevent network transmissions from being intercepted or hacked. In addition, authentication credentials such as passwords must not use vendor-supplied defaults. Clients should be able to change this data easily and frequently.

The second set of control objectives are cardholder data protection requirements. Databases that store vital information such as dates of birth, social security numbers, phone numbers, and mailing addresses should be hack-proof. When cardholder data is transmitted over public networks, it must be efficiently encrypted. Encryption is especially important for e-commerce over the Internet.

The third group of control objectives is protection against loopholes, or system security requirements. In terms of system security, the standard requires regular installation of the operating system (OS) and patches provided by software vendors to ensure the highest level of vulnerability management. At the application level, applications used to process cardholder data should be bug-free to prevent data theft or tampering. The standard also requires the use of up-to-date antivirus, antispyware, or other anti-malware solutions to protect systems from malicious activity.

The fourth group of control objectives are access control requirements. That is, access to and manipulation of system information should be restricted. The first is access control to cardholder data, whether in physical or digital form. For data in physical form, protection measures may include using file shredders and avoiding copying as much as possible. Access control also includes user authentication.

The simplest case is to use a different account and password for each operator.

The fifth group of control objectives are network maintenance and monitoring requirements. Networks require continuous monitoring and periodic testing to ensure that security measures and processes are in place, functioning properly, and up to date. For example, antivirus and antispyware programs should frequently scan all data exchanged, all applications, all random access memory (RAM), and all storage media. Vulnerability scanning and penetration testing also fall under this objective.

The sixth group of control objectives are the requirements for establishing and maintaining an information security policy. All entities involved in processing cardholder data must comply with the security policy at all times. Audits and penalties for non-compliance can be used to ensure policy compliance.

5.5.2 *Medical and health data security regulatory requirements*

Medical and health information touches everyone's tangible interests. As one of the most sensitive types of data, once leaked, it may not be able to be changed. Medical and health information is considered sensitive personal information under the laws of most jurisdictions, and its security is very important.

The earliest protection requirements for medical and health information occurred between physicians and patients, and it is the responsibility of physicians to protect the privacy of patients. As early as the ancient Greek era, the physician Hippocrates issued a proposal on professional ethics for the medical profession (Miles, 2005) that included this part:

"Whatever, in connection with my professional practice or not, in connection with it, I see or hear, in the life of men, which ought not to be spoken of abroad, I will not divulge, as reckoning that all such should be kept secret. While I continue to keep this Oath unviolated, may it be granted to me to enjoy life and the practice of the art, respected by all men, in all times! But should I trespass and violate this Oath, may the reverse be my lot!"

In today's society, the protection of medical data involves not only doctors and patients but also third-party organizations such as insurance companies and payment agencies. In the EU's GDPR or China's relevant laws, medical and health data are classified as sensitive personal data and are protected by uniform standards. However, to date, there are no laws specifically targeting medical and health data.

In the United States, where both medical and legal services are highly developed, there are various laws related to medical and health data. Federal laws that directly restrict medical and health information include the Health Insurance Portability and Accountability Act, the Health Information Technology for Economic and Clinical Health Act, and the Genetic Information Nondiscrimination Act (Hudson *et al.*, 2008). Among them, the Health Insurance Portability and Accountability Act (HIPAA) is relatively early in the legislation and has a relatively wide scope of influence.

HIPAA (Health Insurance Portability and Accountability Act) is a law passed in the United States in 1996. Its purpose is to protect the security and privacy of an individual's medical information and to improve the portability and continuity of health insurance coverage.

HIPAA contains a number of provisions that apply to entities and individuals that handle medical information, including health plans, health care providers, medical information processors, and other organizations that handle medical information. HIPAA establishes the standards and requirements that these entities and individuals must follow when processing, storing, transmitting, and protecting medical information.

Key takeaways from HIPAA include the following:

1. **Privacy rules:** HIPAA ensures that an individual's medical information is properly protected, restricts the use and disclosure of medical information, and requires medical and related organizations to take steps to protect the confidentiality of personal information.
2. **Security rules:** HIPAA requires health care providers and related organizations to take appropriate technical, physical, and administrative measures to protect the security of medical information, including data encryption, access control, disaster recovery plans, and more.

3. **Standardization:** HIPAA has developed a number of standards, including the Electronic Data Interchange (EDI) standard and the Electronic Health Record (EHR) standard, to facilitate interoperability and data exchange of medical information.
4. **Penalties for noncompliance:** HIPAA provides penalties for individuals and organizations that violate its provisions, including fines and criminal penalties.

The implementation of HIPAA has had a significant impact on the healthcare industry, driving the development of medical information technology and strengthening medical information security. It provides individuals with greater control and protection by ensuring the security and privacy of medical information.

The HIPAA Privacy Rule protects all personally identifiable medical and health information that is maintained or transmitted by a covered entity or its business associates in any format, whether electronic, paper, or oral. The Privacy Rule refers to this information as protected health information. Individually identifiable medical and health information includes the following:

(1) Information about an individual's past, present, or future physical or mental health status,
(2) medical information about the individual, and
(3) past, present, or future medical payment information.

The primary purpose of the Privacy Rule is to define and limit the circumstances under which an individual's protected medical and health information may be used and disclosed.

The core principle of the Privacy Rule regarding the use and disclosure of medical and health information is the principle of least restrictive alternative. Covered entities must use, disclose, and request the least amount of protected medical and health information necessary for a particular purpose.

HIPAA gives individuals the right to know, the right to amend, the right to inspect, the right to restrict, the right to complain, and more about medical and health information.

HIPAA requires medical or health care organizations (also known as "entities") to provide privacy notices to ensure an individual's right to know. The Privacy Rule requires that notices describe how protected medical and health information is used and disclosed by the covered entity, state the covered entity's responsibility to protect privacy, and describe individuals' rights, including the right to be heard by HHS and covered entities if they believe their privacy rights have been violated.

Individuals have the right to request an amendment if they believe that the protected health information in the covered entity's designated record set is inaccurate or incomplete. If the Covered Entity accepts the request for amendment, it must make the amendment accordingly. If an individual's request for amendment is denied, the Covered Entity must provide a written explanation of the denial.

Individuals have the right to request an audit of the disclosure of protected personal information by the Covered Entity or its business partners. Once an individual requests an audit, he or she may review up to the previous 6 years of history, unless otherwise specified.

Individuals have the right to ask the covered entity to limit the use or disclosure of protected health information for treatment, payment or health care operations, to limit the disclosure of information to those who are involved in the individual's medical treatment or payment of expenses, or to limit the communication of information such as an individual's status, location or death to family members or others. The Covered Entity is not required to comply with the individual's request for a restriction, and if the parties have agreed to the request for a restriction, the Covered Entity must comply, except for emergency medical treatment.

Covered entities must provide individuals with a complaint procedure for individuals to complain about compliance with the Privacy Policy and Rules and the Privacy Rules, and the complaint procedure must be included in the Notice of Privacy Practices. The Covered Entity must clearly identify to whom an individual may submit a complaint and suggest that the complaint may also be submitted to HHS.

Covered entities may not retaliate against individuals for exercising their rights under the Privacy Rule, and individuals have the right to assist HHS or other relevant authorities in their investigations and to object to conduct that they believe violates the Privacy Rule. Covered entities may not require individuals to waive their rights as a condition of receiving treatment, payment, or enrollment for benefits.

Covered Entities are required to retain such documents and records for a period of 6 years, including documents required by the Privacy Rules and Regulations, Notices of Privacy Practices, Complaint Handling, and other Privacy Rules.

The Security Rule requires covered entities to maintain appropriate administrative, technical, and physical safeguards to protect EPHI. Specifically, covered entities must perform the following:

(1) Ensure the confidentiality, integrity, and availability of all EPHI that is created, received, maintained, or transmitted.
(2) Identify and protect against reasonably anticipated threats to the security and integrity of the information.
(3) Prevent reasonably anticipated inappropriate uses and disclosures.
(4) Ensure that employees comply with security policies.

The confidentiality of the Security Rules requires that EPHI be invisible to unauthorized persons, which supports the inappropriate use of EPHI and the prohibition on disclosure in the Privacy Rules. The Security Rules also require that the integrity and availability of EPHI be maintained. Integrity means that EPHI cannot be altered and destroyed by unauthorized means. Availability means that EPHI is accessible to authorized individuals.

HHS designates entities that can be large and small, even across state lines, so the security rules must be flexible and scalable to allow them to analyze their needs and implement appropriate solutions for their specific contexts. The suitability of the program depends on the size, resources, and business characteristics of the covered entity.

As a result, it is up to the organization to decide which security measures to implement, taking into account the following:

(1) Size, complexity and capacity.
(2) Technology, hardware and software infrastructure.
(3) The cost of the security measures and
(4) The potential impact on potential EPHI risks and the likelihood of their occurrence.

As the environment changes, covered entities must review and modify their security measures to continue to protect EPHI.

The HIPAA Act establishes some common security concepts for healthcare-related industries, clarifies public policy, and establishes codes of practice. Its practical significance lies in truly understanding the importance of information security in the medical industry and regulating it in the form of laws and regulations. The HIPAA Act marks a certain peak in the development of the United States in related aspects, such as medical information system security. China's domestic medical information industry has also started to explore these aspects.

The Health Information Technology for Economic and Clinical Health (HITECH) Act was signed into law by then US President Obama on February 17, 2009 as part of the American Recovery and Reinvestment Act (ARRA). Compared to HIPAA, HITECH strengthens breach notification and minimum dataset requirements, increases penalties, and further promotes the use of electronic health records.

The provisions of HIPAA and HITECH, in addition to the HIPAA Privacy Rule and the HIPAA Security Rule, include breach notification provisions that require notification to individuals and governments in the event of a breach of PHI.

Previously, the penalty was estimated at $100 per violation, capped at $25,000 per year for multiple violations of the same requirement or prohibition. Now, HITECH has adjusted the penalties for violations to range from $100 to $50,000 per violation. There is a cap of $1.5 million per year for multiple violations of the same requirement or prohibition. Individuals held criminally liable for violations are subject to fines of up to $250,000 and imprisonment of up to 10 years. HITECH requires enforcement of breach notification, and service providers will pay for the breach of data related to PHI and will be exposed to the public in a negative light. HITECH states that breach notification applies to "unsecured PHI", which is information that is not rendered readable, usable, or identifiable to unauthorized individuals. In practice, this means encrypting or destroying the data.

HITECH has breach notification requirements similar to the EU GDPR. If 500 or more patients were affected, the law requires healthcare providers to notify individuals affected by the breach and HHS. If the breach affects 500 or more individuals in a particular location, the breached party must also notify the primary media.

Notably, HITECH does not require encryption (it makes encryption an "optional" rather than a "required" control). However, as with state law, breach notification generally need not be enforced if encryption is used. As HHS notes in its guidance, the use of encryption does not change the security requirements for protecting PHI.

Encryption is only one step in ensuring an overall security plan that includes privacy, identity and access management policies enabled and supported by appropriate tools. Today's healthcare information environment, which involves many different institutions and many people playing many different roles and communicating in many different ways, does not lend itself to simple security solutions.

With the development of science and technology, the scope of personal information continues to expand, and genetic information is a brand new sensitive personal information. In response to the resulting invasion of personal privacy, GINA (Hudson *et al.*, 2008) was enacted in 2008. GINA is mainly used to protect individuals from discrimination based on genes and prohibits insurance companies from formulating differentiated insurance plans based on genetic predisposition. GINA also prohibits organizations from discriminating in employment on the basis of genetics. If an organization needs to process genetic information for legal reasons, that information should be recorded and stored separately from sensitive medical information.

The Health Information Trust Alliance (HITRUST) (HITRUST Alliance, 2021) is composed of representatives from the healthcare and security industries. The organization creates and maintains the Common Security Framework (CSF), a certifiable framework that helps healthcare organizations and their vendors demonstrate their security and compliance in a consistent and streamlined manner.

The CSF is based on two U.S. healthcare laws, HIPAA and HITECH, which establish and enforce requirements for the use, disclosure, and protection of individually identifiable health information. HITRUST provides a baseline, standardized compliance framework, assessment, and certification process for providers and covered health care entities to measure compliance. The CSF also incorporates specific healthcare security, privacy, and other regulatory requirements from existing frameworks, such as the Payment Card Industry Data Security Standard (PCI-DSS), the ISO/IEC 27001 Information Security Management Standard, and the Minimum Acceptable Risk Standard (MARS-E).

The CSF is divided into 19 different domains, including endpoint protection, mobile device security, and access control. HITRUST certifies IT products against these controls. HITRUST can also tailor certification requirements based on factors such as organizational, system, and regulatory factors, as well as risk to the organization.

HITRUST offers three levels of certification or assessment: Self-Assessment, CSF Review, and CSF Certification. The levels are strictly sequential. Organizations with the highest level of CSF Certification have met all CSF Certification requirements.

5.5.3 *Children's data supervision requirements*

Many countries around the world have not yet enacted specific laws to protect children's information. The United States passed the Children's Online Privacy Protection Act (Aftab and Savitt, 1999) in 1998 to protect the privacy of children under the age of 13 by regulating the online collection, use, and disclosure of personal information from children by websites and online services. The law details what website operators must include in their privacy policies, when and how to obtain verifiable parental consent, and rules governing operators' responsibilities for protecting children's privacy and safety online. Among these, obtaining "verifiable parental consent" is at the heart of the law. In particular, the requirements, implementation, and exceptions to the application of consent are clearly stipulated in the form of compliance guidelines and other forms of cooperation with the law.

In December 2012, the U.S. Federal Trade Commission issued revisions, effective July 1, 2013, that created additional parental notification and consent requirements for organizations as follows:

(1) Publish a clear and comprehensive online privacy policy describing its information practices for online collection of personal information from children under the age of 13.
(2) Make reasonable efforts, based on available technology, to notify parents directly that the operator is collecting, using, or disclosing personal information from 13-year-olds upon notification of any material changes in the operator's personal information practices regarding children, including prior parental consent.
(3) Obtain verifiable parental consent before collecting, using, and/or disclosing the personal information of children under the age of 13, with limited exceptions.
(4) Provide parents with a reasonable means to review the personal information that the Operator has collected from their children

and to object to the Operator's further use or maintenance of such information.

(5) Establish and maintain reasonable procedures to protect the confidentiality, security and integrity of personal information collected from children under the age of 13, including taking reasonable steps to disclose such information only to parties capable of maintaining the confidentiality and security level of the personal information.
(6) Retain children's personal information collected online only as long as necessary to fulfill the purpose for which it was collected and take reasonable steps to delete such information to prevent unauthorized access or use.
(7) Operators are prohibited from requiring children to provide more personal information than is necessary to participate in an online activity as a condition of participating in the activity.

Verifiable parental consent must be obtained before collecting, using, and disclosing personal information from children. COPPA leaves this issue up to companies, but it is important to properly choose a method, through clear and usable technical design, to ensure that it is the child's parent who is giving consent, not the child.

Parents must choose to allow the collection and use of their children's personal information without bundling consent to disclose that information to third parties. If changes are made to the collection, use, or disclosure practices to which parents have consented, new notices and parental consent must be obtained.

5.5.4 *Manufacturing security supervision requirements*

Security regulatory requirements for the manufacturing industry are complex. The following discussion focuses on two areas: the pharmaceutical industry and the automotive industry.

Using US law as an example, Title 21 of the Code of Federal Regulations (CFR) is reserved for regulations by the Food and Drug Administration (FDA). The purpose of CFR Title 21 Part 11 is to establish the basic rules for technical information systems used by organizations subject to FDA inspection.

Title 21 Part 11 ensures that electronic records and signatures are trustworthy, reliable, and generally equivalent alternatives to paper records and handwritten signatures. It also provides guidance for improving the security of computer systems in FDA-regulated industries. Organizations must demonstrate that their processes and products work as designed, and this demonstration must be revalidated as those processes and products change. The best practice guidelines address the following:

(1) Standard operating procedures and controls that support electronic records and signatures, such as data backup, security, and computer system validation.
(2) Features that ensure the computer system is secure, provides audit trails of data values, and ensures the integrity of electronic signatures.
(3) Validation and documentation that provide evidence that the system does what it is supposed to do and that users can identify when the system isn't working as intended.

To help secure the growing connectivity in the automotive industry, the German Automobile Industry Association (VDA) has developed a catalog of standards for evaluating information security. The VDA Information Security Assessment (German and English) for the automotive industry is based on the international standards ISO/IEC 27001 and ISO/IEC 27002. In 2017, the VDA updated the content to include controls for the use of cloud services.

VDA member companies use this tool for internal security assessments, as well as for suppliers who process sensitive information on their behalf, to assess suppliers, service providers and other partners. However, because these assessments are performed individually by each company, it places a burden on partners and creates duplication of effort for VDA members.

To facilitate assessments, the VDAs have developed a common assessment and exchange standard: the Trusted Information Security Assessment Exchange Mechanism (TISAX). TISAX (Goia, 2019) is an information security standard for the automotive industry based on the ISO 27001 information security management system standard and the VDA-ISA information security assessment checklist. TISAX provides a model for the mutual recognition of the results of information security assessments for different service providers in the automotive industry, and the acceptance of the assessment by the suppliers means that the results are accepted by all participants. The VDA has commissioned a neutral third party, the ENX Association, to implement TISAX. The ENX Association acts as a neutral third party for the qualification of audit providers (auditors), the maintenance of qualification standards and assessment requirements, and the monitoring of the quality of the implementation and the assessment results.

5.5.5 *Media industry safety supervision requirements*

Security regulatory requirements for the media industry include CDSA, MPAA, DPP (UK), FACT (UK). Unlike the mandatory requirements of other industries based on laws and regulations, many of these industry standards come in the form of voluntary compliance and certification.

The Content Delivery & Security Association (CDSA) is a coalition of media industry security executives focused on the secure delivery and storage of entertainment, software, and information content; defending against threats such as piracy and hacking; and identifying and addressing global content compliance risks. In 2018, CDSA and the Motion Picture Association (MPA) launched the Trusted Partner Network (TPN) to help the industry improve content security and enable content owners to measure compliance based on MPA's content protection best practices.

The CDSA's Content Protection and Security (CPS) standard provides guidelines and requirements for protecting media assets in a Content Security Management System (CSMS). The standard establishes a set of controls to ensure the integrity of intellectual property rights and the confidentiality and security of media assets at every step of the digital media supply chain.

The CPS standard has corresponding CPS certifications and includes more than 300 different controls to help secure and manage physical data centers, service hardening, and secure storage facilities. All controls are optimized to handle sensitive and valuable media assets. After the system is validated by a CDSA auditor, the CDSA issues a certificate of compliance to the certified organization.

Since the launch of TPN, both the MPA and the CDSA have discontinued their respective security assessment programs to focus on managing and developing the TPN program and the TPN annual assessment.

The Digital Production Partnership (DPP) and the North American Broadcasters Association had published a joint document, "Broadcaster Cyber Security Requirements for Suppliers", (Michael McEwen, 2016) to assist manufacturers and suppliers in developing products that meet modern cyber security standards and are suitable for integration into broadcast facilities. Working with broadcasters and vendor security experts, the DPP created a self-assessment checklist to help vendors demonstrate to broadcasters their use of cybersecurity best practices. This work evolved into The DPP Committed to Security Program (DPP, 2021), which offers certification and licensing to DPP members and broadcast and content production partners.

Following the launch of the TPN, DPP's media industry security standards were gradually aligned with the best practices published by the TPN. In August 2023, the Trusted Partner Network (TPN) released the MPA Content Security Best Practices Version 5.2. The updates include key requirements for software application vendors to provide hardening guidance, such as application security measures and best practices to prevent cyberattacks.

5.5.6 *Energy industry safety supervision requirements*

The energy industry is part of the country's infrastructure and is also a key target for various attackers. Its security needs are very high.

In May 2021, Colonial Pipeline, the largest fuel pipeline operator in the United States, was hacked, paralyzing nearly half of the fuel supply on the East Coast of the United States. In response, the U.S. Department of Homeland Security (DHS) moved to regulate cybersecurity in the pipeline industry (Manfredi, 2021). The Department of Homeland Security's Transportation Security Administration (TSA) will issue a security directive requiring fuel pipeline companies to report cyber incidents to federal authorities and will issue a binding set of rules, a senior DHS official said. The rules will dictate how pipeline companies must protect their systems from cyberattacks and what they must do if their systems are hacked.

Unlike the U.S. fuel industry, which has complex regulatory requirements and responsibilities for cybersecurity, U.S. electric utilities are explicitly regulated by NERC. The North American Electric Reliability Corporation (NERC) (Nevius, 2020) is a not-for-profit regulatory agency whose mission is to ensure the reliability of the North American bulk power system. NERC is overseen by the U.S. Federal Energy Regulatory Commission (Commission, 1984) and Canadian government agencies. In 2006, FERC granted NERC designation as an Electrical Reliability Organization (ERO) under the Energy Policy Act of 2005 (US Public Law 109 58). NERC develops and enforces a reliability standard known as the NERC Critical Infrastructure Protection (CIP) Standard.

All owners, operators, and users of the bulk power system must comply with the NERC CIP standards. These entities must register with NERC. Cloud service providers and third-party vendors are not subject to the NERC CIP Standards; however, the CIP Standards include objectives that registered entities should consider when using vendors in the operation of a Bulk Electric System (BES).

There is a relationship and mapping between the requirements of the NERC CIP Standards and the security controls of NIST SP 800–53, which together form the basis of the US Federal Risk and Authorization Management Program (FedRAMP). FedRAMP is a U.S. government program that provides a standard approach to the security assessment, authorization, and ongoing monitoring of cloud products and services.

Microsoft Corporation has released the white paper "NERC CIP Standards and Cloud Computing" (Stevan, 2021), which discusses NERC CIP compliance based on established third-party audits applicable to cloud service providers such as FedRAMP. The white paper covers topics such as vetting cloud providers and answers frequently asked questions about logical isolation and multi-tenancy, as well as security considerations for on-premises and cloud deployments.

5.5.7 *Education industry security supervision requirements*

For security oversight laws and regulations of the education industry, especially from a privacy perspective, it is worth mentioning the United States "Family Educational Rights and Privacy Act", referred to as FERPA.

FERPA is a United States federal law designed to protect the privacy of students' educational records, including personally identifiable and confidential information. FERPA was enacted to ensure that parents and students 18 years of age or older can access these educational records, request changes to them, and control the disclosure of information, but in special and limited circumstances, FERPA permits disclosure without consent.

The law applies to schools, school districts, and any other institution that receives funding from the U.S. Department of Education, which includes nearly all public K12 schools and school districts and most public and private postsecondary institutions.

Security is key to FERPA compliance. This law requires that student information be protected from unauthorized disclosure. Using cloud computing, educational institutions need technology vendors to properly manage sensitive student data and ensure this through contracts.

FERPA does not require auditing or certification, so each academic institution subject to FERPA must evaluate for itself whether and how its use of cloud services affects its ability to comply with FERPA requirements.

5.6 Evolution of Data Security and Privacy Protection Supervision in China

China's data security and privacy regulations are designed with the Cybersecurity Law of the People's Republic of China (Qi *et al.*, 2018), the Data Security Law of the People's Republic of China, and the Personal Information Protection Law of the People's Republic of China (Calzada, 2022) as the top-level design.

On June 10, 2021, the 29th Session of the Standing Committee of the 13th National People's Congress enacted the Data Security Law (hereinafter referred to as the "Data Security Law").

On August 20, 2021, the 30th Session of the Standing Committee of the 13th National People's Congress approved the Personal Information Protection Law, which will take effect on November 1, 2021. This is a new milestone in the legal protection of privacy in China. The Personal Information Protection Law of the People's Republic of China is formulated in accordance with the Constitution to protect the rights and interests of personal information, regulate personal information processing activities, and promote the reasonable use of personal information.

Previously, the enactment and implementation of the Cybersecurity Law laid a solid foundation for safeguarding national security and public social interests, protecting the legitimate rights and interests of citizens, legal entities, and other organizations in cyberspace, and ensuring the security of personal information and important data.

This section provides an overview of the key contents of laws such as the Cybersecurity Law, the Data Security Law, and the Personal Information Protection Law, interprets their key contents, and provides recommendations for an organization's legal compliance.

5.6.1 *Legislative trends of data security and privacy protection*

Today's world has entered the digital economy era. The production factors in the era of agricultural economy are land, the production factors in the era of industrial economy are machines and electricity, and the production factors in the era of digital economy are data. As a new type of production factor, data has completely different characteristics from traditional production factors. The convenience of data sharing, the complexity of data use, and the difficulty of data management and control all pose serious challenges to data security.

At the public level, there is also growing concern about how personal data is collected and used, and whether Internet platforms violate personal privacy and interests. The public is actively involved and has high expectations for the protection of personal interests by legislative means.

At the same time, with reference to international legislative practice, legislation on data security and privacy protection must strike a balance in multiple dimensions, such as protecting personal privacy, maintaining national security, promoting economic growth and improving social welfare.

In China's legislative process, the earliest and most fundamental top-level law was the National Security Law promulgated in 2015. In Article 25, it is clearly states that "The State constructs a network and information security protection system, ... it realizes the security and controllability of core network and information technologies, crucial infrastructure and information systems and data in important areas". Subsequently, the Cybersecurity Law, which was implemented on June 1, 2017, is the basic law in the field of network security. The Cybersecurity Law includes the definition of network data in Article 76: "refers to all kinds of electronic data collected, stored, transmitted, processed, and produced through networks". and points out that cybersecurity "refers to taking the necessary measures to prevent cyber attacks, intrusions, interference, destruction, and unlawful use, as well as unexpected accidents, to place networks in a state of stable and reliable operation, as well as ensuring the capacity for network data to be complete, confidential, and usable".

The Data Security Law is a special law in the field of data security, which has its own focus from the Cybersecurity Law, with the legislative goals of "data controllability" and "marco data security", and provides legal guarantees for cyberspace security and data security.

The scope of "data" in the Data Security Law is extremely broad, including "... any record of information in electronic or any other form" (Article 3). In terms of scope, it includes network data as defined in the Cybersecurity Law. Article 3 of the Law also provides a clear definition of "data security": "... ensuring that data is effectively protected and lawfully used through adopting necessary measures, and to possessing the capacity to guarantee the continuous security of data". In terms of legislative purpose, the most central element of the Data Security Law is to adhere to the overall national security concept and ensure national security. Article 4 of the Data Security Law: "In preserving data security, the holistic approach to national security shall be adopted, sound data security governance systems shall be established, and data security and protection capabilities shall be improved".

Unlike the Cybersecurity Law and the Data Security Law, which mainly focus on national security, the Personal Information Protection Law (PIPL) is a special law formulated for the protection of personal information. Personal information is defined in Article 4 of the PIPL: "...all kinds of information, recorded by electronic or other means, related to identified or identifiable natural persons, not including information after anonymization handling". The PIPL also defines the handling of personal information: "includes personal information collection, storage, use, processing, transmission, provision, disclosure, deletion, etc.".

The PIPL clarifies that the processing of personal information: "shall have a clear and reasonable purpose, and shall be directly related to the handling purpose, using a method with the smallest influence on individual rights and interests. The collection of personal information shall be limited to the smallest scope for realizing the handling purpose, and excessive personal information collection is prohibited". The processing of personal information shall comply with the principles of openness, transparency and integrity. As for the scope of application, based on the legislative practice of some countries and regions, such as Europe and the United States, the law establishes the principle of territorial jurisdiction plus extraterritorial application, so as to better protect the rights and interests of natural persons within the territory of China.

In summary, to strengthen the legal protection of data security and personal information, maintain a good ecology of cyberspace, and promote the healthy development of the digital economy, China is rapidly advancing various laws. Enterprises need to gain insight into legal and regulatory requirements, and build appropriate data security management systems and personal information protection systems.

5.6.2 *Interpretation of cybersecurity law*

In the information age, the network has been deeply integrated into all aspects of economic and social life, and network security threats have also penetrated all levels of economic society, and the importance of network security has continued to grow. Network security has become an important issue related to national security and development and the vital interests of the people. Under such circumstances, the enactment of the "Cybersecurity Law" is an objective need to maintain national cyberspace security and an important measure to implement the overall national security concept.

The Cybersecurity Law is China's first basic law to comprehensively regulate the security and governance of cyberspace, providing an important guarantee for the healthy operation of the Internet on the track of the rule of law.

In its basic principles, the Law puts forward the principle of cyberspace sovereignty and the principle of giving equal importance to network security and informatization development.

The first and most important principle is the principle of cyberspace sovereignty. Article 1 of the Cybersecurity Law stipulates: "in order to: ensure cybersecurity; safeguard cyberspace sovereignty and national security, and social and public interests ...", it is clearly stated that the purpose of the legislation is to safeguard national sovereignty in cyberspace. Article 2 of the Cybersecurity Law stipulates that the law applies to the supervision and management of networks and network security within China, and proposes network jurisdiction based on national sovereignty in cyberspace.

The second principle is the principle of giving equal importance to network security and the development of informatization. Article 3 of the Cybersecurity Law stipulates: "...The State advances the construction of network infrastructure and interconnectivity, encourages the innovation and application of network technology, supports the cultivation of qualified cybersecurity personnel, establishes a complete system to safeguard cybersecurity, and raises capacity to protect cybersecurity".

Based on the above basic principles, Article 4 of the Cybersecurity Law proposes to formulate a cybersecurity strategy, clarify the goals of cyberspace governance, and improve the transparency of cybersecurity policies. Article 7 of the Cybersecurity Law stipulates: "...promotes constructing a peaceful, secure, open, and cooperative cyberspace, and establishing a multilateral, democratic, and transparent Internet governance system". This is the cyberspace governance goal declared in legal form. It clearly expresses China's demands for cyberspace governance.

The Cybersecurity Law clarifies the responsibilities and powers of various government departments and further improves the network security supervision system. In particular, it clarifies the division of responsibilities between the Cybersecurity and Informatization Department and other relevant network supervision departments. Article 8 of the Cybersecurity Law stipulates: "State cybersecurity and informatization departments are responsible for comprehensively planning and coordinating cybersecurity efforts and related supervision and management efforts. The State Council departments for telecommunications, public security, and other relevant organs, are responsible for cybersecurity protection, supervision, and management efforts within the scope of their responsibilities, in accordance with the provisions of this Law and relevant laws and administrative regulations".

Regarding the requirements for network service providers, the Cybersecurity Law regulates the relevant duties and responsibilities of network service providers through the following three systems: the classified network security protection system, the user information protection system, and the critical information infrastructure key protection system.

The Cybersecurity Law stipulates the responsibilities and obligations of relevant parties such as national functional departments, industry authorities, and operating companies in terms of the security protection of critical information infrastructure from the three levels of the country, industry, and operators. The Cybersecurity Law emphasizes that on the basis of the network security level protection system, key information protection should be implemented for key information infrastructure, and it is clarified that the operators of key information infrastructure should have more security protection obligations, accompanied by national security review, important Legal measures such as mandatory local storage of data ensure the operational security of critical information infrastructure.

Chapter V of the Cybersecurity Law also defines the network security monitoring, early warning and information notification system, institutionalizing and legalizing monitoring, early warning and emergency response work. This chapter requires network operators and other entities to establish cybersecurity risk assessment and emergency response mechanisms, formulate contingency plans for cybersecurity incidents, and conduct regular drills.

Finally, the Cybersecurity Law clarifies the cybersecurity duties and responsibilities of network operators and other entities, and increases the punishment for violations and crimes. The Cybersecurity Law's chapters on "Network Operation Security", "Network Information Security", and "Monitoring, Early Warning, and Emergency Response" contain relevant requirements, and the chapter on "Legal Responsibility" raises the penalty standard for illegal acts and increases the punishment.

At the level of personal information protection, the "Cybersecurity Law" defines the principles of "lawful, legitimate and necessary", the principle of clear purpose, and the principle of informed consent for the collection and use of personal information by network operators, requires a protection mechanism for personal information, and protects individual rights of deletion and rectification. These principles, mechanisms and protection requirements have been inherited and developed in the subsequent Personal Information Protection Law.

5.6.3 *Interpretation of data security law*

On June 10, 2021, the "Data Security Law of the People's Republic of China" was adopted by the 29th Standing Committee Meeting of the 13th National People's Congress of the People's Republic of China. This is the basic law in the field of data in our country, and it is also an important law in the field of national security.

The full text of the Data Security Law consists of seven chapters and fifty-three articles, and the contents of the seven chapters are General Provisions, Data Security and Development, Data Security System, Data Security Protection Obligations, Security and Openness of Government Data, Legal Liability, and Supplementary Provisions.

The Data Security Law defines data as any record of information, whether in electronic or non-electronic form. Data security refers to the ability to ensure that data is effectively protected and legally utilized by taking necessary measures, and remains in a safe state. The State insists on maintaining data security and promoting data development and utilization, promoting data security through data development and utilization and industrial development, and using data security to ensure data development and utilization and industrial development.

In terms of legislative objectives, the Data Security Law focuses on four dimensions: national sovereignty and data security protection, protection of the legitimate rights and interests of citizens and organizations, healthy development of the digital economy, and e-government and government data openness.

From the system level, the Data Security Law defines the following four systems:

(1) **Data transaction management system (Article 19):** Standardize data transaction behavior and cultivate a data transaction market.

(2) **Data categorized and classified system (Article 21):** Establish a categorized and classified system and carry out data protection according to the importance of the data and the degree of harm caused by the leakage, and propose the concept of "important data".
(3) **Data security review system (Article 24):** Conduct national security reviews of data activities that affect or may affect national security.
(4) **Data security management system (Articles 27 and 39):** Articles 27 and 39 are requirements for legal entities and government agencies that carry out data processing activities, respectively. In addition, for the processors of important data, there are additional requirements: "Processors of important data shall be clear about their persons responsible for data security and the data security management bodies, and fulfill the responsibilities for data security".

The requirements of the data security law are shown in Fig. 5.10.

In addition, the Data Security Law also defines the following three mechanisms:

(1) **Data security collaboration mechanism (Article 9):** Promotes the joint participation by relevant departments, industry organizations, research institutions, enterprises, and individuals in data security protection, so as to create a good

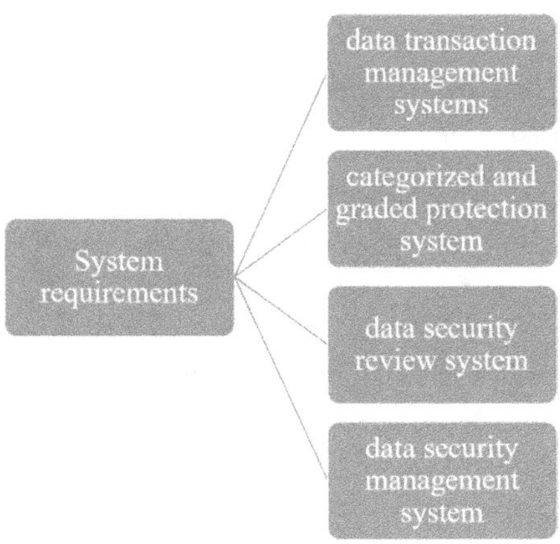

Fig. 5.10. Data security law system requirements.

environment for members of the whole society to jointly protect data, ensure data security and promote the development of relevant industries.

(2) **Data security risk assessment, reporting, information sharing, monitoring and early warning mechanism (Article 22):** Strengthen the acquisition, analysis, research and judgment, and early warning of data security risk information.

(3) **Data security emergency response mechanism (Article 23):** When a data security incident occurs, the relevant competent authorities shall activate emergency plans in accordance with the law, adopt corresponding emergency response measures, prevent the spread of damage, eliminate potential security risks, and promptly release warning information to the public.

The detailed requirements for the systems and mechanisms stipulated above are reflected in other provisions of the Data Security Law. The support of system-related platforms and technical capabilities has also been further clarified and refined through some clauses.

The acceleration of the legislative process of the "Data Security Law" is partly related to the deepening competition and rivalry over data in countries around the world. The security and development of the digital economy also depend on laws that provide institutional guarantees for data security and the protection of personal privacy and personal information.

5.6.4 *Interpretation of personal information protection law*

On August 20, 2021, the 30th Session of the Standing Committee of the 13th National People's Congress adopted the "Personal Information Protection Law of the People's Republic of China" (hereinafter referred to as the "PIPL").

The PIPL consists of 8 chapters and 74 articles, which inherits, draws and absorbs the experience of domestic and foreign legislative practice and international national standards on personal information protection. For example, in terms of the lifecycle of personal information, it relies on the definitions and requirements of the Personal Information Security Specification for Information Security

Technology (GB/T 35273–2020), absorbs the relevant contents of the Cybersecurity Law and the E-Commerce Law in terms of the rights of personal information subjects, and inherits the relevant contents of the Cybersecurity Law and the Data Security Law in terms of the compliance obligations of personal information processors.

Table 5.14. Compliance requirements of personal information lifecycle.

Lifecycle	Compliance requirements
Collection, usage	Informed consent + statutory exceptions (contractual imperatives, statutory duties, public health, news reporting, etc.); the notification is mandatory (including the processing of personal information by state authorities), and must meet the form and content compliance; sensitive personal information must be communicated to the necessity and impact on the individual; general consent must meet the requirements of voluntary consent and expression of will; children's personal information and sensitive personal information must meet special consent rules.
Storage	Minimized storage period; personal data must be stored domestically.
Shared control	Clarify internal responsibilities; joint and several liability.
Delegated processing	Clarify the responsibilities of each party; supervision and management; limited to the specified purpose.
External provision of data	Informed consent (includes scenarios where state organs provide personal data externally); limited to the specified purpose.
Automated decision-making	Transparency; fairness of results; autonomous participation.
Video acquisition and recognition	Only for public safety purposes; distinctive signs and separate consent.
Disclosure	Separate consent to disclosure; The purpose of processing disclosed personal information shall be limited; significant impact notification + consent.

The PIPL details the general principles of personal information protection from the perspective of the lifecycle of personal information. At the same time, special provisions have also been made for sensitive personal information and the processing of personal information by government agencies, the specific contents of which are shown in Table 5.14.

The Personal Information Protection Law does not define the concept of "joint controllers" of personal information as the GDPR does but rather requires joint and several liability for processors of "jointly processed" personal information in Article 20. In addition, the provisions of the Personal Information Protection Law, such as the restrictions on the use of personal image collection and personal identity recognition equipment, and the restrictions on the acquisition and use of disclosed personal information, will have a significant impact on the relevant industries.

The Personal Information Protection Law defines the main rights of personal information subjects, as shown in Table 5.15.

In addition, the "Personal Information Protection Law" also strengthens the provisions related to the quality of personal information, clarifying that "The handling of personal information shall ensure the quality of personal information, and avoid adverse effects on individual rights and interests from inaccurate or incomplete personal information" (Article 8).

For personal information processors, they need to undertake the following compliance obligations (see Table 5.16).

Compared with the previous draft, the Personal Information Protection Law has the following major changes:

The first is the requirement for "withdrawal of consent". Article 15 stipulates that for personal information processing activities based on personal consent, the requirement to provide a convenient way to withdraw consent is added. This clause effectively addresses the problems found in law enforcement practice that the withdrawal of consent is too hidden and the operation is too complicated.

The second is to add new normative requirements for organizations with the function of managing public affairs. A new Article 37 stipulates that "The provisions of this Law regarding personal information handling by State organs apply to the handling of personal information in order to fulfill statutory duties by organizations authorized by laws and regulations to manage public affairs

Table 5.15. The rights of the subject of personal information.

Rights	Compliance requirements	Related laws and regulations
Right to know, right to decide	Protect the rights of individual	Personal Information Protection Law
Right of inquiry	Allow users to query and copy their personal information. Except where otherwise restricted by laws and regulations.	Civil Code, E-Commerce Law, PIPL
Right of rectification, right of supplementation	Premise: Personal information is inaccurate or incomplete; Processing requirements: Verified and processed in a timely manner	Cybersecurity Law, Civil Code, E-Commerce Law, PIPL
Right to erasure	Scenarios where users request deletion; Reason: Expiration or fulfillment of purpose; Service terminated; withdraw consent;	Cybersecurity Law, Civil Code, E-Commerce Law, PIPL
Right to interpretation	Request an explanation of the rules for handling personal information	PIPL

functions", thus strengthening the requirements for those with the function of managing public affairs. Standard requirements for the processing of personal information by organizations. For example, the government's Centers for Disease Control and Prevention may process personal information in accordance with this provision.

The third is the cross-border provision of personal information, which requires stricter requirements. Amend the provisions of Article 41 to prohibitive provisions, that is, from "should apply for approval in accordance with the law" to "Without the approval of

Table 5.16. Requirements for personal information processors.

Compliance obligations	Compliance Requirements	Related laws and regulations
Management system and operating procedures	Organizations should establish management systems and operating procedures covering network security, data security, and personal information protection	Cybersecurity Law, Data Security Law
Data Protection Officer (DPO)	Organizations that process a certain amount of personal data must create this position. Foreign organizations should establish specialized agencies or designate representatives within the territory of China; The information of DPO shall be published and officially reported	Data Security Law, PIPL
Compliance audits	Regular audits; External audits	Article 54, 64 of PIPL
Risk assessment	Applicable situation: sensitive personal information; automated decision-making; Delegate processing, external provision and disclosure; cross-border transfers, etc.; The content of the report is limited and it shall be kept for not less than three years.	Data Security Law
Classification of personal information	According to national standards and industry practices, classify personal data by level.	Data Security Law

(Continued)

Table 5.16. (*Continued*)

Security technical measures	Encryption, backup, de-identification, , etc.	Cybersecurity Law, Cryptography Law
Authority and internal control mechanism	Combined with data classification to promote privacy compliance	PIPL
Education and training	Explain the rules for handling personal information	Cybersecurity Law, PIPL
Event handling	Remediation, reporting and notification (and scenarios for exemption notifications)	Cybersecurity Law, PIPL
Mechanisms for handling requests and complaints	Response should normally be within 30 days; Reasons for not processing should be given.	GB/T 35273 Personal information security specification

the competent authorities of the People's Republic of China, personal information handlers may not provide personal information stored within the mainland territory of the People's Republic of China to foreign judicial or law enforcement agencies".

The fourth is to add the provision that the personal information rights of the deceased should be exercised by close relatives. In legal practice, the subject of personal information is generally a natural person, and the PIPL adds a new Article 49, which stipulates that "When a natural person is deceased, their next of kin may, for the sake of their own lawful, legitimate interests, exercise the rights provided in this chapter to consult, copy, correct, delete, etc., the personal information of the deceased, except where the deceased has arranged otherwise before their death". This clause not only helps protect the rights and interests of the personal information of the deceased but also takes into account the interests of their close relatives.

The fifth is to add the personal information protection obligation requirements of super-large Internet platforms. This clause applies to "Personal information handlers providing important Internet platform services, which have a large number of users, and whose business models are complex". For super-large Internet platforms, Article 58 adds a new provision, requiring the following:

(1) Establish and complete personal information protection compliance systems in accordance with state provisions and establish an independent body composed primarily of external members to conduct oversight of the protection of personal information.
(2) Follow the principles of openness, fairness, and impartiality, draft platform rules, and clarify the norms for the handling of personal information by product or service providers on the platform and their obligations to protect personal information.
(3) Stop providing services to product or service providers on the platform that handle personal information in serious violation of laws or administrative regulations.
(4) Periodically publish reports on social responsibility for the protection of personal information, and accept social oversight.

In October 2021, the State Administration for Market Regulation (SAMR) issued the Guidelines for the Classification and Grading of Internet Platforms (Draft for Comments) and the Guidelines for the Implementation of Entity Responsibilities of Internet Platforms (Draft for Comments), which divides internet platforms into three categories: super platforms, large platforms, and small and medium-sized platforms, which can be used as a reference for enterprises when implementing their personal information protection obligations.

The sixth is the trustee's personal information protection obligations are newly added. Article 59 adds a new provision: "Entrusted persons accepting entrusted handling of personal information shall, according to the provisions of this Law and relevant laws and administrative regulations, take necessary measures to safeguard the security of the personal information they handle, and assist personal information handlers in fulfilling the obligations provided in this Law". The addition of this clause may be related to Article 21.

The seventh is to clarify the powers of the State Cybersecurity and Informatization Department, and to strengthen legislation and research on new technologies and new applications. Article 62 stipulates:

"The State cybersecurity and informatization department coordinates overall the following personal information protection work by the relevant departments:

(1) Formulate concrete personal information protection rules and standards.
(2) Formulate specialized personal information protection rules and standards for small-scale personal information handlers and new technologies and new applications for handling sensitive personal information, facial recognition, artificial intelligence, etc.
(3) Support the research, development, and broad adoption of secure and convenient electronic identity authentication technology, and promote the construction of public online identity authentication services.
(4) Advance the construction of service systems to socialize personal information protection, and support relevant organizations to launch personal information protection evaluation and certification services.
(5) Perfect personal information protection complaint and reporting work mechanisms".

The eighth is to clarify the principle of presumption of fault for personal information violations. Article 69 stipulates that if personal information rights and interests are infringed due to personal information processing activities, and personal information processors cannot prove that they are not at fault, they shall bear tort liability such as damages. This provision compensates for the high cost of protecting the rights of personal information subjects and the difficulty of providing evidence caused by the civil liability principle of "the party who claims shall bear the burden of proof".

5.6.5 *National standards for data security and personal information protection*

In the era of big data, cloud computing and the Internet of Everything, the application of data is becoming more and more extensive, and it also poses great risks to the protection of personal information and data security. National standards for personal information protection play a fundamental, normative and leading role. The implementation of national information security standards in the entire lifecycle of personal information processing is the focus of various organizations in the protection of personal information.

The National Information Security Standardization Technical Committee (SAC/TC260) is a technical working organization engaged in information security standardization work in the professional field of information security technology. The organization was established in April 2002 and is affiliated with the National Standardization Management Committee (hereinafter referred to as "SAC"). The main scope of work of SAC includes technical standardization work in the fields of security technology, security mechanism, security service, security management and security evaluation.

At present, there are six Working Groups (WG) and one Big Data Security Standard Special Working Group (SWG BDS) under the Security Standardization Committee, and the secretariat is set up in China Electronics Standardization Research Institute. The organization chart of SAC/TC260 is shown in Fig. 5.11.

In April 2016, SAC/TC260 established a big data security standardization working group, which is responsible for the development of security standards related to big data and cloud computing.

In the direction of personal information protection, SAC/TC260 focuses on the development of standards for personal information protection requirements and implementation mechanisms. As of the end of June 2020, SAC/TC260 has published two national standards, GB/T 35273 "Personal Information Security Specification" and

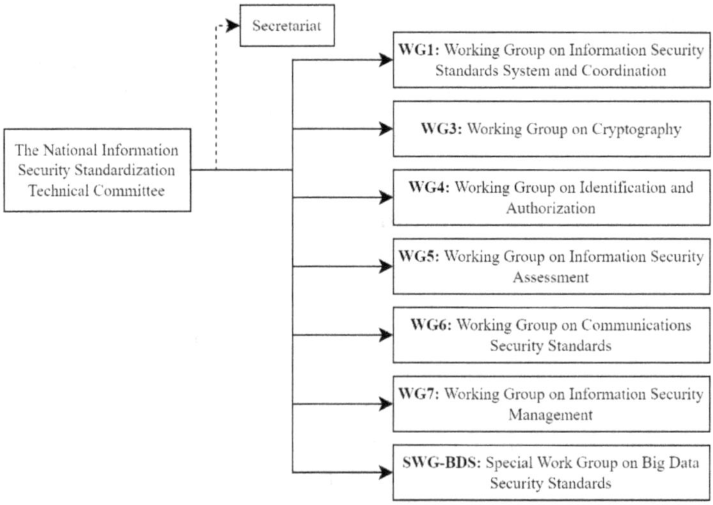

Fig. 5.11. SAC/TC260 organization chart.

Table 5.17. National standards for personal information protection.

No	Title	Standard Number	Contents
1	Personal information security specification	GB/T 35273–2020	Personal information security principles, security requirements for personal information processing activities
2	Guide for de-identifying personal information	GB/T 37964–2019	The management process, technical model, and method of personal information de-identification
3	Guidelines for personal information security engineering	GB/T 41817–2022	Personal information engineering practice in the development, design, testing, and other stages of system engineering
4	Implementation guidelines for notices and consent in personal information processing	GB/T 42574–2023	Notification of personal information processing, and obtaining the consent of the Personal Data Subject
5	Guidance for personal information security impact assessment	GB/T 39335–2020	Personal information impact assessment framework, process, and methodology

GB/T 37964 "Guidelines for the De-identification of Personal Information", five standards under research, and two standards research projects. Table 5.17 lists some of the national standards for personal information protection.

In the direction of data security, focusing on data security capability, data exit assessment, governmental data sharing, data transaction service, health data security, and telecommunication data security, SAC/TC260 has issued four standards: GB/T 35274 "Big Data Service Security Capability Requirements", GB/T 37932 "Data Transaction Service Security Requirements", GB/T 37973 "Big Data Security Management Guidelines", and GB/T 37988 "Data Security Capability Maturity Model". In addition, there are five standards under research and 16 standards research projects. Table 5.18 lists some important national standards for data security.

Table 5.18. National standards for data security.

No	Title	Standard number	Contents
1	Security capability requirements for big data services	GB/T 35274–2017	Security requirements for the big data service lifecycle
2	Security requirements for data transaction service	GB/T 37932–2019	Data transaction object security and transaction activity security, data trading platform security, etc.
3	Government information sharing - Data security technology requirements	GB/T 39477–2020	Technical requirements for data security for government information sharing
4	Big data security management guide	GB/T 37973–2019	Big data activities, roles, responsibilities, risk management, etc.
5	Guide for health data security	GB/T 39725–2020	Security management measures to protect health and medical information, technical measures
6	Data security guidelines for telecom field	GB/T 42447–2023	Construction of big data platform in the field of telecommunications, operational security guide
7	Data security capability maturity model	GB/T 37988–2019	Security controls for the data lifecycle, capability maturity assessment model

5.6.6 *Network security level protection*

Article 21 of the Cybersecurity Law of the People's Republic of China clearly stipulates that "The State implements a cybersecurity multi-level protection system [MLPS]". In order to implement the Cybersecurity Law and adapt to the multi-level protection of network security in new technologies and scenarios such as cloud computing, mobile Internet, Internet of Things, industrial control and big data, in May 2019, the State Administration for Market Regulation and the Standardization Administration of the People's Republic of China officially issued GB/T 22239–2019 "Information Security Technology — Baseline for classified protection of cybersecurity", which officially opened the era of "MLPS 2.0".

The "Terms and Definitions" clearly pointed out that cybersecurity is "The ability to keep the network in a state of stable and reliable operation, as well as to safeguard the integrity, confidentiality, and availability of network data, by taking the necessary measures to prevent attacks, intrusions, interference, destruction, and illegal use of the network, as well as accidents". In other words, the multi-level protection system requires that the object of multi-level protection has the ability to safeguard data security.

In terms of data security, in the general requirements for security, the Baseline for Classified Protection of Cybersecurity has different requirements for different levels of secure communication transmission in networks.

Figure 5.12 shows the multi-level protection requirements for secure transmission.

For the first level, only integrity verification is required. The original text is "6.1.2.1 Communication transmission should use verification technology to ensure the integrity of data during the communication process". Technical measures can adopt the traditional CRC checksum method. The requirements for Level 2 are the same as Level 1. The third level requires the use of cryptographic technology for integrity protection. For its corresponding security

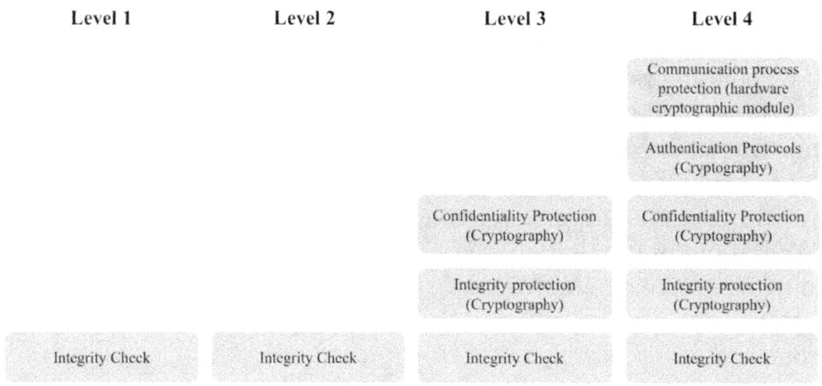

Fig. 5.12. Security transmission level protection requirements.

measures, please refer to the hash function described in Section 4.7 "Hash Algorithm". For scenarios with high security requirements, you can refer to the message authentication mechanism described in Section 4.8 "Message Authentication Code". The third level also requires the use of cryptographic technology to protect the confidentiality of data during the communication process. The original text is "8.1.2.2 This requirement for communication transmission includes: ...

(b) Cryptographic technology should be used to ensure the confidentiality of data during the communication process". For the security measures corresponding to this clause, please refer to the descriptions in Section 4.3 "Key-based Encryption" and Section 4.4 "Public Key Encryption" in this book. The fourth level adds additional requirements for the authentication of both communicating parties and the protection of the communication process on the basis of the third level. The original text is "9.1.2.2 Communication transmission. This requirement includes: ...

(c) Both parties to the communication should be verified or authenticated based on cryptographic technology before communication;

(d) Cryptographic operations and keys for important communication processes should be based on hardware cryptographic modules. manage". For corresponding security measures, in addition to the chapters mentioned above, you can also refer to the description in Section 4.6 "Public Key Infrastructure".

The Classified Protection Requirements also provide requirements for secure computing environments, especially features, such as data integrity, data confidentiality, and data backup and recovery. The corresponding requirements and security measures are similar to the provisions of the secure transmission level protection requirements and will not be described again.

In addition, the Classified Protection Requirements also provide corresponding requirements for data export and personal information protection in the Cloud Computing Security Expansion Requirements. Additional care is required when designing and implementing systems that meet the requirements for classified protection.

5.7 Chapter Summary

Security means that the system must use multiple methods to prevent malicious behavior. Security is primarily concerned with technical aspects such as confidentiality, availability, and integrity. Security also includes attack detection, resilience, and recovery.

From the perspective of purpose, the security protection is the organization's data, to ensure that the organization's secrets will not be leaked, so as to ensure the continuous operation of the organization. Through security protection, the organization can enable users to build trust in the organization.

Privacy refers to the need for users to protect their personal information, such as interactions with others, services, and devices, as well as their own behavioral information. Privacy primarily concerns the user side, which is protected by anonymity and limited processing of personal data. For organizations, privacy can help build user trust.

And trust itself is a broader concept that can encompass both security and privacy. The relationship between security, privacy, and trust is shown in Fig. 5.13. Trust is the belief that people, data, and devices will perform or behave as expected. The infrastructure and services of the next generation converged information and communications network must also be built on trust.

The goal of security and privacy is to securely control systems and data in the "social network-virtual network-physical network". Traditional security systems focus on how to authorize subjects to access data and how to deliver data to authorized subjects. Trust

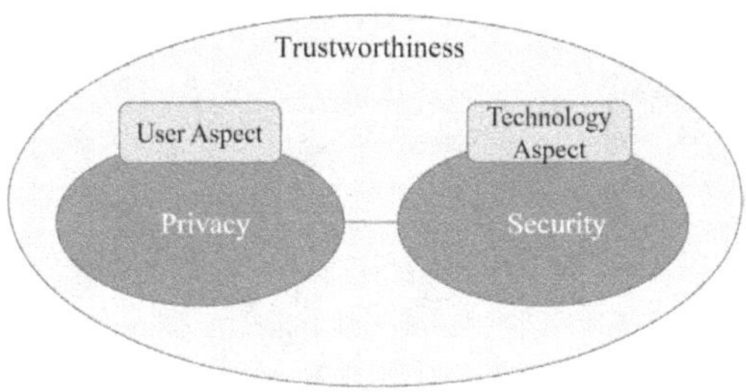

Fig. 5.13. Relationship between security, privacy and trust.

can further measure the reliability of security and privacy as parameters by measuring and observing the difference between objective or subjective expectations of reliable entities and data.

Whether it is security protection or privacy protection, we build trust among users through various channels so that users can use innovative products and services with confidence.

In a sense, the goal of security and privacy is trust and trust building.

Chapter 6

Emerging Technologies and Data Security

The Institute of Electrical and Electronics Engineers (IEEE) has indicated that the Fourth Industrial Revolution (4IR) is upon us, encompassing and integrating a range of new technologies in the physical, digital, and biological worlds that are certain to transform the way people work and live, and enrich the way people connect and communicate. Key areas of the Fourth Industrial Revolution (4IR) include robotics, artificial intelligence, nanotechnology, quantum computing, biotechnology, the Internet of Things, 3D printing, and autonomous vehicles.

This chapter focuses on data security in scenarios such as AI, blockchain, IoT, and 5G/6G mobile communications, providing insights and analyzing the evolutionary trends of data security systems.

6.1 AI and Data Security

With the rapid development of the information technology industry, new services, solutions, and products are constantly being created, and data processing and transmission are at a higher level, greatly improving the convenience and efficiency of people's production and life. Taking human–computer interaction technology as an example, speech recognition technology provides a new field of human–computer interaction and enriches users' information acquisition and experience. Other technologies, such as image recognition technology and natural language translation technology, are also making rapid progress and are gradually becoming popular and widely used in various industries. These technologies are supported by powerful artificial intelligence.

Gartner consulting company expert Costello K. pointed out in the article "Gartner predicts the future of artificial intelligence technology" (Costello, 2020) that from 2018 to 2019, organizations that have deployed artificial intelligence (AI) increased from 4% to 14%. In terms of business domains, conversational AI is a key consideration for organizations, given the huge success of Amazon's Alexa smart speaker and Google Assistant, but other emerging technologies are also being selected, such as augmented intelligence, edge AI, data labeling, and explainable AI. In terms of deployment methods, "AI platform as a service" and AI cloud services have received significant attention. Both Microsoft and Google have launched their own machine learning and AI public cloud service products, such as Microsoft's "Azure Machine Learning" platform (Chappell, 2015).

The rapid development and widespread application of these artificial intelligence technologies heralds the arrival of a "ubiquitous intelligent" society. Especially with the booming and rapid popularity of ChatGPT, generative AI has attracted more and more attention. Research suggests that by 2028, generative AI will affect about 23% of the world's jobs, such as loan underwriters, telemarketers, statistical assistants, cashiers, and journalists, which can be replaced and automated by AI (World Economic Forum, 2023). Goldman Sachs Research predicts that generative AI will drive annual global GDP growth of 7% over the next 10 years (Sachs, 2023). By 2030, generative AI will add $7 trillion to the global economy, with the Chinese market contributing $2 trillion (Chui, 2023).

However, we must also see that behind the rapid development and popularization of AI, there are also great risks. From the analysis of the essence and principle of AI, in the dimension of data security, AI uses a large number of datasets for training, and there is a risk of privacy protection. AI uses data in the real environment, especially personal data for analysis and prediction, and there are huge risks in terms of data confidentiality, integrity and availability. Even AI models themselves, if treated as a form of data, are at risk of confidentiality and integrity protection.

Fig. 6.1. A case where Tesla's autopilot system was attacked.

Ordinary users may be aware of data integrity risks, such as those caused by the introduction of malicious data into road signs in autonomous driving scenarios. According to an editorial selection article "Hackers Made Tesla Cars Autonomously Accelerate Up to 85 In A 35 Zone" (Winder, 2020) published on the Forbes website in February 2020, McAfee security researchers used machine learning model confrontation to make the machine learning image classification system make a wrong decision and recognize the sign with the speed limit of 35 as the speed limit of 85. Figure 6.1 shows a case where Tesla's Autopilot system was attacked.

This case vividly illustrates the dire consequences of attacks on AI systems. Therefore, it is very necessary for AI systems to take appropriate protective measures based on the lifecycle process of their system, model, and data, and based on the risk assessment and response mechanism described in Section 2.2 "Data Security Response Mechanism".

6.1.1 *Artificial intelligence, machine learning, and deep learning*

The concept of artificial intelligence was proposed in the 1950s and refers to programs that can simulate "intelligence". In computer science, the study of artificial intelligence is defined as the study of "intelligent agents": any device that can perceive its environment and take actions to maximize the achievement of its goals. A more detailed definition describes artificial intelligence as "the ability of a system to correctly interpret external data, learn from that data, and use those learnings to achieve specific goals and tasks through flexible adaptation". In practice, artificial intelligence can refer to algorithms that reason or learn given the necessary input and knowledge. Envisaged application scenarios include prediction (e.g., weather forecasting) and planning, identification, and autonomous decision-making. Machine learning (ML) is a specialized branch of AI that uses algorithms to understand models of phenomena from examples (statistical machine learning) or experience (reinforcement learning).

According to a common understanding, the relationship between artificial intelligence, machine learning, and deep learning is an inclusive relationship. Machine learning is a field of artificial intelligence, and deep learning is a common method in machine learning. The relationship between artificial intelligence, machine learning, and deep learning can be represented in Fig. 6.2.

Machine learning is a method of achieving artificial intelligence. It focuses on algorithms that automatically improve through experience and the use of data. That is, building a model from sample data and being able to make predictions or decisions without explicitly implementing business logic code.

In the early days of artificial intelligence as a specialized field of study, researchers were interested in machines learning from data and proposed several applications based on symbolic methods and "neural networks". However, the boom in machine learning began in the 1990s. The goal of machine learning is no longer to "realize artificial intelligence" but to solve practical problems. Since then,

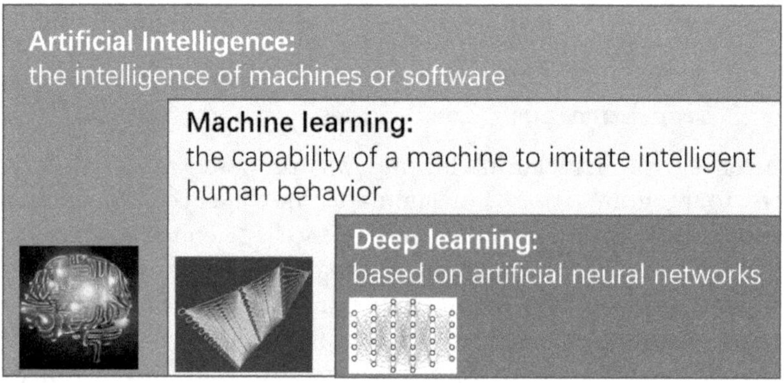

Fig. 6.2. Relationship among artificial intelligence, machine learning and deep learning.

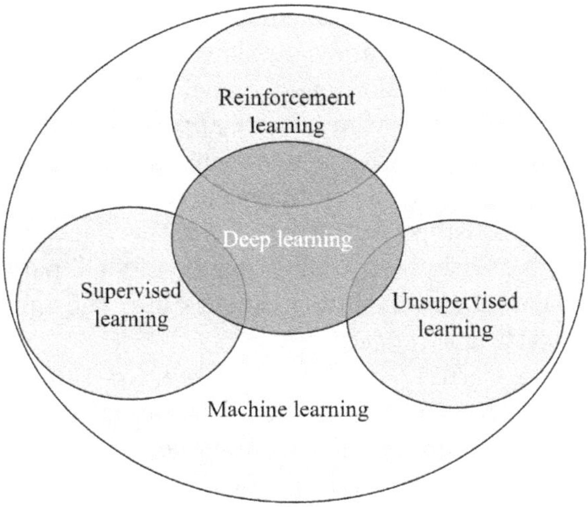

Fig. 6.3. Common categories of machine learning.

methods and models borrowed from statistics and probability theory have been proposed. Common classifications of machine learning are supervised learning, unsupervised learning, semi-supervised learning, deep learning, and reinforcement learning, as shown in Fig. 6.3.

Supervised learning is a machine learning method whose main feature is the use of labeled datasets. These datasets can be used to train, or "supervise" algorithms to produce accurate classification or prediction results. Using labeled inputs and outputs, users can measure the accuracy of a supervised learning model and the model learns over time. Supervised learning can be divided into two categories: classification and regression algorithms.

Classification refers to the precise assignment of data into specific categories, such as separating apples from oranges. In the real world, supervised learning algorithms can be used to classify spam into a separate folder. Linear classifiers, support vector machines, decision trees, and random forest algorithms are all common classification algorithms. In the abstract, if the type of output is limited or the value is a relatively fixed set of values, it is more appropriate to use a classification algorithm.

Regression is another type of supervised learning algorithm that uses an algorithm to understand the relationship between dependent and independent variables. Regression models can predict values based on different data points, such as predicting a company's sales revenue. Some popular regression algorithms include linear regression, logistic regression, and polynomial regression. In the abstract, a regression algorithm is more appropriate when the output value can be any number within a certain range.

Unsupervised learning uses unlabeled training data to describe and extract relationships or structure in the data, i.e., unsupervised learning algorithms try to discover hidden patterns in the data without human intervention. This is also the literal meaning of "unsupervised".

Unsupervised learning models are used for three main tasks: clustering, association, and dimensionality reduction.

Clustering is a data mining technique used to group unlabeled data based on their similarity or difference. For example, the K-means clustering algorithm assigns similar data points into groups, where the K value represents the size and granularity of the grouping. Clustering algorithms are used in scenarios such as market segmentation, customer segmentation, and image compression.

Association is another type of unsupervised learning method that uses various rules to find relationships between variables in a given dataset. Personalized recommendation engines often use this algorithm, such as the association recommendation of "Customers who bought this product also bought" on shopping websites.

Dimensionality reduction is a learning technique used when the number of features (or dimensions) in a given dataset is too large. It reduces the number of data entries to a manageable size while maintaining data integrity. Typically, dimensionality reduction techniques are used in the data preprocessing stage, such as removing noise from visual data to improve image quality.

Semi-supervised learning lies between unsupervised learning (without any labeled training data) and supervised learning (with fully labeled training data), and is a special case of weak supervision. Some training instances lack training labels, but many machine learning researchers have found that unlabeled data, when used with a small amount of labeled data, can yield significant gains in learning accuracy. When machine learning is applied to real-world scenarios, the high human or resource cost of labeling may make large, fully labeled training sets infeasible, while obtaining unlabeled data is relatively inexpensive. In such cases, semi-supervised learning can be of great practical value. Some studies suggest that for automatic analysis of medical images (such as CT scans), a small amount of training data labeled by professional radiologists can significantly improve the accuracy of disease assessment.

Deep learning is a subfield of machine learning, while artificial neural networks and representational learning are the backbone of deep learning algorithms. Artificial neural networks (ANNs) are inspired by the information processing mechanisms of biological systems but are quite different.

"Deep" in deep learning refers to the use of multiple layers in an artificial neural network to progressively extract higher-level features from the original input. In deep learning, each layer learns to transform its input data into a slightly more abstract, comprehensive representation. For example, in a face recognition scenario, the original input image can be represented as a two-dimensional matrix of pixels. The deep learning model contains several layers of convolutional neural networks. The first layer can abstract pixels and encode edges, the second layer can synthesize and encode the arrangement of edges, the third layer can encode noses and eyes, and the fourth layer can identify faces in the image. The four layers here are just a representation of the image description, and the actual deep learning algorithm for face recognition can be much more complex than this description.

Deep learning has been applied to areas such as computer vision, speech recognition, natural language processing, machine translation, drug design, medical image analysis, and board games. It has been shown to produce results that match, and in some cases exceed, the performance of human experts in certain scenarios. As a result, deep learning is becoming increasingly popular in a wide variety of scenarios.

Reinforcement learning is one of the three basic machine learning paradigms, along with supervised and unsupervised learning. Reinforcement learning differs from supervised learning in that in supervised learning the training data has the answer, so the model is trained to have the correct answer itself, whereas in reinforcement learning the training data does not have the answer, but the reinforcement agent decides how to perform the given task. Reinforcement learning is the study of decision making in complex environments with potentially different consequences over time.

Reinforcement learning systems can progressively optimize decisions based on external and potentially delayed feedback. Reinforcement learning is therefore particularly well suited to problems where long-term and short-term rewards must be weighed, including robot control, elevator scheduling, telecommunications scheduling, and various chess games (e.g., AlphaGo in Go).

6.1.2 *Artificial intelligence and data security*

Artificial intelligence is based on the analysis and understanding of data, and realizes intelligent and automatic decision making in the application domain. Modern artificial intelligence is usually deeply integrated with robotics and the Internet of Things and is applied to scenarios with ever-increasing data types and huge data volumes. Data is the most valuable asset in the entire artificial intelligence scene. Throughout the entire lifecycle of artificial intelligence, data is stored, transformed, transmitted, and processed in various forms.

The two fields of artificial intelligence and data security are closely related, and their intersection exists in at least the following dimensions:

(1) **AI data security:** Focus on the protection of various types of data involved in the entire lifecycle of artificial intelligence. For example, the confidentiality and integrity protection of the AI model needs to be considered at the deployment stage, and the security protection of the original data and training set needs to be considered at the training stage.
(2) **AI privacy protection:** Focus on the application of the principle of data collection minimization, the realization of AI privacy data access control, and the combination of various privacy enhancing technologies such as differential privacy and AI.
(3) **The security of AI for personal data processing:** Focus on the realization of personal rights of private data in artificial intelligence.

First, AI systems require data security. That is, the integrity, confidentiality and usability dimensions of artificial intelligence systems, as well as the protection of privacy and the prevention of misuse, require the support of security functions. The lack of such security features will lead to security risks of the artificial intelligence system itself.

Second, AI systems must comply with legal requirements for privacy protection. This is especially true when personal data is included in the raw and training data. The combination of privacy-enhancing technologies, such as anonymization and differential privacy, with artificial intelligence technology is still a hotspot of research and practice in academia and industry.

Finally, the use of AI to process personal data poses a number of security, privacy, and trustworthiness risks that need to be carefully assessed and addressed. For example, the right to erasure and the right to rectification required by the GDPR may be more complicated to address in AI data flow scenarios. In addition, when AI is used to process personal data, the fairness, ethics, and accountability of automated decisions involving individuals also need to be considered.

AI systems introduce complexity that traditional information technology systems do not have. At the compliance level, AI systems introduce borderline complexity for personal data controllers and processors. At the organizational level, there are far more roles and personnel involved in the design and deployment of artificial intelligence systems than in traditional information technology systems, such as data scientists, statisticians, and algorithm engineers. At the business level, there are intricate interactions between artificial intelligence systems and existing business flows, data flows, and workflows. At the technical level, how AI systems can securely and reliably process personal data is still in the exploratory stage, and there are no clear processing principles and best practices. At the implementation level, AI systems may also rely on third-party software or code, adding complexity to the supply chain.

Overall, the impact of artificial intelligence on data security depends on the following factors:

(1) How AI systems are built and deployed.
(2) The complexity of the organization deploying the AI system.
(3) The maturity of existing risk management capabilities.
(4) The nature, purpose, scope, and context of the personal data processed by the AI system.

When formulating and implementing security control measures, the above factors should also be fully analyzed and considered.

6.1.3 *AI data threat model*

The problem of data security in artificial intelligence, or machine learning, is a horizontal problem that spans all stages of the artificial intelligence lifecycle, so a systemic perspective is required.

The premise of building a data threat model for artificial intelligence is to systematically identify and define the "data" involved in artificial intelligence. The narrow data processed by the artificial intelligence system includes training data, raw data, test data, etc., and processed datasets, such as labeled datasets, also belong to this category. In addition, models and algorithms are the two most direct types of data involved in AI systems. Algorithms include preprocessing algorithms, training algorithms, prediction algorithms, and so on. The model includes model parameters, model tuning, etc. These data have corresponding data security protection requirements, especially based on the environment and context, there are different confidentiality, integrity and availability requirements. The environmental dependency of the artificial intelligence system, such as the artificial intelligence framework adopted, the software and platform relied on, and even the storage environment, can be defined as data in a broad sense.

From the lifecycle dimension, the steps for modeling artificial intelligence data threats are as follows:

The first step is to define the target system. Determine the security attribute requirements of the target system. The security attribute requirements of artificial intelligence for different targets may be completely different, such as artificial intelligence systems for autonomous driving and artificial intelligence systems for automatic classification of photo libraries. In this step, it is also necessary to sort out the overall architecture of the system, key components or subsystems, and their internal and external dependencies and interaction patterns.

The second step is data identification. Combined with the lifecycle of the target artificial intelligence system, key data at each stage should be identified. Generally speaking, the lifecycle of an artificial intelligence system includes several interdependent phases, such as the design and development phase (including requirements analysis, data collection and processing, training, testing, integration, and other subphases), installation phase, deployment phase, operation phase, maintenance phase, destruction phase, and so on. For example, preprocessing datasets and preprocessing algorithms are key data in the data collection and processing phase. It should be noted that due to the complexity of business, models and algorithms, the lifecycle stages of artificial intelligence systems in different scenarios will be different, and the key data they contain will also vary widely.

The third step is threat identification. Identify the sources of threats to critical data and their specific impact on the data. Threats can come from many sources and have varying effects. Broadly speaking, threat sources can include malicious or non-malicious people within the organization, external professional researchers, hackers or hobbyists, competitors or cybercriminals, etc.

The fourth step is to identify vulnerabilities. Determine the vulnerability of critical data based on known attack patterns. That is, which types of critical data are vulnerable to which threat sources, the consequences, and the level of risk involved. Attack modes can include misuse, destruction, theft, interception, physical attack, accidental damage, denial of service (DoS), and so on, and each attack mode includes different attack behaviors. For example, stealing behavior can include different attack behaviors such as stealing raw data and stealing model parameters. For vulnerability analysis and assessment, refer to Section 2.2.1 "Risk Assessment Method".

The lifecycle of the AI training phase is shown in Fig. 6.4, which is also a summary of the previous four main steps.

It is not difficult to see that the steps above are similar to those described in Section 2.2.7 "Data-Centric Threat Modeling" and are part of the same model.

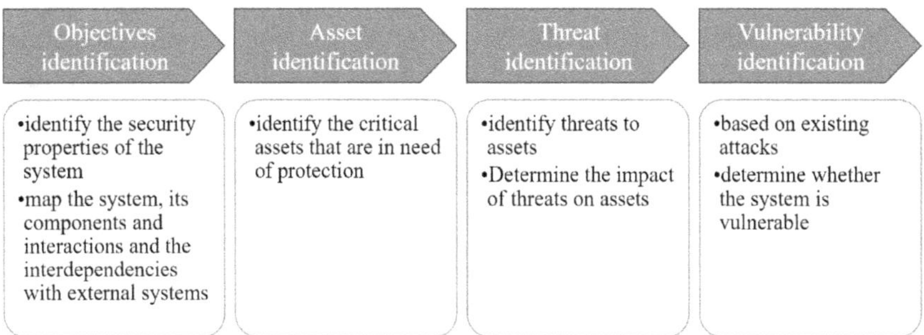

Fig. 6.4. Lifecycle of AI training phase.

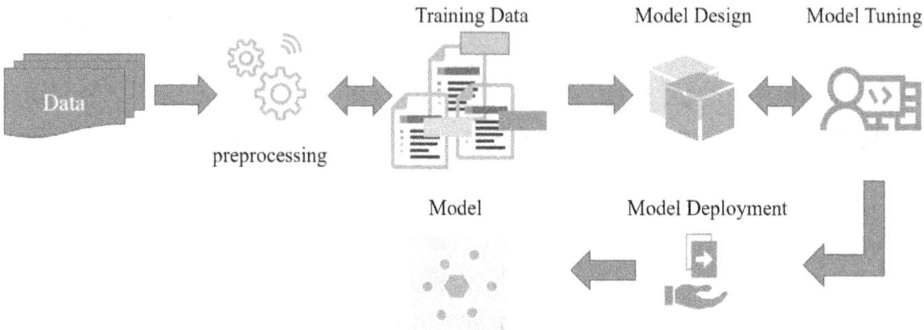

Fig. 6.5. Process of the model training phase.

For ease of further analysis, the various stages of artificial intelligence development and deployment can be summarized into two distinct stages of machine learning model design and development, deployment, and operation, generally referred to as two stages of training and prediction. It can also be analyzed and expanded on the basis of each substage according to the actual scene.

The process of model training phase is shown in Fig. 6.5.

The model training phase is divided into the following steps:

(1) Preprocessing based on raw data may include data cleaning, data labeling, and data format conversion. This step can be repeated to obtain the most accurate training set data.
(2) Select the appropriate machine learning algorithm.
(3) Perform model training and tuning based on the training set obtained after preprocessing.
(4) Deploy the tuned model in the test or production environment.

For each link in the above steps, apply the threat modeling method to obtain the typical threats faced by each link, and then formulate mitigation measures. The threats present in the "raw data" are as follows:

(1) Raw data contains information that identifies natural persons, which may violate personal privacy.
(2) The manipulation of the original data, such as "poisoning attack", may cause the training set to be contaminated and indirectly lead to inaccurate models.

(3) A large amount of garbage data is used as input, similar to the DoS (Denial of Service) attack in network security, which makes the model unusable.

For example, if a machine learning system is designed to find characteristic gene fragments that can cause thalassemia, collecting a large number of detailed medical records of each patient from multiple hospitals will pose a privacy risk to the original data.

Threats that exist in "training set data" are that the training set (possibly including the test set) is stored, managed, managed and used, even shared with third parties. Therefore, data security and privacy are at risk.

The security measures that can be selected are as follows:

(1) Adopt data lifecycle process security measures to record and archive data movement and storage behaviors.
(2) Implement logging and auditing mechanisms that meet accountability and evidence retention requirements.
(3) Apply privacy-enhancing techniques at the source or prior to disclosure.

Accordingly, the prediction phase is divided into the following steps:

(1) The application receives real data.
(2) The application processes the real data and passes it to the model in the deployment environment. The model is based on a pre-defined learning algorithm and computes its results.
(3) The result output of the model.

Figure 6.6 shows the process of AI prediction stage.

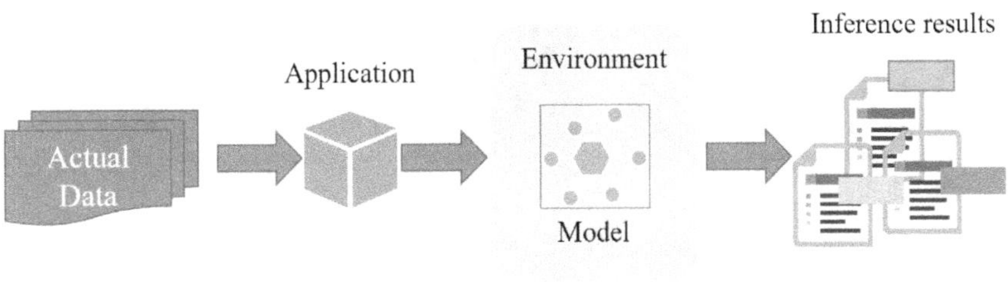

Fig. 6.6. Process of AI prediction stage.

From the "real data" dimension, the types of attacks that affect the correctness of the results include adversarial sampling attacks, physical world attacks, etc. There are other attack methods, such as model inversion attacks to obtain personal information present in the model, and membership inference attacks.

Adversarial sample attack refers to the addition of small perturbations, which are difficult for humans to detect or recognize in normal samples, to well-designed input samples in order to fool or deceive machine learning models, especially deep learning models, so that they produce incorrect results.

The most well-known case of an adversarial example attack is presented by Ian J. Goodfellow, a security expert at Google, in his 2015 paper "Interpreting and Exploiting Adversarial Examples" (Goodfellow *et al.*, 2014). An example of an adversarial example attack is shown in Fig. 6.7. By adding distortions to the image that are barely noticeable to the human eye, it tricks the AI image recognition algorithm into recognizing the panda in the image as a gibbon.

The physical world attack refers to not tampering with the digital sample form but directly modifying the data source in the physical world to cause the machine learning model to misidentify and produce incorrect results. Compared to adversarial sampling attacks, physical world attacks are more complex and must account for the loss and distortion caused by data sampling and conversion.

A model inversion attack assumes that the attacker has obtained some personal information about specific subjects in the training set, after which the attacker can further infer the personal information

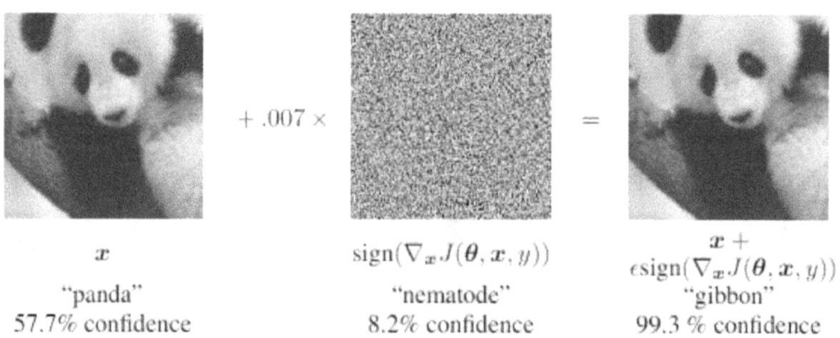

Fig. 6.7. Example of adversarial sample attack.

of these specific subjects by observing the inputs and outputs of the machine learning algorithm model, and the information it can know is not limited to general inferences about similarly characterized subjects. Recent cases have shown that model inversion attacks against facial recognition techniques can be used to reconstruct facial images that were previously recognized during training (Fredrikson *et al.*, 2015).

Membership inference attacks allow attackers to infer whether a particular subject is present in the training data of a machine learning algorithm model. The difference with model inversion attacks is that the attacker in a membership inference attack may not necessarily be able to extract other personal information about a specific data subject. For example, health insurance companies have deployed AI systems in some hospitals that predict when patients can be discharged based on their medical records. However, if an attacker is able to gain access to the system, it may be possible to determine whether someone has been to those hospitals. If the training data comes from vulnerable groups, such as Alzheimer's patients or the mentally ill, this could pose a serious privacy risk.

In short, based on the characteristics of the model itself and the different deployment environments of the model, there are different types of attack methods for the confidentiality, integrity, and availability of the model itself.

If the attacker does not have direct access to the model, e.g., if the model is deployed in the cloud, this scenario belongs to the "black box attack", where the attacker has limited knowledge of the model and is therefore severely limited in the generation of adversarial samples. There are several typical attack patterns in this scenario. Among them, Nicolas Papernot et al. in their paper "Mobility in Machine Learning: From Phenomena to Blackbox Attacks with Adversarial Samples" (Papernot *et al.*, 2016) described such a model transfer attack: the attacker obtains the query results by querying the target model several times, and uses them as inputs to train the "alternative model". After obtaining the alternate model, the attacker searches for adversarial samples based on the alternate model, and the adversarial samples obtained are also effective in attacking the target model, thus successfully fooling the target model.

Models are not static. If the model is retrained with run-time data input to improve the accuracy of the model and cope with changes in the data, it can be affected by the "bait attack". "Bait" is a vivid metaphor, meaning that the attacker "contaminates" the training data with carefully designed samples. After a model is "poisoned", its accuracy can be significantly affected.

The other two types of model attacks are backdoor attacks and model stealing attacks.

A traditional backdoor refers to source or binary codes embedded in a target system. Machine learning models typically consist of a set of parameters and do not have source code, so their backdoors are more hidden. For most data inputs, the model with the backdoor can produce normal results. But for special inputs, it is controlled by the attacker. When the backdoor is triggered, the attacker gets the result he intended.

Model stealing attack means that the attacker analyzes the input and output of the target model and combines it with other relevant information to infer the parameters of the model and even the training data of the model. This type of attack requires special attention to the use of cloud services to provide AI model training and identification, prediction, and services, collectively known as "AI as a Service" (AIaaS). For example, an attacker guesses the type, characteristics, and important parameters of a model by making multiple calls to a cloud-provided image recognition service. Model stealing attacks can lead to the theft of important intellectual property, or the stolen model information can be used in further transitive attacks against the model.

Of course, the attacker's goal in the attack process is often not single, such as stealing the target model by exploiting the confidentiality weakness and then attacking it by analyzing the vulnerability of the target model, thus threatening the availability and integrity of the model.

There may also be privacy risks in the model itself. Some models may unintentionally contain personal data, and some types of AI algorithm models may have some original training data at the time of design. For example, Support Vector Machine (SVM) and K-Nearest Neighbors (KNN) algorithm models themselves contain some of the training data. In these cases, the storage and use of these models is subject to relevant laws and regulations regarding privacy and personal data protection.

6.1.4 *Privacy protection of artificial intelligence*

Our lives are increasingly digitized, and the collection and storage of personal data is almost exclusively digital. At the same time, the adoption of more and more sensor technologies has generated a wide range of data covering personal health, sports, and physiological dimensions. At the same time, whether in the data collection dimension or in the data analysis dimension, artificial intelligence technology has been increasingly applied.

In the digital world, data has the characteristics of extremely low replication cost and extremely fast transmission speed, so it is much more difficult to control data collection and data access. Artificial intelligence exacerbates the existing problems of data security and privacy protection. For example, the application of the face recognition function based on artificial intelligence technology in business scenarios has raised many concerns about the infringement of consumers' privacy and interests; user portraits and personalized recommendation technologies based on "device fingerprints" have also attracted some consumers' disgust.

While people enjoy the free and rich services provided by the Internet, the "digital traces" they leave behind are being targeted and analyzed for use in a variety of scenarios, including advertising. The proliferation and popularization of the Smart Internet of Things and Smart Cities has created an even larger data collection and analysis system that can provide various types of detailed data and increasingly informative related information in real time, raising even greater concerns about data security and privacy.

Looking back at the content mentioned in Section 5.2 "OECD Eight Principles of Privacy Protection", there is great uncertainty whether the application of AI technology can fully meet several principles. For example, the development of artificial intelligence models requires a large amount of data for training, which conflicts with the Collection Limitation Principle. The Purpose Specification Principle requires that data be collected for its intended purpose, and that the individual's consent be obtained again if the purpose changes. However, the data sources required by AI are diverse and may be collected from multiple channels. It is difficult to be sure that all data has been notified and approved. It is even difficult to describe exactly what an AI model "learns" from data provided by an individual. The model of artificial intelligence as a kind of "black box" poses a great challenge to the implementation of the Openness Principle and the Individual Participation Principle. The Accountability Principle is the "sword of Damocles" hanging over the heads of data controllers and data processors using artificial intelligence technology.

A number of technologies, tools, and solutions are gradually being proposed to mitigate and respond to privacy protection issues in artificial intelligence to comply with the requirements of privacy protection laws and regulations.

When it comes to personal data and privacy, the most basic requirements must be met for any type of system to be designed and implemented. For example, Article 25 of the General Data Protection Regulation (GDPR) calls for the principles of "privacy by design" and "privacy by default". Article 35 of the GDPR requires a Data Protection Impact Assessment (DPIA) when the processing of data, in particular when the processing is carried out using new technologies, is likely to pose a high risk to the rights and freedoms of natural persons.

For AI scenarios, privacy-preserving technologies or solutions can be divided into the following types:

(1) Minimize or replace personal data from the source of the data.
(2) Introduce privacy-enhancing technologies in data processing.
(3) From the data output, pay attention to the interpretability of the results.

The technologies that minimize or replace personal data mainly include Federated Learning, Generative Adversarial Network (GAN), Matrix capsules, etc.

Federated learning is a distributed machine learning technique. The basic idea is to download the model to the local computer, and through local data training, the model is improved or merged on the server side. Federated learning can save the complex and risky process of transferring private data to the server, and protect the privacy of participants to a certain extent. A figurative analogy is that machine learning (specifically centralized) "feeds" data to the model, and federated learning "feeds" the model to the data.

Generative Adversarial Networks (GANs) can be used to synthesize training data for artificial intelligence algorithms. Existing research includes the use of generative adversarial networks to generate large numbers of face images that do not actually exist. These face images can be used to calibrate and improve face recognition algorithms. Generative adversarial networks have the potential to generate large amounts of high-quality synthetic training data to meet the demand for labeled datasets and training samples, rather than relying on massive amounts of personal data.

Matrix capsules are a variant of neural networks that can use less training data to achieve similar effects as deep learning.

In computing, privacy enhancing technologies include differential privacy, homomorphic encryption, and transfer learning.

Differential privacy adds "noise" to personal data or query results, preventing personal information from being retrieved. Differential privacy is introduced in detail in Section 5.3.3 "Differential Privacy".

Homomorphic encryption is a method of processing or computing encrypted data. With this method, the confidentiality of the dataset can be guaranteed while the required query, statistical, or computational functions can be realized. The efficiency and versatility of homomorphic encryption are issues that need to be addressed in applications. Microsoft has published a white paper on the use of homomorphic encryption in image recognition.

Transfer learning is an approach in machine learning that aims to improve learning performance by applying learned knowledge and experience from one task or domain to another. In traditional machine learning methods, a model must be trained from scratch for each task, while transfer learning leverages existing knowledge and reduces the time and resources required to train new tasks. Specifically, transfer learning involves fine-tuning a model pre-trained for one task to a new, related task.

The interpretability of results mainly refers to the research and application of Explainable Artificial Intelligence (XAI).

Explainable AI refers to explaining how artificial intelligence works and how it makes decisions in a way that is understandable to humans. Explainable AI focuses on the interpretation of input variables and the decision-making phase of the model. It also includes the structure of the model itself.

LIME is an XAI solution. It provides an explanation of artificial intelligence that ordinary people can understand without relying on specific models. Take image recognition, for example, which is able to show which parts of an image are relevant to what it thinks it is. This makes it easy for anyone to understand the basis for decisions.

6.1.5 *Artificial intelligence and network security*

With the rapid development of mobile Internet, cloud computing, big data, Internet of Things, machine learning, and other technologies that support people's work and life, the network security environment is becoming more complex and the security threats are escalating. In this complex environment, the traditional research methods of analyzing security issues and setting fixed rules have become increasingly inefficient and even impotent. For example, relying on the manual repair method of security experts can no longer solve the problem of zero-day vulnerabilities. The traditional network intrusion detection method that relies on fixed rules has a large number of false-positive alerts or too long detection time and other problems in the face of ever-increasing data volume and complex network behavior. Spam filtering systems that rely on fixed rules have problems such as low detection efficiency or untimely rule updates.

The analysis of the above scenarios shows that network security is mainly related to "scale". Security control measures that are suitable for single machines, small networks, and single services are often not as effective when faced with large networks, complex networks, and dynamic services. The dynamism and adaptability of machine learning can meet the requirements of many new scenarios. Therefore, the application of machine learning in network security defense has also received sufficient attention.

Machine learning has shown great advantages in natural language processing, medical data analysis, etc., and machine learning methods also offer the possibility of solving complex network security

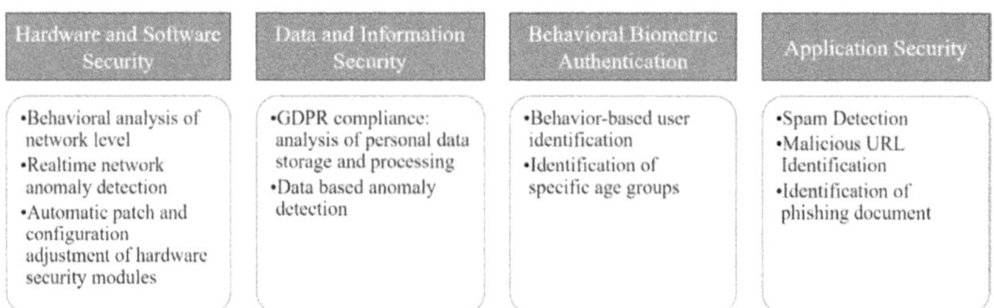

Fig. 6.8. Example of the application of AI in the field of network security.

problems. Figure 6.8 shows an example of the application of artificial intelligence in network security.

Of course, there are some other scenarios that are not within the scope of application described above, such as abnormal account and behavior detection in the field of network risk control and network public opinion analysis based on natural language processing.

6.2 Emerging Applications Represented by Blockchain

Blockchain is an ingenious invention that uses a lot of cryptography, computer algorithms and classical data structures. It was developed by Satoshi Nakamoto, who proposed it in the article "Bitcoin: A Peer-to-Peer Electronic Cash System" in 2008. Some people believe that Satoshi Nakamoto may be a pseudonym and his real identity has not been revealed.

Satoshi Nakamoto did not directly define the concept of blockchain in the white paper but proposed the explanation and definition of the data structure required by Bitcoin, including the concepts of transactions, blocks, chains, etc., and later formed the definition of blockchain. Blockchain was originally designed to be used as a digital currency (Bitcoin), and other potential uses have since been discovered and are booming.

According to Don and Alex Tapscott, authors of "Blockchain Revolution: How the Technology Behind Bitcoin is Changing Money, Business, and the World" (Tapscott and Tapscott, 2016), stated: "The blockchain is an incorruptible digital ledger of economic transactions that can be programmed to record not just financial transactions but virtually everything of value" (Bawa, 2019). Blockchain uses technology to solve the problem of fraud in financial and commercial transactions. The World Economic Forum (WEF) predicts that by 2025, 10% of GDP will be tracked and traded using distributed ledgers based on blockchain technology.

From a business scenario perspective, the three outstanding advantages of blockchain-decentralization, transparency, and immutability-make it suitable for recording transaction information, supply chain provenance, medical records, etc.

From a technical perspective, blockchain is a master of innovative application consensus mechanisms, encryption algorithms, digital signatures, distributed networks, and hash tree data structures.

Although the concept of blockchain was first proposed by Satoshi Nakamoto in 2008, the concept did not come out of nowhere, and many pioneers made early explorations.

In his 1982 paper, "Computer Systems Established, Maintained and Trusted by Mutually Suspicious Groups" (Chaum, 1979), cryptographer David L. Chaum first proposed a blockchain-like protocol, describing a distributed "treasury system" that encompasses technologies such as consensus algorithms, member consensus and node authentication, and private transaction computation. Subsequent work on chains of cryptographically secure blocks also included "How to time-stamp a digital document" (Haber and Stornetta, 1991), proposed by Stuart Haber and W. Scott Stornetta in 1991. In 1992, Haber, Stornetta, Dave Bayer, and others introduced Merkle trees into their design and demonstrated their efficiency and effectiveness for including multiple documents in a block. Cryptographer Nick Szabo conducted research in 1998 on the mechanisms of a distributed electronic currency, which he called Bit Gold (Szabo, 1998).

In Satoshi Nakamoto's whitepaper, the words "block" and "chain" are used separately, and when they are widely used, they are collectively referred to as "block-chain", and it was not until 2016 that they became a single word: "blockchain". Due to the proliferation of blockchain transaction data, in August 2014, the blockchain file size of bitcoin was 20 GB, and in the third quarter of 2019, the blockchain file size of bitcoin reached 242 GB.

In 2014, the concept of "Blockchain 2.0" based on "Smart Contracts" rather than pure "transactions" was proposed, and it has been widely used in digital currencies such as Ethereum. The exact meaning of "Blockchain 2.0" is also debated. There is a view that the application of blockchain outside the digital currency occasion, especially the generation and application of blockchain as a service (BaaS), is "Blockchain 2.0".

6.2.1 *Basic concepts of blockchain*

Blockchain was first proposed as the underlying technology for bitcoin. It is essentially a decentralized list of records that can only be added to, not deleted or tampered with. Records are stored in blocks and linked together using cryptographic methods. In the original design, each block of data not only contains a batch of bitcoin transaction information but also contains the validity verification of its information and is also used to generate the next block.

In the white paper written by Satoshi Nakamoto, it also includes the characteristics of distributed network, distributed data storage, multi-node consensus mechanism and other characteristics, and uses a lot of cryptography and data structure concepts such as digital signatures, encryption algorithms and hash trees.

In general logic, the blockchain can be divided into the following levels:

(1) Infrastructure or hardware.
(2) Distributed network (including node discovery, information dissemination, and verification).
(3) Consensus (consensus among multiple nodes, such as proof of work).
(4) Data (such as block data structure, transaction records).
(5) Applications (such as smart contracts, distributed applications).

At the consensus, data, and application levels, blockchain has relatively many innovations, while at the infrastructure or hardware and distributed network levels, it uses a large number of existing technologies.

Blockchain is generally a decentralized peer-to-peer network and is used as an open distributed ledger. Each participant (called a node) follows a communication protocol to communicate with each other and verify new blocks. Nodes may choose to maintain their own copy of the distributed ledger.

If there is an authoritative and credible entity in the system, such as the national driver's license registration system, the responsible authority and its employees or authorized personnel are responsible for updating and maintaining the contents of the system's database as trusted authoritative subjects. As a distributed system without an authoritative central entity, the blockchain trust protocol is implemented using a consensus mechanism among nodes.

A consensus mechanism is a fault-tolerant mechanism for a distributed system that achieves the necessary agreement on data values or states among distributed nodes and potentially resists unintentional or malicious node behavior. For example, in the context of cryptocurrencies, consensus mechanisms are very useful for keeping records in ledgers.

In the case of bitcoin, the earliest application of blockchain, the publicly shared ledger requires an efficient, fair, practical, reliable, and secure mechanism to ensure that all transactions occurring on the network are authentic and that all participants have a consensus on the state of the ledger.

Different types of consensus algorithms are based on different principles. Proof of Work (PoW) is a common consensus algorithm used by the most popular cryptocurrency networks, such as Bitcoin and Litecoin. It requires participating nodes to prove that the work they have performed and submitted qualifies for the right to add new transaction records to the blockchain. However, this "mining" mechanism of bitcoin requires high energy consumption and long processing time.

Proof of Stake (PoS) is another common consensus algorithm and a low-cost, low-energy alternative to the PoW algorithm. PoS distributes the responsibility for maintaining the public ledger among participating nodes in proportion to the virtual currency held by the node. However, it also has the disadvantage of encouraging cryptocurrency savings rather than consumption.

Similarly, there are other consensus algorithms, such as Proof of Capacity (PoC). The PoC algorithm encourages nodes to contribute memory or hard disk space (capacity) to a shared blockchain network. The more memory or disk space a node owns or contributes, the more rights it is granted to maintain the public ledger.

In the Bitcoin network, blocks record all transaction information within a certain period of time. The data structure of each block includes the hash value (Pre-hash) of the previous block, the timestamp, and the transaction data recorded in that block in the form of a Merkle tree.

With the exception of the Genesis block (the first block on the blockchain), each block contains the hash of the previous block, forming a chain. The blockchain data structure is shown in Fig. 6.9. The characteristics of the hash algorithm ensure that each block is a security enhancement of the previous block.

As a result, blockchain is resistant to modifications to the data on the chain. Once recorded, any modification to a block must be modified in all subsequent blocks, otherwise it is easy to detect and trace. If a node attempts to modify a block and all subsequent blocks, it will not be accepted by the majority of other nodes in the blockchain network.

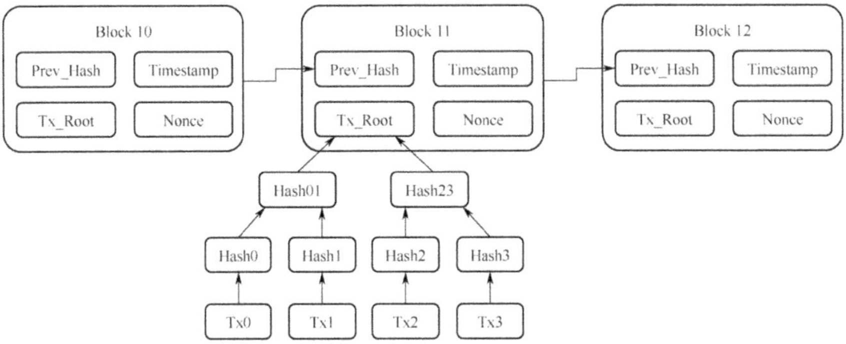

Fig. 6.9. Blockchain data structure.

While blockchain records are not theoretically immutable because of the possibility of forks, blockchain implements a distributed computing system with Byzantine style high fault tolerance (Castro *et al.*, 1999) and has no known serious security flaws in its design.

The application layer of the blockchain has a wide range of content. There is an opinion that the blockchain platform that includes the "smart contract" function is the "blockchain 2.0" version. The term "smart contract" was first coined by Nick Szabo in the early 90s of the 20th century, and the meaning is "a set of commitments, specified in digital form, including an agreement by the parties to fulfill those commitments". Since the launch of the Ethereum in 2015, the term "smart contract" has often been used to refer to a concept more specifically applied to general computing on blockchains and distributed ledgers. The National Institute of Standards and Technology (NIST), in its Blockchain Technology Overview (Yaga *et al.*, 2018), describes a "smart contract" as "a collection of code and data (sometimes referred to as functions and state) that is deployed using cryptographically signed transactions on the blockchain network".

Similar to transactions on the blockchain, deployment of smart contracts on the blockchain is accomplished by sending transactions from the blockchain wallet. The transaction contains not only the transfer information like a traditional transaction but also the compiled code of the smart contract and a special receiving address. This transaction must be included in a block added to the blockchain, at which point the smart contract code is executed to establish the initial state of the smart contract. The blockchain protects smart contracts in a Byzantine, fault-tolerant algorithm and in a decentralized manner to prevent tampering. Once a smart contract is deployed, it cannot be updated.

The earliest and most popular blockchain running smart contracts is Ethereum. On Ethereum, smart contracts are typically written in a Turing-complete programming language called Solidity, compiled into low-level bytecode, and executed by the Ethereum Virtual Machine.

A key feature of smart contracts is that they do not require a trusted third party (such as a middleman or arbitrator) to act as an intermediary between the contracting parties. In business scenarios, they can provide provable transaction transparency to promote trust and provide insight into business decisions, thereby reducing the reconciliation costs that exist in traditional enterprises.

In short, blockchain innovatively integrates the application of network, encryption, and recording technologies, and continues to innovate and promote applications. Its context-specific strengths and weaknesses should be recognized and used only where appropriate.

6.2.2 *Consensus algorithm and its application*

In a centralized system, all activities are controlled by a single authoritative entity and must have a unified charter. The charter must be followed by all members but does not require consensus among the members. However, this centralized system can also be vulnerable to malicious behavior by the authoritative entity or an attack that results in the destruction of the entire system.

In a completely decentralized network such as blockchain, how do you ensure the privacy and security of transaction information? In other words, how to ensure that participants jointly hold the same ledger, and the increment of the block (transaction record) reaches a consensus among the majority of participants?

By definition, a consensus algorithm is a group decision-making process in which individuals construct and support decisions that are most effective for other members. It is a form of decision making in which individuals must support the decision of the majority, whether they like it or not. Popularly, but inaccurately, the consensus algorithm embodies the principle that the minority obeys the majority. The "minority obeys the majority" here does not necessarily refer to participants or nodes but can also be computing power, shares or inputs, etc.

The consensus model of blockchain tries to achieve the following goals:

(1) **Common goal:** Agreements are collected among team members, and a majority agreement is reached as the common goal.
(2) **Collaboration:** Everyone can participate, and agreements based only on their own interests cannot be approved by the majority.
(3) **Equality:** All participants can vote, and each participant's vote is important.
(4) **Participation:** Every participant must vote, and abstentions or invalid votes should be designed and implemented as little as possible.

The consensus model attempts to solve the problem of "untrusted nodes" and "untrusted behavior" in distributed systems, the Byzantine General Problem. This problem was proposed by Leslie Lamport and others in 1982 in the paper "The Byzantine Generals' Problem" to explain a metaphor for distributed fault-tolerance technology. The inventor originally called it the Albanian Generals Problem, but to avoid potential offense, it was renamed the Byzantine Generals Problem.

This problem presents an easy to understand scenario. A group of Byzantine generals each have their own armies and deploy those armies in different locations. The generals can only communicate with each other through messengers. To attack and destroy a city, the generals must coordinate an attack plan. Suppose there are some traitor generals or traitor couriers who will try to sabotage the entire attack. The traitor couriers may not deliver at all, or they may deliver the wrong attack plan. The overall goal is that all loyal generals agree on the plan, and in the case of a certain number of traitor generals, the entire offensive mission will not fail.

The mathematical analysis of this problem shows that if there are n traitor generals, $3n + 1$ generals are needed to have a consensus plan.

The Byzantine Generals Problem is a metaphor for distributed computer systems. In a distributed system composed of computers, due to reasons such as network congestion, hardware problems, or malicious attacks, unpredictable behavior may occur in the distributed system, and the business goals of the distributed system cannot be achieved. The proposed solution to this problem is called Byzantine Fault Tolerance (BFT) (Castro *et al.*, 1999).

The blockchain consensus model is not to solve the Byzantine Generals problem perfectly but to achieve the goal of majority consensus so that some more reliable and fault-tolerant solutions can be adopted.

Proof of Work (PoW) is the earliest consensus algorithm introduced in the blockchain network. The Bitcoin network uses this consensus algorithm to confirm all of its transactions and add the resulting blocks to the network chain.

In layman's terms, the PoW consensus protocol involves "miners" who process blockchain transactions through "mining".

Miners on the network compete against each other, solving complex mathematical problems. The miner who solves the problem first receives the corresponding amount of cryptocurrency as a "block reward". Solving complex mathematical problems is equivalent to validating transactions. Successfully verified transactions result in the creation of new blocks. In essence, the miner's job is to create new blocks and add them to the blockchain.

The advantage of the PoW consensus algorithm is that it is clear and easy to implement. It does not require a large amount of information interaction between nodes, and it can resist malicious behavior of nodes and DDoS attacks to some extent. The disadvantage of PoW consensus algorithm is extremely high energy consumption and long transaction processing time.

Proof of Stake (PoS) is another popular consensus algorithm. The basic idea is that the more stake users have in the system, the more they want the system to succeed and the less likely they are to disrupt or undermine it. Stake generally refers to the amount of cryptocurrency invested in the system by users of a blockchain network. Staking can be done in a variety of ways, such as locking cryptocurrency through special transaction types, sending cryptocurrency to specific addresses, or holding cryptocurrency in special wallet software. Once staked, cryptocurrencies generally cannot be spent again. A blockchain network that uses the Proof of Stake (PoS) consensus algorithm uses the amount of equity owned or invested by a user as the determining factor for publishing a new block. That is, the possibility of a user publishing a new block is positively related to the proportion of the user's stake in the total stake of the blockchain network.

This consensus algorithm eliminates the need for resource-intensive computations such as proof-of-work, thereby reducing energy consumption.

Some proof-of-stake algorithms can suffer from a problem known as "nothing at stake". If, at some point, there are multiple competing blockchains (perhaps due to ledger conflicts), users may take the same action on each competing chain to increase their probability of receiving a reward. This can lead to blockchain forks rather than consensus as a single branch.

Other common consensus models applied to distributed systems include the Round Robin Consensus Model, the Proof-of-Authority Consensus Model, and the Proof-of-Time Consensus Model.

Round Robin Consensus Model is a consensus model used by some permissioned blockchain networks. In this consensus model, nodes take turns creating blocks. The Round Robin Consensus Model is based on distributed system architecture and has a long history. The model uses a timeout mechanism to allow available nodes to release blocks without unavailable nodes causing a block release to stall. The model ensures that no single node can create a majority of blocks. The round-robin consensus model is simple in its approach and does not require computation, but it requires trust between nodes and therefore does not work well in the permissionless blockchain networks used by most cryptocurrencies. Malicious actors can continue to add other nodes to increase the probability of publishing new blocks and, in the worst case, disrupt the proper functioning of the blockchain network.

The Proof of Authority (also known as Proof of Identity) consensus model relies on the partial trust that a publishing node gains through its known connections to real-world identities. The publishing node must have a verified and verifiable identity in the blockchain network (e.g., an identifying document that has been verified and notarized and included in the blockchain). Publishing nodes use their identity or reputation to publish new blocks, and the behavior of publishing nodes directly affects the reputation of publishing nodes. Publishing nodes that act in a way that the users of the blockchain network do not agree with will lose some of their reputation. The lower the reputation, the lower the probability of publishing a block. Therefore, it is in the publishing node's interest to maintain a higher reputation. The model is only applicable to permissioned blockchain networks with high trust.

In the Proof of Elapsed Time (PoET) consensus model, each publishing node requests a wait time from a secure hardware time source within its computer system. The publishing node is randomly assigned a period of time and is in an idle state for that period of time. Once a publishing node awakens from its idle state, it creates a block and publishes it to the blockchain network, notifying other nodes of the new block. Any publishing node that is still idle will stop waiting and the whole publishing process will start again.

The PoET consensus model requires ensuring that random times are used, otherwise a malicious publishing node will set the minimum wait time allowed. This model also requires ensuring that the publishing node does not start early. These requirements can be met by running software in trusted execution environments (such as Intel SGX or ARM TrustZone) on specific computer processors. Validated and trusted software can run in these secure execution environments and cannot be modified by external programs.

6.2.3 *Merkle tree data structure*

A hash tree is a binary tree or multi-tree data structure in which each leaf node contains the hash value of its data block, and each non-leaf node (root and branch nodes) contains the hashes of its child nodes. The hash tree is named after Ralph Merkle, who first described and patented it in 1979. Therefore, it is often referred to as a Merkle tree.

Most implementations of Merkle trees are binary trees, and their structure is shown in Fig. 6.10. To facilitate drawing, "$Hash()$" is abbreviated to "$H()$". In this scenario, for non-leaf nodes,

$$Hash(x) = Hash(Hash(x_0) + Hash(x_1))$$

Among them, x_0 and x_1 are two child nodes of x, respectively.

Merkle trees allow efficient and secure verification of large data contents. Its steps are to divide large data into blocks, calculate a hash for each block, and generate a Merkle tree data structure. Since the structural characteristics of the Merkle tree ensure that

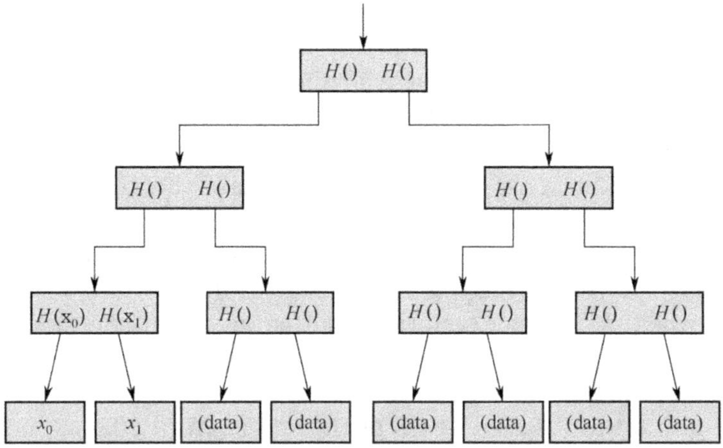

Fig. 6.10. Merkle tree structure.

any change to a data block is propagated to its parent node and then to the root node, the integrity of the data can be effectively verified. The data structure of the Merkle tree also guarantees the time complexity of querying whether a given leaf node exists in the tree ($\log n$, where n is the number of nodes). The query performance has advantages over the hash list, and the query complexity of the hash list is generally n.

Traditionally, Merkle trees are used to verify that stored and transmitted data has not been tampered with. For example, file transfers over a peer-to-peer network can quickly verify the integrity of blocks of data received from multiple endpoints. Some file systems such as ZFS, Btrfs, etc. also use Merkle trees. In addition, version control systems such as Git, etc. also use Merkle tree implementations when committing and comparing files and directories.

Comparing the consistency of the data can be simplified to comparing its Merkle tree. Starting from the root node of the Merkle tree, if the root node is consistent, it means that the two replicas are consistent, if not, you just need to compare the child nodes with different hash values until the data is inconsistent with the leaf node. In the data synchronization scenario, only inconsistent data blocks can be synchronized, which can significantly reduce the comparison time and the amount of data transferred.

The tamper protection and fast consistency comparison features of the Merkle tree have been directly applied to the blockchain.

The first application scenario of the Merkle tree in blockchain is to ensure the immutability of historical transactions. The data records in the block (usually containing the information of multiple transactions) form a hash value of the root node through hash calculation and recursion, put it in the block header, and record the hash of the previous block in a chain structure to ensure that the block data cannot be tampered with.

The second application scenario of the Merkle tree in the blockchain is Simple Payment Verification (SPV). As mentioned above, when the verifier confirms whether a certain transaction has been verified and confirmed by the blockchain network, he only needs to store all the block headers and does not need to synchronize all the block data up to 329.31 GB (as of the end of April 2021). The verifier only needs to provide the hash value of the transaction, the hash value of the associated leaf node, and the hash path to the root node to quickly verify that the transaction is actually stored on the blockchain. Some people call this "zero knowledge proof" of transactions. The process does not reveal the content of the transaction itself, such as the time of the transaction, the object of the transaction, or the amount of the transaction.

6.2.4 *Comprehensive application of cryptographic algorithms*

Blockchain innovatively uses a variety of cryptographic algorithms, the most prominent of which are cryptographic hash functions, asymmetric cryptography, and digital signatures.

Cryptographic hash functions are widely used in blockchain. For example, only implementation mechanisms for immutable ledgers are allowed to be added.

Review the three properties of cryptographic hash functions explained in Section 4.7 "Hash Algorithm": The first property is preimage resistance, also called one-way property. It is difficult to find x for a digest m, so that $Hash(x) = m$; The second property is the second preimage resistance. Given the plaintext x, it is mathematically difficult to find a plaintext y except by enumeration, so that $Hash(x) = Hash(y)$; The third property is collision resistance. For any two plaintexts x and y, it is difficult to obtain $Hash(x) = Hash(y)$.

Due to the above three properties, cryptographic hash functions are used in many scenarios in the blockchain:

(1) The security of the block data is written into the block header by the hash of the block data to achieve integrity (tamper-proof) protection.
(2) The security of the block header data, including the hash of the block header data and the hash of the previous block, to achieve integrity (non-tampering) protection.
(3) Creation of a unique identification number or string.
(4) Derivation of the blockchain address.

Some blockchain networks use "addresses" to represent the originator and recipient of network transactions. An address is typically a short string of letters and numbers derived from a blockchain network user's public key using a cryptographic hash function, also known as a blockchain address derivation. Some additional data fields may also participate and serve as input to the hash calculation, such as version number, checksum, etc. The blockchain network address is not a secret and can be made public. For permissionless blockchain networks that allow anonymous account creation, blockchain network users can generate as many asymmetric key pairs as possible to generate the required addresses, achieving varying degrees of pseudo-anonymity. The address can serve as a user's public identifier in the blockchain network. Addresses can even be represented as QR codes for easy use on mobile devices, similar to the "receipt QR" of mobile payment apps.

A "user" of a blockchain network is typically represented by a set of public and private key pairs. This is where asymmetric cryptography is used. Asymmetric key cryptography uses a pair of keys: a public key and a private key that are mathematically related. The public key can be made public without compromising security, but the private key must be kept secret. Even if there is some correlation between the two keys, asymmetric cryptography guarantees that it is difficult to deduce the private key from the public key alone. Depending on the usage scenario, you can encrypt with the private key and then decrypt with the public key, or you can encrypt with the public key and then decrypt with the private key.

Users of blockchain networks must securely manage and store their private keys. The software or hardware used to securely store private keys is often referred to as a wallet. The wallet can store private keys, public keys, and the user's associated addresses. Some wallets also include other functions, such as calculating the total amount of digital assets owned by a user.

Blockchain transactions exist between users who do not know or trust each other. The authenticity and integrity of transactions is based on digital signature technology. The sender of a transaction signs the transaction using his or her private key, and anyone on the network with the public key can verify the authenticity and integrity of the transaction.

A widely used public key cryptographic algorithm in blockchain is the elliptic curve cryptographic algorithm. The security of the elliptic curve cryptographic algorithm depends on the difficulty of the elliptic curve discrete logarithm problem. Compared to the RSA public and private key algorithm, the elliptic curve cryptographic algorithm has a shorter key length and requires less network bandwidth and storage. For example, Bitcoin uses the secp256k1 elliptic curve and the corresponding ECDSA digital signature algorithm.

The secp256k1 elliptic curve is an elliptic curve based on the $\mathbb{F}p$ finite field and has the characteristics of high performance. Compared to similar NIST-recommended elliptic curves, secp256k1's constants are predictable, preventing designers from implanting backdoors.

The following example can be used to illustrate the transaction process supported by bitcoin using signature and verification algorithms. Suppose Alice decides to pay 1 bitcoin to Bob, and the miner Mina is responsible for recording the transaction.

Alice withdraws 1 bitcoin from her wallet and generates a transaction message: Alice pays 1 bitcoin to Bob. To ensure the authenticity of the transaction, Alice must digitally sign the transaction to confirm that the transaction was actually issued by her. Alice uses her own private key to sign based on the secp256k1 elliptic curve signature algorithm. Given the size of the message and the efficiency of the public key cryptographic algorithm, signing the transaction message is actually signing the hash value of the transaction message. In addition to the transaction message itself, the content that Alice broadcasts to the entire network also includes Alice's signature on the message and Alice's public key information.

After the transaction message sent by Alice is sent along with the signature, it is received by Mina. In order to record this transaction on the blockchain, Mina must first verify the source of the transaction, i.e., perform signature verification. Mina also uses the same secp256k1 elliptic curve to verify Alice's signature, which is the decryption process using Alice's public key. If the verification passes, Mina records the transaction information "Alice pays 1 bitcoin to Bob" in the block. If the verification fails, it means that there is something wrong with the message Mina received, and he will not record the corresponding transaction in the block.

In short, the use of elliptic curve signature and verification algorithms can not only ensure that accounts are not counterfeited but also ensure that users cannot deny their signed transactions.

6.2.5 *Data structure of blockchain*

Blockchain is a one-way chain data structure formed by chaining blocks. Except for the first block (called the genesis block), each block contains the hash of the previous block (the unique identifier of the block), thus forming a chain relationship between blocks. The chained data structure of the blockchain is shown in Fig. 6.11.

Users of the blockchain network submit candidate transactions to the blockchain network. A user sends a transaction to one or more nodes in the blockchain network, and the submitted transaction is broadcast to other nodes in the network, but this action does not place the transaction on the blockchain. In many blockchain implementations, after a pending transaction is distributed to a node, it must wait in a queue until it is added to the blockchain by a publishing node.

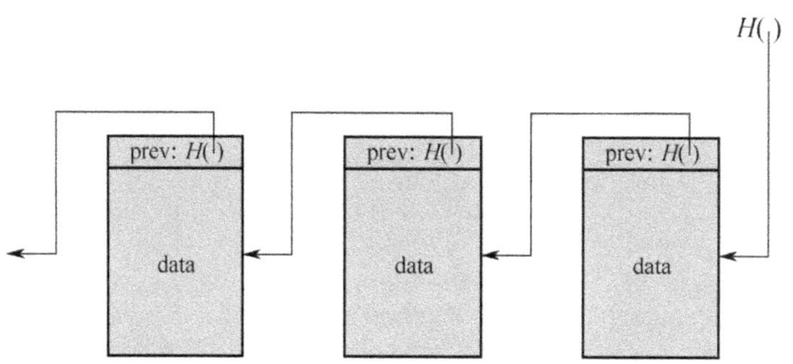

Fig. 6.11. Chain data structure of blockchain.

When a publishing node successfully publishes a block, the transaction is added to the blockchain. A block contains a block header and the block data body. The block header contains metadata about the block. The data body of a block contains the list of valid and real transactions that have been submitted to the blockchain network. Validity and authenticity are ensured by verifying that the transactions are properly formatted and that the originator of each transaction has cryptographically signed the transaction. Full nodes check the validity and authenticity of all transactions in published blocks, and if a block contains invalid transactions, it will not be accepted.

Each blockchain implementation may define its own data fields. A typical blockchain implementation includes the following data fields.

The block header typically contains the following:

(1) Block number, also called block height in some blockchain networks.
(2) The hash value of the previous block header.
(3) Hash representation of the block data.
(4) Timestamp.
(5) Block size.
(6) One-time random value (nonce).

The block body typically contains a list of the transactions contained in the block.

Using bitcoin's block structure as an example, bitcoin transaction records are stored in blocks of data. In the bitcoin system, a block

Table 6.1. Block structure.

Field	Description	Size/Byte
Magic No.	0xD9B4BEF9	4
Blocksize	number of bytes following up to end of block	4
Blockheader	consists of 6 items	80
Transaction Counter	A positive integer	1 9
Transactions	the (non empty) list of transactions	< Transaction counter > -many transactions

Table 6.2. Block header structure.

Field	Purpose	Size/Byte
Version	Block version number	4
Prev Block Hash	256-bit hash of the previous block header	32
Merkle Tree Root Hash	256-bit hash based on all of the transactions in the block	32
Timestamp	Current block timestamp as seconds since 1970-01-01T00:00 UTC	4
Difficulty Target	Current target in compact format	4
Nonce	32-bit number (starts at 0)	4

is generated approximately every 10 minutes. The data structure of the block is shown in Table 6.1.

The data structure of the block header is shown in Table 6.2.

6.2.6 *Bitcoin mining algorithm*

Bitcoin uses a consensus mechanism based on the Proof of Work (PoW) algorithm. The PoW algorithm requires users to publish the next block by being the first to solve a computationally intensive puzzle. The amount of work put into solving a puzzle is a testament to that user's investment. Puzzles must be difficult to solve and easy to verify. This prevents spoof attacks, where the attacker claims to have solved the puzzle but does not.

There are few puzzles with this characteristic, and cryptographic hash functions can provide one.

The blockchain's proof-of-work consensus algorithm uses a one-time random number (Nonce) combined with a hash function to calculate. That is, a one-time random number is combined with the data, and then a hash calculation is performed:

$$Hash(data + Nonce) = Digest$$

By changing the value of the one-time random number, different digests can be generated while keeping the data unchanged. The preimage resistance of cryptographic hash functions ensures that it is very difficult to find a hash value that conforms to certain rules, such as a hash value that starts with 8 0x00 bytes (commonly known as "leading zeros").

Applied to the blockchain scenario, the publishing node frequently modifies the one-time random number of the block header, and then performs hash calculation on the block header, trying to obtain a hash value that conforms to the target format, and successfully publishes it. This is a computationally intensive process, colloquially known as "mining". The target format can be modified over time to dynamically adjust the difficulty (up or down) to affect the frequency of block issuance.

For example, Bitcoin, which uses the proof-of-work model, adjusts the difficulty of the puzzle every 2016 blocks to affect the frequency of block issuance, ensuring an approximately fixed frequency of about once every 10 minutes. This puzzle difficulty adjustment essentially increases or decreases the hash format requirements, such as the number of leading zeros required. Increasing the number of leading zeros reduces the number of possible hash values that meet the requirements and increases the difficulty of solving the puzzle. Conversely, decreasing the number of leading zeros makes the solution less difficult. In fact, as the computing power available to the Bitcoin network has increased over time, as has the number of publishing nodes, the difficulty of puzzles has generally increased.

Adjusting the difficulty level ensures that it is very difficult for someone to control the majority of nodes in the Bitcoin network, thereby reducing the possibility of malicious behavior by the controller. However, the large amount of computation caused by the proof of work will cause a large amount of resource consumption.

The following example considers the computational difficulty of the SHA-256 algorithm as a puzzle. Find a hash that meets the format requirements (also called difficulty level):

SHA256 ("DataSec" + Nonce) starts with N leading 0s.

In the example, the text string "DataSec" and an appended one-time random number (Nonce) are combined to calculate the hash value. For ease of understanding, set Nonce to start incrementing from 0. Assume that the target starts with 6 leading 0s, i.e., "000000".

SHA256("DataSec0") = 0x63 9b 27 bf 9f 1f 72 57... (unresolved)

SHA256("DataSec1") = 0x86 2c 9a 25 e4 c1 60 11... (unresolved)

...

SHA256("DataSec309872") = 0x00 00 00 26 b9 db 81 4c... (Solved)

After more than 300,000 attempts, a hash value that met the requirements was finally found. On a decent laptop, results are available in less than 0.5 seconds.

Each additional leading 0 increases the difficulty significantly. If the number is increased to 7 leading 0s, using the same laptop, after about 560 million calculations in nearly 4 minutes, the result is obtained:

SHA256("DataSec566083470") = 0x 00 00 00 02 d4 da 4e... (Solved)

And if the number is increased to 8 leading 0s, using the same laptop, it will take about 5.18 billion calculations and nearly 30 minutes.

As of now, there are no known shortcuts to this computational process, other than exhaustive or random attempts. The publishing node must spend computational time and resources to find the correct nonce value for the target. Typically, in a blockchain network, publishing nodes that successfully solve computational puzzles receive some sort of reward, which can also take the form of cryptocurrency.

Once a publishing node reaches its goal first, it sends the block along with a valid nonce to the full nodes in the blockchain network. Receiving full nodes verify that the new block meets the puzzle requirements, then add the block to their copy of the blockchain and send the block out again to their peers. In this way, new blocks are quickly distributed throughout the network, including all participating nodes. Verifying the nonce is simple, as you only need to perform the same hash calculation repeatedly to confirm that the puzzle has been solved.

In many proof-of-work-based blockchain networks, publishing nodes are typically organized into "pools" or "groups" that work together to solve puzzles and share rewards. Since work can be distributed across multiple nodes, this possibility exists and is the dominant form of some blockchain networks.

Still the above example. Assuming that 4 nodes form a "pool" (generally called a "mining pool"), each node can calculate within the pre-allocated Nonce value range:

(1) Node 1: 5 000 000 6 000 000.
(2) Node 2: 6 000 000 7 000 000.

(3) Node 3: 7 000 000 8 000 000.
(4) Node 4: 8 000 000 9 000 000.

If 6 leading 0s are required, the result will be quickly obtained by Node 4:

SHA256("DataSec8483761") = 0x 00 00 00 25 36 62 b9 7b...... (solved)

Note that this is different from "DataSec309872" which also met the target last time. In fact, nodes can choose different Nonce values for calculation and compete with each other to try to achieve the goal and obtain rewards as early as possible.

The PoW algorithm comprehensively utilizes the advantages of the number of network nodes (the cost is small), the computing power (the cost is large), and the random lottery system (just like a lottery ticket, the more you buy, the higher the probability of winning, but there is no guarantee of winning), which can resist the Sybil attack to a certain extent. The Sybil attack attempts to create a large number of nodes to gain influence or control over a distributed network.

The leading 0 target in the example above is just an indication of the difficulty of the computation. In fact, using the bitcoin network as an example, the 600,000th block generated on October 19, 2019 had 19 leading zeros. The 200,000th block generated on September 22, 2012 had only 13 leading zeros. This means that an enormous amount of computing power has gone into bitcoin's calculations.

6.3 IoT Data Security

With the development of the economy and the progress of society, the Internet of Things has emerged as the need of the hour. According to statistics and forecasts from Statista, there were 22 billion IoT devices in the world at the end of 2018, and it is expected that there will be 50 billion IoT devices in the world by 2030, which can form a huge device interconnection network covering all fields and scenarios, from smartphones to kitchen appliances.

This growth is driven by the promise of increased business insight, improved customer satisfaction, and increased efficiency. These benefits are made possible by the functional fusion of sensor data from devices and Internet-based cloud services. Critical to the widespread adoption of the IoT is whether there are sufficiently strong security mechanisms built into the ecosystem to mitigate the security risks of connecting devices to the Internet.

IoT security primarily includes data security, network security, and node security. Among them, network security and node security are mainly for data security. Data integrity, comprehensive collection, secure transmission, and effective protection are the foundation of IoT security.

6.3.1 *Concept of internet of things*

There are several definitions of IoT. In general, the Internet of Things (IoT) refers to the network of connected physical objects, that is, "things". The Internet of Things uses technologies such as sensors, software, and network connectivity to connect devices to other devices and systems and to exchange data.

The combination of computer technology and communications technology constitutes a computer network, that is, the Internet. The combination of Internet and mobile communication technology constitutes the mobile Internet. The superposition of mobile Internet and sensing technology constitutes the Internet of Things. Figure 6.12 shows the composing technologies of the Internet of Things.

The function of IoT can be abstractly summarized as capturing data, transmitting data, processing data, and taking action:

(1) **Capturing data:** Through sensors, IoT devices capture data from their environment. Data can be simple, such as temperature, or complex, such as real-time video, audio. Simple data pre-processing or data analysis can also be performed on IoT devices, such as data sampling, generating alarm information after data exceeds a threshold, and so on.

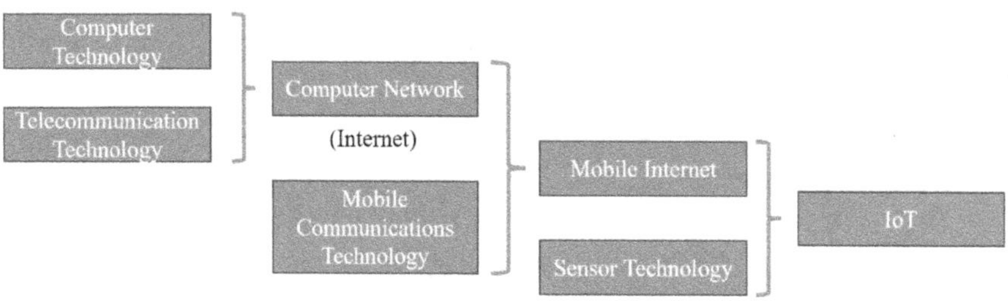

Fig. 6.12. IoT technologies.

(2) **Transmitting data:** IoT devices utilize available network connections to transmit captured or pre-processed data over public or private networks, depending on the configuration.
(3) **Processing data:** Analyze aggregated data from devices within the IoT network, extract the information expressed in the data, and analyze the action plan based on the information.
(4) **Take action:** Based on the analysis and insight of data and information, support the next action or business decision.

The network architecture of IoT can be simply divided into three layers: perception layer, network layer, and application layer. At the perception layer, ubiquitous IoT sensors capture data about objects in their environment. At the network layer, IoT devices transmit data to the cloud directly or through IoT gateways over wired networks and wireless networks. At the application layer, based on the processing and analysis of the data, cloud services generate data-based results, trends, and insights, and take the next step autonomously or manually as needed. The schematic diagram of the IoT architecture is shown in Fig. 6.13.

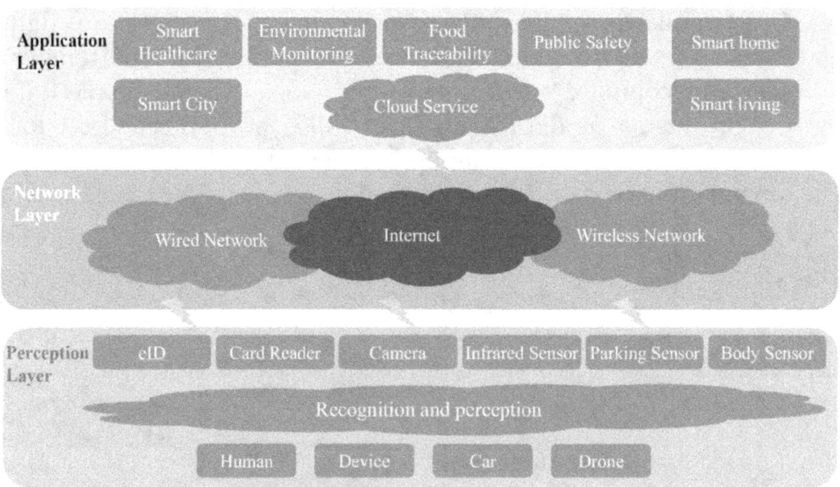

Fig. 6.13. Schematic diagram of IoT architecture.

6.3.2 *Data security risks of the internet of things*

Internet of Things security means that the data in the hardware, software, and systems of the Internet of Things is protected from being damaged, altered, or leaked for accidental or malicious reasons. The Internet of Things system can operate continuously, reliably, and normally so that Internet of Things services are not interrupted. IoT security includes all technical or management means to resolve or mitigate security threats in the application process of IoT technology, as well as the security threats themselves and related activities.

In the article "Best Practices of Internet of Things Security" (Corser *et al.*, 2017) published by IEEE in February 2017, it is pointed out that the security of the Internet of Things is divided into three levels of issues:

(1) Device security issues.
(2) Network security issues.
(3) Data security issues.

The most important data security issue is the most important IoT security issue and the core of IoT security. On the Internet of Things, where everything is connected, information is directly connected to the physical world, which brings great convenience but also great risk. As great as national information security, production security, and even personal private information, because it runs on the Internet of Things, it increases the risk of being stolen, intercepted, and modified. Based on these risks, data protection is essential.

6.3.2.1 *Security attributes of the internet of things*

IoT security also has the following three basic attributes: confidentiality, integrity, and availability. These are three key goals at the heart of security but with slightly different meanings:

(1) Confidentiality mainly refers to the confidentiality of data, ensuring that private or confidential information cannot be used by or disclosed to unauthorized individuals.
(2) Integrity mainly refers to data integrity and system integrity.

    Data Integrity: Ensuring that information and procedures can be changed only in designated and authorized ways.

System integrity: Ensuring that the system performs its intended functions in an uncompromised manner and prevents intentional or unintentional unauthorized operations on the system.

(3) Availability, also known as effectiveness, refers to the characteristic that information resources can be accessed by authorized entities as needed, used normally, or restored to use under abnormal circumstances (system security features for user services). The required information can be accessed correctly when the system is running. If the system is accidentally attacked or damaged, it can be quickly restored and used. This is a measure of the user-oriented security performance of network information systems to ensure the provision of services to users.

In addition, IoT security requirements include privacy, which refers to ensuring that individuals control the collection and storage of their information, as well as control over who can share that information or with whom.

6.3.2.2 *IoT security risk analysis*

A study released by Hewlett Packard in 2014 (Rawlinson, 2019) found that 70% of the most commonly used Internet of Things (IoT) devices have high-risk vulnerabilities, including password security, encryption, and a general lack of proper user access settings.

HP researchers scanned more than a dozen popular IoT devices, including TVs, webcams, home thermostats, remote power outlets, sprinkler controllers, hubs that control multiple devices, door locks, home alarms, scales and garage door openers, and found an average of 25 vulnerabilities per device.

The top five categories of vulnerabilities are as follows:

(1) Eighty percent of IoT devices or their applications/cloud services have issues with collecting consumer privacy data. The data collected includes names, email addresses, home addresses, birth dates, credit card information, and health information.
(2) Eighty percent of IoT devices or their applications/cloud services do not require passwords of sufficient complexity and length.
(3) Seventy percent of IoT devices communicate with the Internet or local networks without encryption, while 50% of devices have

mobile applications that communicate with the cloud, Internet, or local networks without encryption. Since the data collected and transmitted by IoT devices contains sensitive information, encryption during transmission is critical.

(4) The web interface of 60% of IoT devices has security vulnerabilities such as stored XSS, poor session management, weak default credentials, and credentials transmitted in clear text. The cloud and mobile components of 70% of devices are vulnerable to potential attackers identifying valid user accounts through account enumeration or password reset functions.

(5) Sixty percent of IoT devices do not use encryption when downloading software updates, which can lead to manipulation of device functions. Some downloaded software can even be intercepted, extracted, and mounted as a file system in Linux for easy viewing or modification.

The combination of the number of IoT devices, the heterogeneity of protocols, and the complexity of the network create daunting challenges for IoT security. From an IoT device perspective alone, while security is basically controllable, there are still issues such as misconfiguration and lack of updates. There are some challenges with consumer IoT devices because consumers often lack security awareness. From the perspective of the IoT network, or in the industrial IoT scenario, security issues are more complex.

In April 2021, statistics on IoT security (Segal, 2021) showed the following:

(1) Less than 20% of IoT security risk professionals understand all or most of the IoT devices used by their organization.
(2) Seventy-six percent of IoT risk professionals believe their organization's IoT security posture makes them vulnerable to cyberattacks.
(3) Fifty-six percent of organizations do not have a complete list of the IoT devices they use; 64% of organizations do not have a complete list of their IoT applications.
(4) In 2020, IoT malware attacks increased by 30% on the premise that comprehensive factors such as the COVID-19 epidemic led to a 39% decrease in the total amount of malware.

Based on the essential characteristics of the Internet of Things, there are several prominent security risks as follows:

First, the complexity of deployment increases the remote exposure and external attack surface. IoT devices are not deployed in physically separate computer rooms like most IT equipment but may be placed in open spaces in offices, factories, homes, and other locations beyond the control of the owners of various devices. As a result, both the device itself and the network connection present a particularly large attack surface, giving attackers the ability to interact with the device remotely. IoT security must address a large number of entry points to protect assets.

Second, the resource constraints of IoT devices result in a lack of processing power to integrate security features (such as encryption in transit) or security products (such as firewalls). Some devices even have little ability to connect to other devices, such as IoT devices that only support Bluetooth technology. In 2020, a cybersecurity expert exploited a Bluetooth vulnerability to hack into the system of a Tesla Model X in less than 90 seconds. Other cars that rely on over-the-air (FOB) keys to open and start may also be vulnerable to attack for similar reasons.

Third, the complexity of upgrades and maintenance. IoT device manufacturers are typically slower to release patches and updates than companies that make general-purpose operating systems and other general-purpose software. It's common for IoT devices to remain unpatched for weeks or months, running outdated software. Even in some organizations, some devices never receive software or firmware upgrades.

Fourth, end users or operators of IoT devices lack security knowledge or awareness. IoT devices themselves have many security measures in place, but the end user is often the weakest link in an IoT security system. IoT devices are small and easy to carry, and have become inseparable partners in consumers' daily lives and work. If you include smartphones, most people have their own IoT device. The problem is that most people are not aware of the security risks, such as frequently connecting their devices to unknown networks.

Fifth, the industry lacks insight and foresight into IoT security. In the process of digital transformation, industries such as automotive, logistics, and healthcare are using IoT devices on a large scale to improve productivity and efficiency. At the same time, the risk of data leakage from IoT has not yet received sufficient attention from these industries nor has it received sufficient and necessary investment of funds and resources.

6.3.3 *IoT data security response*

On the Internet of Things, the form of data has changed dramatically, and the speed of data is not the same as in the past. In addition, the complexity of IoT device forms and deployment methods has created great challenges for data security and privacy protection.

Currently, IoT devices have multiple connection scenarios, such as local, wireless, and wide area network connections, to transmit various data. In the future, IoT will be more independent of human intervention, and through the deep application of artificial intelligence in IoT, things and things may communicate more and more actively. In such complex interconnection and interaction scenarios, the issues of data confidentiality, integrity, and privacy are becoming more complicated.

To identify and respond to IoT data security challenges, we should also start with the most basic data identification and classification. Questions to consider include: What are the IoT devices in the system? For example, endpoints, gateways, etc. What data needs to be protected on these devices and during transmission? For example, sensitive data includes audio, video, geographic location, sports and health data, etc., which require confidentiality and integrity protection. For physical world data, such as ambient temperature collected by sensors, confidentiality protection is not necessarily required, but integrity (anti-tampering) and availability protection may be.

After clarifying the protected data, from the security requirements analysis and solution design dimensions, the following solutions can be analyzed and implemented based on different scenarios:

First, provide physical security for the device, which is the most basic step. Physical security includes device hardware security, sensor security, sensor anti-jamming and transmission anti-interception. Qualified devices can also consider features such as security chips, encryption and key storage, and device identity authentication. Without the support of physical security, other security measures will be very difficult to implement.

Second, implement the authentication mechanism. By deploying multiple authentication and access control mechanisms for appropriate IoT devices, such as two-factor authentication and fingerprint scanners, the security of human–device and device–device interactions will be enhanced to ensure data security.

Third, system security. System security includes the security of the operating system of IoT devices, the hardening of the operating system, and the security of the APIs provided by the operating system. For example, operating system code should be digitally signed to prevent tampering. For the interface used to transmit data, it is necessary to analyze whether it provides data encryption function and integrity protection function.

Fourth, transport security. Transport layer protocol design must systematically analyze threats and risks in terms of security and privacy, and derive security requirements. For example, personal wearable devices track and record an individual's location in time and space. How to prevent the risk of invasion of privacy by tracking the individual's location without the individual's knowledge is one of the issues to be considered in protocol design.

Fifth, network security. Some devices or functions of traditional network security can still be used as means or complements of Internet of Things network security or network perimeter security, such as gateway security, firewall, antivirus and anti-malware products, intrusion detection system (IDS), intrusion prevention system (IPS).

Sixth, application security. IoT application developers need to pay attention to the security and privacy issues of their IoT applications. When designing an IoT system, it should conduct a comprehensive analysis of its security, and try to find an appropriate balance between user experience and protection of IoT applications.

Seventh, cloud security. Due to the limited processing power of IoT devices, the analysis and processing of IoT data is often done in the cloud. Cloud data security solutions need to take into account the unique characteristics of IoT data, such as massive endpoints and fast flow.

The above solutions are considered based on the design and implementation of the product. In addition to technical solutions, organizational security policies cannot be ignored, especially the security policies of organizations such as IoT device manufacturers, IoT software developers, and IoT operation service providers.

In addition, the "human" factor must be fully considered. Owners of IoT devices need to be aware of all possible security threats to keep their data safe. IoT operators and maintainers need to understand their security responsibilities to avoid data leakage and privacy violations.

6.4 5G and Data Security

The latest generation of cellular mobile communication networks (5G) is gradually being used by people, and organizations and consumers are looking forward to faster transmission rates and richer application scenarios. 5G has ushered in a new round of technological and industrial changes, and has gradually become an important support and driving force for global economic development.

5G offers higher data rates by using higher frequency radio waves than previous cellular networks. Other benefits of 5G include reduced latency, energy savings, lower costs, and increased system capacity and device connectivity at scale — the Internet of Things (IoT).

In addition to IoT, 5G's high data rate and low latency will support technologies such as virtual reality (VR) and augmented reality (AR) and other application scenarios, as well as handle the large amounts of data required for the safe operation of autonomous vehicles.

The first phase of the 5G specification (Release 15) was completed in April 2019 to support early commercial deployments (3GPP, 2019), and the second phase (Release 16) was completed in July 2020 (Baek *et al.*, 2021).

According to the GSMA's Mobile Economy Report 2024 GSMA 2024, 5G is expected to add more than $930 billion to the global economy by 2030. Most of this will be achieved in developed regions, including Asia-Pacific, North America and Europe, which are expected to see strong growth over the next five years. By 2040, low- and middle-income countries are expected to capture an increasing share of the 5G economy as 5G technology begins to be scaled up and widely adopted. From 2023 to 2030, 36% of revenues are expected to come from manufacturing, 15% from government, and 10% from services, driven by smart factory, smart city, and smart grid applications.

Most people in the telecom industry believe that 2019 is the "first year" of the 5G mobile network industry, as the first batch of experimental networks for 5G services were launched in the fourth quarter of 2018.

In fact, major operators around the world are accelerating the deployment of 5G networks, and various industries, university and research institutions are also actively exploring 5G industry application scenarios. In China, the three major operators (China Mobile, China Telecom, and China Unicom) simultaneously announced the launch of 5G services for commercial use in October 2019. NTT DOCOMO, Japan's largest telecommunications operator, also launched 5G services for pre-commercial use in September 2019, and will officially launch services for commercial use in 2020. In Europe, Norway, Germany, the United Kingdom, and other countries have also launched commercial use services for 5G networks in 2020.

5G defines a new mobile network architecture and introduces key technologies such as Network Functions Virtualization (NFV), network slicing, and edge computing to support Enhanced Mobile Broadband (eMBB), Massive Machine-Type Communications (mMTC), Ultra-Reliable Low Latency Communications (URLLC), and other application scenarios. The 5G application scenario is shown in Fig. 6.14.

As networks and services continue to innovate and evolve rapidly, it is very important for 5G standard-setters, industry participants and consumers to identify 5G-related security risks in a timely manner and respond effectively. From a security perspective, new network architectures, new key technologies and new application scenarios

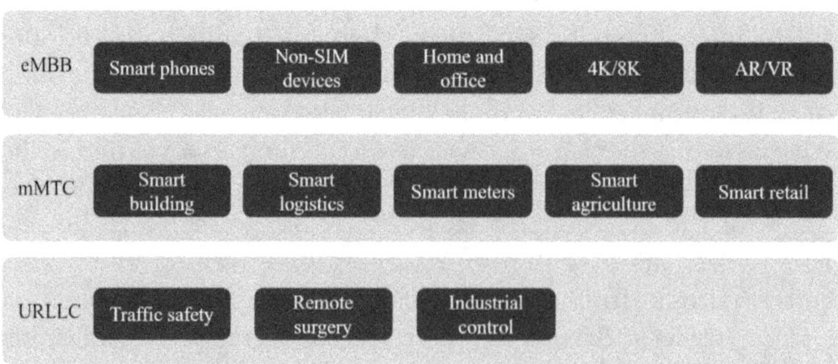

Fig. 6.14. 5G application scenarios.

introduce security risks at different levels and dimensions, which are described in detail in the following.

6.4.1 *5G data security standards*

The 3rd Generation Partnership Project (3GPP), a global communications technology organization, is responsible for formulating 3G, 4G, and 5G mobile communications standards. The organization has established an independent Security and Privacy Working Group (SA WG3) to develop a set of 5G security standards (Prasad *et al.*, 2018). This working group is responsible for identifying the security and privacy requirements of 5G, and defining the security architecture and associated protocols to meet these requirements. The formulation of encryption algorithm specifications is also within the scope of this working group.

The overall definition specification of 5G security technology 3GPP TS 33.501 "Security architecture and procedures for 5G system" has two versions for Release 15 and Release 16. The specification was released in version 16.4.0 in September 2020.

5G is an evolution of 4G, but compared to 4G, 5G network is more heterogeneous, deployment is more complex, form is more virtualized, and services are more open. The security steps for 5G are as follows:

First, focus on whether it can provide the same security capabilities as 4G networks and network elements.

Second, based on the vulnerability analysis and security threat assessment of 4G networks and network elements, analyze the corresponding security enhancement and security evolution requirements and define the security architecture and security protocols based on the attack scenarios such as wireless interface, signaling plane, user plane, counterfeiting, privacy stealing, and man-in-the-middle attacks.

Again, based on the new network architecture and deployment environment of 5G, define the corresponding security standards.

Finally, based on the new types of services and new capabilities provided by 5G, analyze the security and privacy threats and generate corresponding security requirements and security standards.

The steps described above summarize the different dimensions of 5G security protocols and are not done in chronological order.

In short, 3GPP's security standards are divided into three main categories: scenario-oriented service security and industry

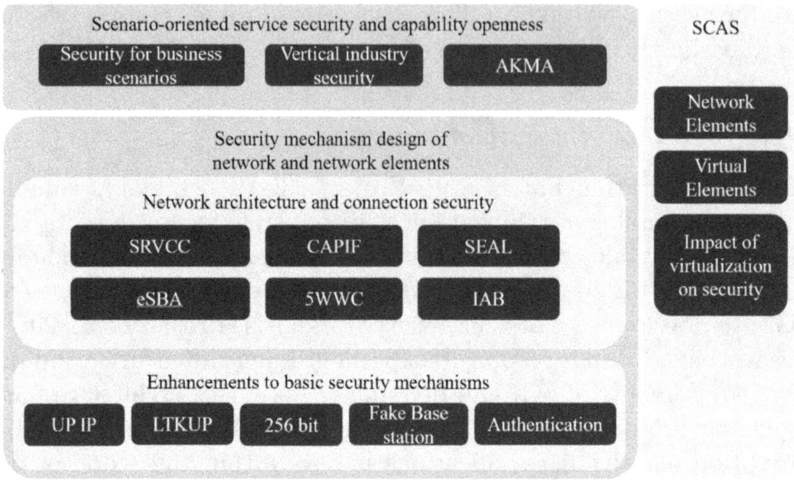

Fig. 6.15. 5G network security standards.

empowerment, 5G network and NE's own security mechanism design, and security assurance requirements. Figure 6.15 shows the 5G network security standards.

Scenario-oriented service security and industry empowerment include the following:

(1) Security of business scenarios such as Cellular Internet of Things (cIoT), URLLC, and Internet of Vehicles.
(2) Local area network security for slicing and vertical industries.
(3) Security open architecture (Authentication and Key Agreement for Applications, AKMA): Provides authentication and key management for applications.

Network architecture and connection security include the following:

(1) SRVCC (Single Radio Voice Call Continuity, a VoLTE voice service continuity solution proposed by 3GPP) security.
(2) Common API Framework (CAPIF; used for open architecture) security.
(3) Vertical Service Enabler Architecture Layer (SEAL) security.
(4) Enhanced Service-Based Architecture (enhanced SBA, eSBA) security;
(5) 5G Wireless Wireline Convergence (5WWC) security.
(6) Integrated Access and Backhaul (IAB) security.

Basic security mechanism enhancements include the following:

(1) **UP IP:** Enhancement of uplink IP link security.
(2) Long Term Key Update Process (LTKUP).
(3) **256bit:** Support 256-bit key length and encryption algorithm to prevent quantum computers from attacking cryptography.
(4) Enhanced ability to detect rogue base stations.
(5) **Authentication enhancement:** The defense against user traceability attacks is standardized by the TS33.846 series of protocols.

Security assurance includes three dimensions: security assurance of network equipment, security assurance of virtualization equipments, and the impact of virtualization on security.

From a data security perspective, 5G data security requirements cover three dimensions: data regulatory requirements, data protection technology, and user privacy protection. These requirements are mapped to different business scenarios, and there are different standards, such as vertical industry standards and open capabilities, and data lifecycle protection requirements. At the network security layer and terminal security layer, the network and terminal provide appropriate capabilities and security control measures to support and protect data security. 5G relies on infrastructure such as virtualization, cloud platforms, and SDN (software-defined networking). Infrastructure security is a key cornerstone. This part may not be fully defined in 3GPP international standards. It needs to work with the IETF, which coordinates the ISO family of international standards. Domestically, standards organizations such as China Communications Standards Association (CCSA) and TC260 have also taken the lead in formulating a series of 5G standards and security standards. The framework, capabilities and control measures provided by the basic capability layer include other layers, such as the IPSec security transmission technology defined by the IETF standards organization. In addition, technical technologies such as access control technology, sandbox and isolation technology, and encryption technology provide solid security support for terminals, clouds, devices, and services.

In addition, domestic 5G network security operations, especially related business and processes of security operation and maintenance, emergency response, and content control, also have corresponding standard definitions.

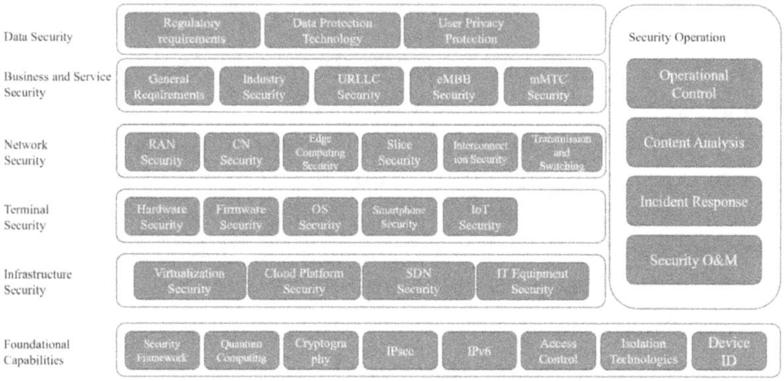

Fig. 6.16. Layered architecture of 5G data security standards.

Figure 6.16 shows the layered architecture of 5G data security standards.

From a layered architecture perspective, the security of new network architectures such as edge computing and slicing in the network security layer, and the security of new business scenarios such as vertical industries, URLLC, eMBB, and mMTC in the business and service security layer are the key changes in 5G data security.

6.4.2 *Security of 5G new network architecture*

With the development of 5G networks, cloud virtualization technologies such as software-defined networking (SDN) and network functions virtualization (NFV) are becoming more popular, but they also bring new security risks. Due to the open, flexible, and programmable nature of cloud virtualization technologies, SDN and NFV open up a new avenue of security threats. For example, network elements of SDN, such as the management interface, can be used to attack the SDN controller or management system and shut down the system.

The 3GPP 5G Security paper in the *Journal of ICT Standardization* (Prasad *et al.*, 2018) points out that a multi-pronged approach to 5G security includes trust models, authentication and key agreement (AKA), and secondary authentication based on Extensible Authentication Protocol (EAP).

The security of the 5G network infrastructure needs to evolve with the standards. For example, since a 5G network can be divided into slices for different purposes, each virtual network slice may require unique security features based on the needs of different usage scenarios. Similarly, infected 5G equipment on the radio access network (RAN) side may pose a greater distributed denial of service (DDoS) threat.

6.4.3 *Security of 5G new service scenarios*

5G networks have the potential to spark explosive growth in vertical industries, enabling the creation of a wide variety of new services, all of which will require new and different levels of security.

For example, with the proliferation of self-driving cars, the threat of automotive cyberattacks will increase. To address this, the National Highway Traffic Safety Administration (NHTSA) is taking a layered approach to cybersecurity when approving driver-assist technologies.

In healthcare, 5G capabilities will facilitate faster transfer of large image files of patients through IoT devices for remote surgery and remote monitoring of patients. However, these technological advances have also brought with them the need for stricter security. Potential security risks include medical identity theft, health privacy breaches, and disruptions in medical data management, among others. According to Verizon's "2019 Data Breach Investigation Report" (Verizon Enterprise Solutions, 2019), the healthcare industry is the second most targeted industry for attacks. There were 466 security incidents in this industry in 2019, of which 304 incidents were confirmed to have been caused by a data breach. The growing adoption of IoT devices will make it even more challenging to address the increasing data security risks.

Smart homes also require stronger authentication methods. For example, biometric authentication methods are increasingly used in smart homes. Smart speakers support voice-based identification, and smart door locks are starting to support fingerprint unlocking.

In the 5G scenario, IoT devices and sensors will require more complex authentication to prevent unauthorized access.

6.4.4 *Formal analysis of 5G authentication protocol*

A 2018 study by a team from ETH Zurich, the University of Lorraine, INRIA, and the University of Dundee describes some of the concerns surrounding 5G communication protocols.

In the paper, "Formal Analysis of 5G Authentication" (Basin *et al.*, 2018), the researchers performed a comprehensive security analysis of the Rel-15 version of the 5G standard. They concluded that 5G has improved privacy compared to previous 3G and 4G standards but still has security vulnerabilities.

The researchers systematically examined the 5G Authentication and Key Agreement (AKA) security protocols using a security protocol verification tool for analyzing cryptographic protocols, taking into account the specified security goals.

The security tool automatically identifies the minimum security standards required to meet the security goals set by the 3rd Generation Partnership Program (3GPP). The 3rd Generation Partnership Program is a collaboration between the Telecommunications Standards Association (TSI) community and the developer of 5G communications standards.

The analysis shows that this security standard is insufficient to meet all of the key security objectives of the 5G AKA protocol.

The researchers also found that 5G authentication and key security protocols allow for certain types of traceability attacks. In this attack, the phone does not send the full identity of the user to the tracking device, and the researchers are still able to determine the approximate location of the phone.

The researchers believe that more sophisticated tracking devices could also pose a threat to 5G users in the future. The research team believes that as new mobile technologies are introduced with the Rel-15 standard, it could lead to a large number of network security attacks.

An article published by security news site SDxCentral, "What Are the Key Challenges to 5G Security?" (SDxCentral Studios, 2020) outlines some of the key security challenges, threats, and vulnerabilities associated with 5G, such as the threat of automotive cyberattacks as self-driving cars gain popularity, medical identity theft associated with new healthcare and health industry application scenarios, and unauthorized access to IoT devices.

According to the article, new cloud and virtualization technologies, such as software-defined networking (SDN) and network functions virtualization (NFV), are gradually becoming more popular in 5G networks, but at the same time, they bring new security risks. Due to their open, flexible and programmable nature, SDN and NFV can be insecure technology isolation. The security of 5G network infrastructure needs to evolve with the standards.

SDxCentral cites Cisco's recommendations for security protections that many organizations should be concerned about. One of the recommendations is that by using firewalls to protect the network, access controls to minimize user-based risks, and intrusion detection and prevention tools to block basic 5G security threats, the fundamental problem with most security incidents can be minimized.

Another recommendation is to detect and block advanced malware attacks that evade basic filters through signature verification tools and other means. It is important to conduct behavior-based inspections (possibly using sandbox inspections) of endpoints. Once a threat is detected, organizations should be able to remove all instances of the threat from the network.

Other suggestions include using packet captures, big data, and machine learning to identify threats that are not detected by basic filters. These technologies will be even more effective when embedded in network switches and routers, turning the devices into 5G security sensors.

6.5 Outlook for 6G and Data Security

Predictions show that since the first generation of mobile communication systems appeared in 1982, a new generation of mobile communication systems has appeared every ten years. Therefore, although 5G has just started commercial services in most parts of the world, or has not even started commercial services yet, Beyond 5G (B5G)/6G is already on the research road.

If we look back at previous generations of wireless systems, 2G brought voice, 3G brought Internet access and video calling, and 4G brought a rich application ecosystem. Wireless technology and applications are evolving rapidly, from smartphones to tablets to various Internet of Things devices to autonomous driving. 4G can no longer meet the needs of these scenarios. 5G offers high data rates, low latency, and large-scale connectivity services, with the vision of enabling a truly connected society. The proposed 6G will provide further improvements for bandwidth-intensive needs, while enabling further penetration of artificial intelligence technology and providing higher levels of availability for specific industrial environments such as robotic control, remote surgery or connected cars.

This section outlines the research directions of 6G, the evolution of 6G security, and the attack and defense of terahertz networks.

6.5.1 *6G research direction*

The mobile communications industry is evolving rapidly. As 5G has just started commercial use in China, Japan, South Korea and other East Asian regions in 2019, 6G has entered the research process. The current general view is that terahertz communication will become the new radio technology that constitutes 6G.

It is expected that 6G will bring at least 100GB/s transmission rate and lower latency. In contrast, the International Telecommunication Union (ITU) IMT-2020 predicts that 5G speeds will evolve to 20 Gbps and beyond with the adoption of high mmWave frequencies (M Series, 2015). In 2018 field tests by Verizon (US operator) and Nokia, based on the 28 GHz band, 5G achieved a delay of 1.5 ms and a throughput of 1.8 Gbps. In contrast, 4G often uses bands of a few gigahertz, with typical download speeds of less than 20 Mbps.

"6G is expected to use high frequencies in the range of 100 GHz to 1 THz (terahertz) for 100 Gbps transmission rates", UC Santa Barbara scientists said in a 2018 press release. By contrast, Verizon, Qualcomm, and Novatel Wireless's 5G mmWave experiment was conducted in the 39 GHz spectrum (Patrick, 2018).

"Extreme densification of communications systems, enabling hundreds and even thousands of simultaneous wireless connections, ... with 10 to 1,000 times higher capacity than the nearer-term 5G systems and network". the researchers said.

Medical imaging, augmented reality and the Internet of Things are expected to benefit from the higher data rate and lower latency of 6G. (IoT) sensors are some of the target application scenarios for 6G.

In terms of key technologies for 6G networks, the research hotspots are spatial multiplexing technology and multiple-input multiple-output (MIMO) technology used in Wi-Fi and 5G networks. Spatial multiplexing allows individual data signals to be sent in streams to ensure efficient, continuous reuse of bandwidth. MIMO is a way to take advantage of multi-path to maximize the antenna, which also improves efficiency. Overall, terahertz communications should require less power and more capacity.

Based on the propagation characteristics of radio waves, in the frequency bands of terahertz communications, the wavelength is shorter, and obstacles will become an obvious problem. In this scenario, the study of the reflection and diffraction of radio waves in this frequency band will be more important.

6.5.2 *Evolution of 6G security*

Although 5G has not yet been rolled out globally, technological breakthroughs will not stop. 6G research needs to consider and design security earlier. Specifically, 6G network architecture, application scenarios, security of transmitted data (confidentiality, integrity, availability), and consumer privacy are key issues to be considered.

Looking ahead, the security goal of 6G should be based on the security of 6G application scenarios. Although it has not been formally defined yet, based on the development direction of 5G, at least the following key points need to be considered. The concept of 6G security architecture is shown in Fig. 6.17.

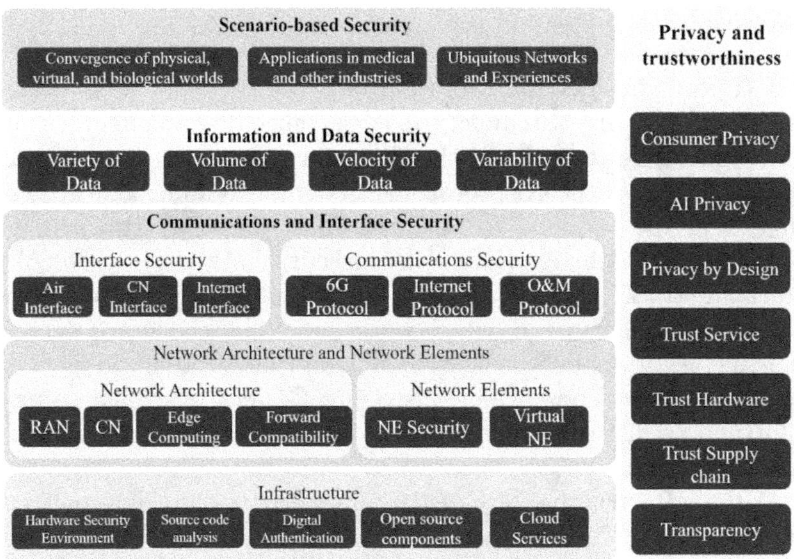

Fig. 6.17. 6G security architecture assumptions.

In the target scenarios of 6G, the physical, virtual and biological worlds will be further integrated; ubiquitous networks and ubiquitous experiences will be further popularized; in vertical industry fields, a variety of new applications will appear in medical, education, transportation and other industries. Even aerospace networks, satellite Internet and ocean networks are included in the design of 6G target scenarios.

These scenarios directly affect the level of information and data security, including the further expansion of the form, connotation, and category of data, the rapid increase of data volume, the order of magnitude increase of data transmission speed, and the change of data processing methods (cloud computing, edge computing, and on-device computing).

As for communication and interface security, it is more complicated due to the further integration of heterogeneous networks, including the complexity of the air interface in the terahertz communication frequency band, the evolution of the core network, and the interface security with the Internet. The evolution of communication security itself must take into account not only the 6G communication protocol but also the IETF Internet protocol and the Operation and Maintenance (O&M) protocol, and even some disruptive technologies, such as the impact of post-quantum cryptography.

At the level of network architecture and network elements, it is necessary to consider the evolution of access network architecture, core network security architecture, edge computing architecture, and even connectivity security with the historical architecture. The 2G fallback feature of the 4G protocol is a counterexample. The attacker can trigger the terminal to fall back from the 4G network to the 2G network, thereby exploiting the insecurity of the 2G protocol to launch an attack. At the same time, traditional network elements still need to be compatible in 6G networks, and the security of network elements needs to be considered; in addition, virtualized network elements will be more widely used in 6G networks. The security of virtualized network elements and virtualized networks must be considered.

At the infrastructure level, existing security technologies will still play a role in the 6G era, such as hardware and chip-based security environments, source code security, certificates and authentication, etc. It is worth noting that the network and network elements in the 6G era will become even more software-based, generalized, and cloud-based. Whether it is the implementation of network protocols or basic computing and storage functions, they all rely heavily on open source code. Open source security, vulnerability prevention, and security integration are key points of infrastructure that are often overlooked.

There is another dimension to the 6G security panorama: privacy and trustworthiness. Privacy-preserving design, or the integration of privacy-preserving requirements into the design of 6G products, is critical. Privacy here refers not only to consumers' personal data processed, transmitted, and stored in the network but also to different privacy requirements for each scenario, business characteristics, and even end-to-end privacy protection for enabling technologies such as AI. In the trustworthiness dimension, how to deliver trustworthy products to operators and industry through hardware trustworthiness and supply chain trustworthiness, realize trustworthy operation and maintenance of the 6G network through service trustworthiness, and be transparent to government regulators, upstream and downstream, media and industry partners is the focus of every equipment manufacturer.

The EU 5G PPP organization's "Smart Connectivity beyond 5G" (European Union, 2020) project aims to provide "personalized, multitenant and permanent protection based on security, privacy and trust mechanisms in highly virtualized and software environments, considering including hardware security functions and the end-to-end dimension of software processes. Blockchain technology can be explored in this context". This project shows the possible direction of data security and privacy protection after 5G.

6.5.3 *Attack and defense of terahertz network*

Currently, the 6G security research direction in the academic community is focused on the feasibility study of attacks on terahertz networks, and papers have been published.

In October 2018, researchers from Brown University, Rice University, and the University at Buffalo published the paper "Security and eavesdropping in terahertz wireless links" (Ma *et al.*, 2018) in the journal Nature, describing serious vulnerabilities in 5G's successor, terahertz data communication networks.

Terahertz, the extremely high wavelength between microwaves and infrared, could form the basis of 6G networks, which could be deployed by 2025 or even 2030. 6G may use sub-millimeter to terahertz frequencies, much higher than those used by 5G (millimeter spectrum). If feasible, 6G should offer higher reliability and lower latency than 5G.

"The conventional wisdom in the terahertz community is that it's almost impossible to monitor a terahertz data link unnoticed", said Daniel Mittleman, a professor at Brown University's School of Engineering and a co-author of the paper. "But we demonstrate that undetected eavesdropping in the terahertz space is easier than most imagine, and that we need to consider security when designing network architectures".

Due to its high frequency, terahertz communications can carry 100 times more data than the microwaves used in wireless communications today, making terahertz an attractive option for 6G wireless networks. As bandwidth increases, it is also widely believed that the way high-frequency radio waves travel naturally enhances security. Unlike microwaves, which travel in wide beams, terahertz waves travel in narrow and highly directional beams.

The extremely short wavelengths of terahertz waves have led many to believe that they are too small to intercept. Man-in-the-middle receivers placed in narrow, directional terahertz beams to avoid interference would block entire transmissions and be immediately detected. But research now shows that this assumption is wrong.

In microwave communications, the signal is slightly conical and covers a wide area. Eavesdroppers can place the receiver almost anywhere and pick up the signal without interfering with other receivers in the coverage area. Assuming an attacker can decode the signal, they can eavesdrop undetected. But in terahertz networks, the narrow beam means that the eavesdropper must place the antenna between the transmitter and receiver. The previous assumption was that eavesdropping would cause some or all of the signal to be blocked, making the eavesdropping attempt easily detectable by the intended receiver.

Daniel Mittelman and his team at Brown University, Rice University, and the University at Buffalo set out to test this hypothesis, as shown in Fig. 6.18. They established a direct terahertz data

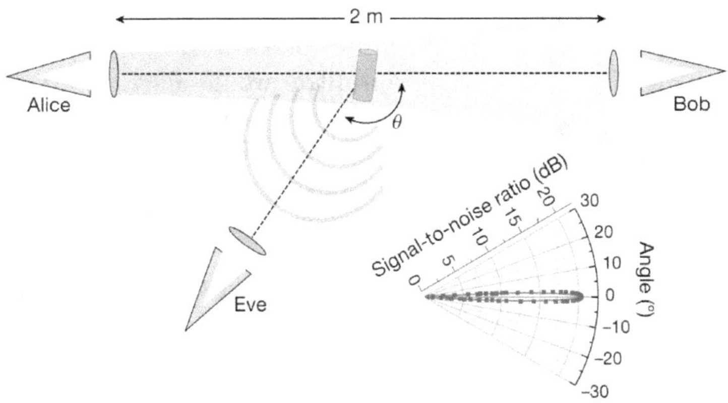

Fig. 6.18. Schematic diagram of terahertz transmission network monitoring.

link between the transmitter and receiver and tried to install devices to intercept the signal. They were able to experiment with several strategies that allowed them to steal the signal undetected, even when the signal carried a beam of light that was highly directional, with a cone angle of less than 2° (cellular networks often use 120° sectors).

One set of strategies involves placing objects at the edge of the beam, which scatters a small portion of the beam. For a reliable data link, the diameter of the beam must be slightly larger than the aperture of the receiver. This allows an attacker to work with a small segment of the signal without leaving a detectable shadow on the receiver.

The researchers found that a flat piece of metal can redirect a portion of the beam to a secondary receiver operated by an attacker. The researchers were able to obtain a usable signal at the secondary receiver without significant power loss at the primary receiver.

The research team also demonstrated a more flexible approach (from an attacker's perspective) by using metal cylinders in the beam instead of flat plates.

Study co-author Josep Jornet, an assistant professor of electrical engineering at the University at Buffalo, also noted that cylinders have the advantage of scattering light in all directions, giving attackers more options for mounting receivers. Due to the physics of terahertz wave propagation, even small cylinders can significantly scatter the signal without blocking the line-of-sight path.

The researchers then demonstrated another attack using lossless beam splitters that is difficult, if not impossible, to detect. A beam splitter placed in front of the transmitter would allow an attacker to steal enough to be useful without alerting network administrators.

Most importantly, the researchers note that there are inherent security enhancements in terahertz links compared to lower frequencies, but these security enhancements are far from foolproof.

6.6 Chapter Summary

In the era of convergence of all things, the further convergence of the physical world, the digital world, and even the biological world, and the continuous emergence of new scenarios, new businesses, and new processing methods have brought great challenges to data security and privacy protection. Focusing on the data itself, as mentioned in Chapter 1, with the rapid popularization of 5G, the Internet of Things, and artificial intelligence, the scope, speed, quantity, and quality of data have increased by orders of magnitude, and the risks to data security have grown accordingly.

Focusing on the level of protection, the goals of protection (data confidentiality, integrity, and availability) remain unchanged, and the principles of protection (such as defense-in-depth architecture, data-centric security, zero-trust architecture, etc.) can still be followed. It is only in the dimension of specific control measures, based on threat modeling in different business scenarios, that different security control measures need to be designed and implemented to mitigate or eliminate negative impacts such as threats.

Security is a control measure in the technical dimension, and it also depends on the deep understanding and participation of everyone in every link of the whole chain. A good data security ecosystem requires upstream and downstream collaboration. Security is the defense against threats and harm from potential attackers, while privacy is an organization's agreement and commitment to its users, customers, or consumers.

Let's work together to make connections more secure, data more protected, and privacy more credible, and welcome a brighter future of integrating it all together!

Bibliography

3GPP (2019). Release 15, https://www.3gpp.org/release-15.

Abdalla, M., Haase, B., and Hesse, J. (2021). Security analysis of cpace, in *International Conference on the Theory and Application of Cryptology and Information Security* (Springer), pp. 711–741.

Aftab, P. and Savitt, N. L. (1999). The children's online privacy protection act of 1998, *Preventive L. Rep.* **18**, 32.

Alwen, J., Coretti, S., Dodis, Y., and Tselekounis, Y. (2020). Security analysis and improvements for the ietf mls standard for group messaging, in *Annual International Cryptology Conference* (Springer), pp. 248–277.

Anderson, R. (1994). A5 (was: Hacking digital phones), *Newsgroup Communication.*

Anina Ot (2022). The cloud access security broker (CASB) market, https://www.datamation.com/applications/cloud-access-security-broker-market/.

Apple Inc. (2019). ios security (ios 12.3), https://css.csail.mit.edu/6.858/2023/readings/ios-security-may19.pdf.

Apple Inc. (2022). Apple platform security, https://help.apple.com/pdf/security/en_US/apple-platform-security-guide.pdf.

Armour, W. W., Bukhari, N., & Butler, W. (2013). *Nist Cloud Computing Security Reference Architecture*, Tech. rep., National Institute of Standards and Technology.

Arute, F., Arya, K., Babbush, R., Bacon, D., Bardin, J. C., Barends, R., Biswas, R., Boixo, S., Brandao, F. G., Buell, D. A., *et al.* (2019). Quantum supremacy using a programmable superconducting processor, *Nature* **574**, 7779, 505–510.

Baek, S., Kim, D., Tesanovic, M., and Agiwal, A. (2021). 3gpp new radio release 16: Evolution of 5g for industrial internet of things, *IEEE Communications Magazine* **59**, 1, 41–47.

Ball, P. (2016). Google moves closer to a universal quantum computer, *Nature News.*

Bambysheva, N. (2021). Satoshi & company: The 10 most important scientific white papers in development of cryptocurrencies, https://www.forbes.com/sites/ninabambysheva/2021/02/13/satoshi-company-the-10-most-important-scientific-white-papers-in-development-of-cryptocurrencies/.

Barbaschow, A. (2019). Public transport victoria in breach of privacy act after re-identifiable data on over 15m myki cards released, https://www.zdnet.com/article/public-transport-victoria-in-breach-of-privacy-act-after-re-identifiable-data-on-over-15m-myki-cards-released/.

Barkan, E., Biham, E., and Keller, N. (2008). Instant ciphertext-only cryptanalysis of gsm encrypted communication, *Journal of Cryptology*, **21**, 392–429.

Barker, E., Smid, M., Branstad, D., and Chokhani, S. (2013). *Nist Special PubliCation 800-130: A Framework for Designing Cryptographic Key Management Systems,* National Institute of Standards and Technology.

Basin, D., Dreier, J., Hirschi, L., Radomirovic, S., Sasse, R., and Stettler, V. (2018). A formal analysis of 5g authentication, in *Proceedings of the 2018 ACM SIGSAC Conference on Computer and Communications Security*, pp. 1383–1396.

Bawa, N. (2019). It's time to explain blockchain, https://www.forbes.com/sites/forbestechcouncil/2019/01/31/its-time-to-explain-blockchain/.

Ben-Sasson, E., Chiesa, A., Tromer, E., and Virza, M. (2014). Succinct {Non-Interactive} zero knowledge for a von neumann architecture, in *23rd USENIX Security Symposium (USENIX Security 14)*, pp. 781–796.

Bernstein, D. J. *et al.* (2008). Chacha, a variant of salsa20, in *Workshop Record of SASC*, Vol. 8 (Citeseer), pp. 3–5.

Bernstein, D. J. (2008). The salsa20 family of stream ciphers, in *New Stream Cipher Designs: The eSTREAM Finalists* (Springer), pp. 84–97.

Bertoni, G., Daemen, J., Peeters, M., and Van Assche, G. (2009). Keccak sponge function family main document, *Submission to NIST (Round 2)* **3**, 30, 320–337.

Big Data Working Group (2013a). Expanded top ten big data security and privacy challenges, *Cloud Security Alliance.*

Biryukov, A., Shamir, A., and Wagner, D. (2001). Real time cryptanalysis of a5/1 on a pc, in *Fast Software Encryption: 7th International Workshop, FSE 2000 New York, NY, USA, April 10–12, 2000 Proceedings 7* (Springer), pp. 1–18.

Bitansky, N., Canetti, R., Chiesa, A., and Tromer, E. (2012). From extractable collision resistance to succinct non-interactive arguments of knowledge, and back again, in *Proceedings of the 3rd Innovations in Theoretical Computer Science Conference*, pp. 326–349.

Bok, S. (1982). *Secrets: On the Ethics of Concealment and Revelation* (Oxford University Press, New York).

Briceno, M., Goldberg, I., and Wagner, D. (1999). A pedagogical implementation of the gsm a5/1 and a5/2, *Internet Security Applications, Authentication and Cryptography Group (ISAAC) at the University of California at Berkeley and the smartcard developers association.*

CAICT (2019). White paper on personal information protection in the smart terminal industry (2018) [zh-cn], http://www.caict.ac.cn/kxyj/qwfb/bps/201812/P020181203675524292768.pdf.

Calzada, I. (2022). Citizens' data privacy in China: The state of the art of the personal information protection law (PIPL), *Smart Cities* **5**, 3, 1129–1150.

Castro, M., Liskov, B., *et al.* (1999). Practical byzantine fault tolerance, in *OsDI*, **99**, 173–186.

Catalin, C. (2019). Chinese companies have leaked over 590 million resumes via open databases, https://www.zdnet.com/article/chinese-companies-have-leaked-over-590-million-resumes-via-open-databases/.

CCPA, C. (2018). California consumer privacy act of 2018.

Center for Internet Security (2022). CIS controls, https://www.cisecurity.org/controls.

Center for Internet Security (2023). CIS benchmarks list, https://www.cisecurity.org/cis-benchmarks.

CEPAL Electronic Engineers (2000). Recommended practice for architectural description of software intensive systems, *IEEE Standards Department, Piscataway, New Jersey, USA.*

Chappell, D. (2015). Introducing azure machine learning, *A Guide for Technical Professionals, Sponsored by Microsoft Corporation.*

Chaum, D. L. (1979). *Computer Systems Established, Maintained and Trusted by Mutually Suspicious Groups* (Electronics Research Laboratory, University of California).

Chui, M. (2023). The economic potential of generative AI: The next productivity frontier, https://www.mckinsey.com/capabilities/mckinsey-digital/our-insights/the-economic-potential-of-generative-ai-the-next-productivity-frontier.

Churchhouse, R. F. (2002). *Codes and Ciphers: Julius Caesar, the Enigma, and the Internet* (Cambridge University Press).

Cloud Security Alliance (2013). Cloud security alliance security, trust & assurance registry, https://cloudsecurityalliance.org/star.

Cohn-Gordon, K., Cremers, C., Dowling, B., Garratt, L., and Stebila, D. (2020). A formal security analysis of the signal messaging protocol, *Journal of Cryptology* **33**, 1914–1983.

Corser, G., Fink, G., and Bielby, J. (2017). Internet of things (IoT) security best practices; IEEE internet technology policy community; white paper, *IEEE: Piscataway, NJ, USA.*

Costello, K. (2020). Gartner predicts the future of ai technologies, https://www.gartner.com/smarterwithgartner/gartner-predicts-the-future-of-ai-technologies/.

Dalenius, T. (1977). Towards a methodology for statistical disclosure control, *Statistics Sweden* **15**, 2–1.

Data Governance Institute (2024). Dgi data governance framework, https://datagovernance.com/the-dgi-data-governance-framework/.

Dempsey, K., Chawla, N., Johnson, L., Johnston, R., Jones, A., Orebaugh, A., Scholl, M., and Stine, K. (2011). *Information Security Continuous Monitoring (ISCM) for Federal Information Systems and Organizations,* Tech. rep., National Institute of Standards and Technology.

Den Boer, B. and Bosselaers, A. (1993). Collisions for the compression function of md5, in *Workshop on the Theory and Application of Cryptographic Techniques* (Springer), pp. 293–304.

Dobbertin, H. (1996). Cryptanalysis of md5 compress, *Rump Session of Eurocrypt* **96**, 71–82.

Doyle, C. and Bagaric, M. (2005). *Privacy Law in Australia* (Federation Press).

DPP (2021). Committed to security, https://www.thedpp.com/security.

Dwork, C., McSherry, F., Nissim, K., and Smith, A. (2006). Calibrating noise to sensitivity in private data analysis, in *Theory of Cryptography: Third Theory of Cryptography Conference, TCC 2006, New York, NY, USA, March 4-7, 2006. Proceedings 3* (Springer), pp. 265–284.

Dwork, C. (2008). Differential privacy: A survey of results, in *International Conference on Theory and Applications of Models of Computation* (Springer), pp. 1–19.

Dworkin, M. *et al.* (2016). Recommendation for block cipher modes of operation: Methods for format-preserving encryption, *NIST Special Publication 800-38G.* https://doi.org/10.6028/nist.sp.800-38g.

Dworkin, M. (2007). *Nist Special Publication 800-38d: Recommendation for Block Cipher Modes of Operation: Galois/Counter Mode (GCM) and gmac,* US National Institute of Standards and Technology http://csrc.nist.gov/publications/nistpubs/800-38D/SP-800-38D.pdf.

Ehrsam, W. F., Meyer, C. H., Smith, J. L., and Tuchman, W. L. (1978). Message verification and transmission error detection by block chaining, US Patent 4074066.

ElGamal, T. (1985). A public key cryptosystem and a signature scheme based on discrete logarithms, *IEEE Transactions on Information Theory,* **31**, 4, 469–472.

Erlingsson, Ú., Pihur, V., and Korolova, A. (2014). Rappor: Randomized aggregatable privacy-preserving ordinal response, in *Proceedings of the 2014 ACM SIGSAC Conference on Computer and Communications Security,* pp. 1054–1067.

European Union (2020). 5g ppp — smart connectivity beyond 5g, https://cordis.europa.eu /programme/id/H2020_ICT-52-2020.

Feynman, R. P. *et al.* (2018). Simulating physics with computers, *International Journal of Theoretical Physics* **21**, 6/7.

FIPS Publication (2000). Digital signature standard (DSS), *FIPS PUB*, pp. 186–192.

FireEye Inc. (2019). M-trends 2019, *M-Trends 2019.*

Firstbrook, P., Perkins, E., Wheatman, J., Mahdi, D., Zhang, J., and Olyaei, S. (2018). Top security and risk management trends.

Force and Initiative (2013). Security and privacy controls for federal information systems and organizations, *NIST Special Publication* 800, **53**, pp. 8–13.

Fredrikson, M., Jha, S., and Ristenpart, T. (2015). Model inversion attacks that exploit confidence information and basic countermeasures, in *the 22nd ACM SIGSAC Conference.*

Gartner. (2014, June 4). Gartner says big data needs a data-centric security focus.

Gennaro, R., Gentry, C., and Parno, B. (2010). Non-interactive verifiable computing: Outsourcing computation to untrusted workers, in *Advances in Cryptology–CRYPTO 2010: 30th Annual Cryptology Conference, Santa Barbara, CA, USA, August 15-19, 2010. Proceedings 30* (Springer), pp. 465–482.

Gentry, C. (2009). Fully homomorphic encryption using ideal lattices, in *Proceedings of the Forty-First Annual ACM Symposium on Theory of Computing*, pp. 169–178.

Goia, O.-S. (2019). Tisax assessment for information security in the automotive industry, *Tisax*.

Girard, J. and Dulaney, K. (2010). Four architectural approaches to limit business risk on consumer smartphones and tablets.

Goldreich, O., Goldwasser, S., and Halevi, S. (1997). Public-key cryptosystems from lattice reduction problems, in *Advances in Cryptology — CRYPTO'97: 17th Annual International Cryptology Conference Santa Barbara, California, USA August 17–21, 1997 Proceedings 17* (Springer), pp. 112–131. Springer Berlin Heidelberg.

Goldreich, O. (1998). Secure multi-party computation, *Manuscript. Preliminary Version* **78**, 110.

Goldwasser, S., Micali, S., and Rackoff, C. (1989). The knowledge complexity of interactive proof systems, *SIAM Journal on Computing* **18**, 1, 186–208.

Golić, J. D. (1997). Cryptanalysis of alleged a5 stream cipher, in *International Conference on the Theory and Applications of Cryptographic Techniques* (Springer), pp. 239–255.

Goodfellow, I. J., Shlens, J., and Szegedy, C. (2014). Explaining and harnessing adversarial examples, arXiv preprint arXiv:1412.6572.

GSMA (2024). The mobile economy 2024, https://www.gsma.com/solutions-and-impact/connectivity-for-good/mobile-economy/.

Grandison, T., Bilger, M., O'Connor, L., Graf, M., Swimmer, M., Schunter, M., Wespi, A., and Zunic, N. (2007). Elevating the discussion on security management: The data centric paradigm, in *2007 2nd IEEE/IFIP International Workshop on Business-Driven IT Management* (IEEE), pp. 84–93.

Haber, S. and Stornetta, W. S. (1991). *How to Time-Stamp a Digital Document* (pp. 437–455). Springer Berlin Heidelberg.

Hartnett, K. (2019). A new law to describe quantum computing's rise, *Quanta Magazine* **18**.

Hellman, M. (1976). New directions in cryptography, *IEEE Transactions on Information Theory* **22**, 6, 644–654.

Hermann, M., Pentek, T., and Otto, B., *et al.* (2015). Design principles for industrie 4.0 scenarios: A literature review, *Technische Universität Dortmund, Dortmund* **45**, 1–15.

Hoffstein, J., Howgrave-Graham, N., Pipher, J., Silverman, J. H., and Whyte, W. (2003). Ntrusign: Digital signatures using the ntru lattice, in *Cryptographers' Track at the RSA Conference* (Springer), pp. 122–140.

Holdings, T. (2014). Trustwave global security report, *Retrieved June* **15**, p. 15.

Howard, M. and Lipner, S. (2006). *The Security Development Lifecycle*, **8** (Microsoft Press Redmond).

Hu, V. C., Ferraiolo, D., Kuhn, R., Friedman, A. R., Lang, A. J., Cogdell, M. M., Schnitzer, A., Sandlin, K., Miller, R., Scarfone, K., *et al.* (2013). Guide to attribute based access control (abac) definition and considerations (draft), *NIST Special Publication 800*, **162**, 1–54.

Hudson, K. L., Holohan, M., and Collins, F. S. (2008). Keeping pace with the times — the genetic information nondiscrimination act of 2008, *New England Journal of Medicine* **358**, 25, 2661–2663.

Huelsing, A., Butin, D., Gazdag, S.-L., Rijneveld, J., and Mohaisen, A. (2018). XMSS: eXtended Merkle Signature Scheme, RFC 8391, doi:10.17487/RFC8391, https://www.rfc-editor.org/info/rfc8391.

IBM Security (2019). IBM: Cost of a data breach report, *Computer Fraud & Security* **2019**, 8, 4.

IBM Security (2021). Cost of a data breach report 2021, *Risk Quantification* **73**.

IBM (2018). Ibm study: Responding to cybersecurity incidents still a major challenge for businesses, https://www.prnewswire.com/news-releases/ibm-study-responding-to-cybersecurity-incidents-still-a-major-challenge-for-businesses-300613590.html.

International Organization for Standardization (2011). Iso/iec 29100:2011 – information technology — security techniques — privacy framework.

International Organization for Standardization (2015). Iso/iec/ieee 15288: 2015, *IEEE Standard for Systems and Software Engineering — System Life Cycle Processes.*

International Organization for Standardization (2018). Iso 27005: 2018, information technology–security techniques–information security risk management.

International Organization for Standardization (2022). Iec 42010: 2022, "iso/iec 42010: Software, systems and enterprise — architecture description," *Int. Organ. Stand. Geneva Switz.*

International Organization for Standardization (2022). Iso/iec 27001:2022 information security, cybersecurity and privacy protection information security management systems requirements.

International Organization for Standardization (2022). Iec 27002: 2022 information security, cybersecurity and privacy protection — information security controls.

ISACA (2021). Control objectives for information and related technologies (COBIT), www.isaca.org.

Jarecki, S., Krawczyk, H., and Xu, J. (2018). Opaque: An asymmetric pake protocol secure against pre-computation attacks, in *Advances in Cryptology–EUROCRYPT 2018: 37th Annual International Conference on the Theory and Applications of Cryptographic Techniques, Tel Aviv, Israel, April 29-May 3, 2018 Proceedings, Part III 37* (Springer), pp. 456–486.

Katz, J. and Lindell, Y. (2014). *Introduction to Modern Cryptography* (Chapman and Hall/CRC).

Kaur, G. *et al.* (2017). A review on database security, *International Journal of Engineering and Management Research (IJEMR)* **7**, 3, pp. 269–272.

Khovratovich, D., Rechberger, C., and Savelieva, A. (2012). Bicliques for preimages: Attacks on skein-512 and the sha-2 family, in *International Workshop on Fast Software Encryption* (Springer), pp. 244–263.

Kindervag, J., Shey, H., and Mak, K. (2015). *The Future of Data Security and Privacy: Growth and Competitive Differentiation*, Cambridge, MA.

Krawczyk, D. H., Bellare, M., and Canetti, R. (1997). HMAC: Keyed-Hashing for Message Authentication, RFC 2104, doi:10.17487/RFC2104, https://www.rfc-editor.org/info/rfc2104.

Krebs, B. (2018). Google: Security keys neutralized employee phishing, https://krebsonsecurity.com/2018/07/google-security-keys-neutralized-employee-phishing/.

Kruchten, P. B. (1995). The 4 + 1 view model of architecture, *IEEE Software* **12**, 6, 42–50.

Lamberger, M. and Mendel, F. (2011). Higher-order differential attack on reduced sha-256, *Cryptology ePrint Archive.*

Lenstra, A., Wang, X., and de Weger, B. (2005). Colliding x. 509 certificates, *Cryptology ePrint Archive.*

Lenzner, R. (2013). Ibm ceo Ginni Rometty crowns data as the globe's next natural resource.

Li, N., Li, T., and Venkatasubramanian, S. (2006). T-closeness: Privacy beyond k-anonymity and l-diversity, in *2007 IEEE 23rd International Conference on Data Engineering* (IEEE), pp. 106–115.

Lipner, S. (2004). The trustworthy computing security development lifecycle, in *20th Annual Computer Security Applications Conference* (IEEE), pp. 2–13.

Lopez, J. and Dahab, R. (2000). An overview of elliptic curve cryptography, *Technical Report* Institute of Computing, Sate University of Campinas, Sao Paulo, Brazil.

M Series (2015). Imt vision–framework and overall objectives of the future development of imt for 2020 and beyond, *Recommendation ITU* **2083**, 0.

Ma, J., Shrestha, R., Adelberg, J., Yeh, C.-Y., Hossain, Z., Knightly, E., Jornet, J. M., and Mittleman, D. M. (2018). Security and eavesdropping in terahertz wireless links, *Nature* **563**, 7729, 89–93.

Machanavajjhala, A., Kifer, D., Gehrke, J., and Venkitasubramaniam, M. (2007). l-diversity: Privacy beyond k-anonymity, *ACM Transactions on Knowledge Discovery from Data (TKDD)* **1**, 1, 3–es.

Manfredi, L. (2021). Colonial pipeline hack prompts dhs to issue new cybersecurity regulations, http://a.msn.com/00/en-us/AAKnc3b?ocid=se.

Manin, I. I. (1980). Vychislimoe i nevychislimoe, *Soviet Radio.*

Mariantoni, M. (2014). Building a superconducting quantum computer-invited talk in pqcrypto 2014. (October 2014).

Market Search (2021). Database encryption market with report in depth industry analysis on trends, growth, opportunities and forecast till 2026, https://www.marketwatch.com/press-release/database-encryption-market-with-report-in-depth-industry-analysis-on-trends-growth-opportunities-and-forecast-till-2026-2021-05-18.

Mayer, H. (2016). Ecdsa security in bitcoin and ethereum: A research survey, *CoinFaabrik*, June 28, **126,** 50.

McAfee (2019). Cloud adoption and risk report, https://www.mcafee.com/enterprise/en-us/assets/skyhigh/white-papers/cloud-adoption-risk-report-2019.pdf.

McGrew, D., Curcio, M., and Fluhrer, S. (2019). Leighton-Micali Hash-Based Signatures, RFC 8554, doi:10.17487/RFC8554, https://www.rfc-editor.org/info/rfc8554.

McMillan, R. and Proctor, P. (2018). Cybersecurity and digital risk management: Cios must engage and prepare, *Gartner Research.*

Michael McEwen (2016). Naba and dpp broadcasters unite to promote cyber security requirements for suppliers, https://nabanet.com/wp-content/uploads/2017/08/NABAcaster-Issue_26.pdf.

Microsoft Corporation (2020). Password-less protection, https://www.microsoft.com/zh-cn/security/business/identity/passwordless.

Miles, S. H. (2005). *The Hippocratic Oath and the Ethics of Medicine* (Oxford University Press).

Mogull, R. (2006). Database activity monitoring is a viable stopgap to database encryption for the payment card industry data security standard (and beyond), July 2006, *Gartner.*

Moody, D. and Chen, L. (2019). The 2nd round of the nist pqc standardization process, Online: https://csrc.nist.gov/CSRC/media/Presentations/the-2nd-round-of-the-nist-pqc-standardization-proc/images-media/moody-opening-remarks.pdf[accessed: April 2021].

Mouha, N. and Preneel, B. (2013). A proof that the arx cipher salsa20 is secure against differential cryptanalysis. *IACR Cryptol. ePrint Arch.* **2013**, p. 328.

Murthy, P., Bharadwaj, A., Subrahmanyam, P. A., Roy, A., & Rajan, S. (2014). Big data taxonomy, *Cloud Security Alliance.*

Narayanan, A. and Shmatikov, V. (2008). Robust de-anonymization of large sparse datasets, in *2008 IEEE Symposium on Security and Privacy (sp 2008)* (IEEE), pp. 111–125.

NCSC (2020). Quantum security technologies, https://www.ncsc.gov.uk/white paper/quantum-security-technologies.

Netter, E. (2019). Sanction à 50 millions d'euros: au-delà de google, la cnil s' attaque aux politiques de confidentialité obscures et aux consentements creux, *Dalloz IP/IT* **3**, 3, p. 165.

Nevius, D. (2020). The history of the north american electric reliability corporation, *North American Electric Reliability Corporation, Atlanta, GA.*

Nguyen, P. Q. and Regev, O. (2006). Learning a parallelepiped: Cryptanalysis of ggh and ntru signatures, in *Annual International Conference on the Theory and Applications of Cryptographic Techniques* (Springer), pp. 271–288.

Olovsson, T. and Bodley-Scott, J. (2005). The jericho forum: De-perimeterisation of network resources, in *Black Hat Conference 2005/Jericho Challenge Winner*, pp. 1–34.

Organization for Economic Co-operation & Development (1980). OECD guidelines on the protection of privacy and transborder flows of personal data, *Paris, France.*

Paganini, P. (2020, March 23). 538 million Weibo users' records being sold on Dark web. Security Affairs. Retrieved December 19, 2024, from https://securityaffairs.com/100243/data-breach/weibo-data-dark-web.html.

Palacios-Berraquero, C., Mueck, L., and Persaud, D. M. (2019). Instead of 'supremacy' use 'quantum advantage', *Nature* **576**, 7786, 213–214.

Papernot, N., McDaniel, P., and Goodfellow, I. (2016). Transferability in machine learning: From phenomena to black-box attacks using adversarial samples, *arXiv preprint arXiv:1605.07277.*

Parno, B., Howell, J., Gentry, C., and Raykova, M. (2016). Pinocchio: Nearly practical verifiable computation, *Communications of the ACM* **59**, 2, 103–112.

Patrick, N. (2018). Get ready for upcoming 6g wireless too, https://www.networkworld.com/article/3285112/get-ready-for-upcoming-6g-wireless-too.html.

Payment Card Industry (2010). Data security standard, *Requirements and Security Assessment Version* **3**.

Pednault, E. (2019). On quantum supremacy, https://research.ibm.com/blog/on-quantum-supremacy.

Petitcolas, F. A. (2023). Kerckhoffs' principle, in *Encyclopedia of Cryptography, Security and Privacy* (Springer), pp. 1–2.

Popov, A. (2015). Prohibiting RC4 Cipher Suites, RFC 7465, doi:10.17487/RFC7465, https://www.rfc-editor.org/info/rfc7465.

Porter, J. (2019). Google confirms quantum supremacy breakthrough, https://www.theverge.com/2019/10/23/20928294/google-quantum-supremacy-sycamore-computer-qubit-milestone.

Positive Technologies (2019). Cybersecurity threatscape 2018: Trends and forecasts, https://www.ptsecurity.com/ww-en/analytics/cybersecurity-threatscape-2018/.

Prasad, A. R., Arumugam, S., Sheeba, B., and Zugenmaier, A. (2018). 3gpp 5g security, *Journal of ICT Standardization* **6**, 1–2, 137–158.

Preimesberger, C. (2016). Rsa conference 2016: What's old security is new again, https://www.eweek.com/security/rsa-conference-2016-what-s-old-security-is-new-again/.

Preskill, J. (2012). Quantum computing and the entanglement frontier, *arXiv preprint arXiv:1203.5813.*

Qi, A., Shao, G., and Zheng, W. (2018). Assessing china's cybersecurity law, *Computer Law & Security Review* **34**, 6, pp. 1342–1354.

Radack, S. M. (2004). Federal information processing standard (FIPS) 199, standards for security, *Gaithersburg, MD:[US] National Institute of Standards and Technology, US Department of Commerce.*

Rawlinson, K. (2019). Internet of things research study, https://www8.hp.com/us/en/hp-news/press-release.html?id=1744676.

Rivest, R. L., Shamir, A., and Adleman, L. (1978). A method for obtaining digital signatures and public-key cryptosystems, *Communications of the ACM* **21**, 2, 120–126.

Rivest, R. L. (1992). The MD5 Message-Digest Algorithm, RFC 1321, doi:10.17487/RFC1321, https://www.rfc-editor.org/info/rfc1321.

Rogaway, P. (2002). Authenticated-encryption with associated-data, in *Proceedings of the 9th ACM Conference on Computer and Communications Security*, pp. 98–107.

Ross, R. *et al.* (2010). Nist sp 800-37, revision 1, *Guide for Applying the Risk Management Framework to Federal Information Systems.*

Ross, R., McEvilley, M., and Oren, J. (2016). Nist special publication 800-160: systems security engineering considerations for a multidisciplinary approach in the engineering of trustworthy secure systems, *Gaithersburg: National Institute of Standards and Technology.*

Rowley, J. (2007). The wisdom hierarchy: Representations of the dikw hierarchy, *Journal of Information Science* **33**, 2, pp. 163–180.

Sachs, G. (2023). Generative ai could raise global gdp by 7%, https://www.goldmansachs.com/intelligence/pages/generative-ai-could-raise-global-gdp-by-7-percent.html.

Sakai, M. and Oshima, Y. (2020). The enforcement of personal data protection law in Japan, *Global Privacy Law Review* **1**, 3.

Saltzer, J. H. and Schroeder, M. D. (1975). The protection of information in computer systems, *Proceedings of the IEEE* **63**, 9, 1278–1308.

Sarbanes, P. (2002). Sarbanes-oxley act of 2002, in *The Public Company Accounting Reform and Investor Protection Act. Washington DC: US Congress*, **55**, 1–66.

Schwab, K. (2016). The fourth industrial revolution: What it means, how to respond, https://www.weforum.org/agenda/2016/01/the-fourth-industrial-revolution-what-it-means-and-how-to-respond/.

Schwab, K. (2017). *The Fourth Industrial Revolution* (Currency).

Scoping Special Interest Group and Tokenization Taskforce (2011). Information supplement: PCI-DSS tokenization guidelines, *Standard: PCI Data Security Standard (PCI DSS)* **24**.

Scoping Special Interest Group (2011). Information supplement: Pci dss tokenization guidelines, https://www.pcisecuritystandards.org/documents/Tokenization_Guidelines_Info_Supplement.pdf.

SDxCentral Studios (2020). What are the top 5g security challenges? https://www.sdxcentral. com/5g/definitions/top-5g-security-challenges/.

Secretary's Advisory Committee on Automated Personal Data Systems (1973). *Records, Computers, and the Rights of Citizens: Report*, Vol. 10 (US Department of Health, Education & Welfare).

Segal, B. (2021). Breaking down iot security challenges and solutions, https://telnyx. com/resources/iot-security-solutions.

Shannon, C. E. (1949). Communication theory of secrecy systems, *The Bell System Technical Journal* **28**, 4, pp. 656–715.

Shaw, M. and Garlan, D. (1996). *Software Architecture: Perspectives on An Emerging Discipline* (Prentice-Hall, Inc.).

Shor, P. W. (1994). Algorithms for quantum computation: Discrete logarithms and factoring, in *Proceedings 35th Annual Symposium on Foundations of Computer Science* (Ieee), pp. 124–134.

Shostack, A. (2008). Experiences threat modeling at microsoft. *MODSEC@ MoDELS* **2008**, p. 35.

Souppaya, M. and Scarfone, K. (2016). Guide to data-centric system threat modeling, Tech. rep., *National Institute of Standards and Technology*.

Srinivas, S., Balfanz, D., Tiffany, E., Czeskis, A., and Alliance, F. (2015). Universal 2nd factor (u2f) overview, *FIDO Alliance Proposed Standard* **15**.

Stütz, T. and Uhl, A. (2010). Efficient format-compliant encryption of regular languages: block-based cycle-walking, in *IFIP International Conference on Communications and Multimedia Security* (Springer), pp. 81–92.

Stalla-Bourdillon, S. and Knight, A. (2016). Anonymous data v. personal data-false debate: an eu perspective on anonymization, pseudonymization and personal data, *Wisconsin International Law Journal* **34**, 284.

Statista Inc. (2023). Number of smartphone users worldwide from 2013 to 2028, https://www.statista.com/forecasts/1143723/smartphone-users-in-the-world.

Stefik, M. (1997). Shifting the possible: How trusted systems and digital property rights challenge us to rethink digital publishing, *Berkeley Technology Law Journal* **12**, p. 137.

Stevan, V. (2021). Nerc cip standards and cloud computing, https://aka.ms/AzureNERC.

Stine, K. M., Kissel, R., Barker, W. C., Lee, A., Fahlsing, J., and Gulick, J. (2008). *SP 800-60 Rev. 1. Volume I: Guide for Mapping Types of Information and Information Systems to Security Categories; Volume II: Appendices to Guide for Mapping Types of Information and Information Systems to Security Categories* (National Institute of Standards & Technology).

Stoneburner, G., Goguen, A., Feringa, A., *et al.* (2002). Risk management guide for information technology systems, *Nist Special Publication* **800**, 30, 800–30.

Sweeney, L. (2002). k-anonymity: A model for protecting privacy, *International Journal of Uncertainty, Fuzziness and Knowledge-Based Systems* **10**, 05, 557–570.

Szabo, N. (1998). Secure property titles with owner authority, p. 1, http://szabo.best.vwh.net/securetitle.html.

Tapscott, D. and Tapscott, A. (2016). *Blockchain Revolution: How the Technology behind Bitcoin is Changing Money, Business, and the World* (Penguin).

Tracey, S. (2015). The fall of the data retention directive, *Communications Law* **20**, 2, 53.

Transforma Insights (2020). Global iot market to grow to 24.1 billion devices in 2030 generating 1.5 trillion annual revenue, https://transformainsights.com/news/iot-market-24-billion-usd15-trillion-revenue-2030.

Ucedavélez, T. and Morana, M. M. (2015). *Risk Centric Threat Modeling: process for attack simulation and threat analysis* (John Wiley & Sons).

Valsorda, F. (2013). The ecb penguin, https://words.filippo.io/the-ecb-penguin/.

Verizon (2012). Data breach Investigations Report' now available as a free Ibook. (2012, June 18). Retrieved December 19, 2024, from https://www.verizon.com/about/news/press-releases/verizon-2012-data-breach-investigations-report-now-available-free-ibook.

Verizon (2018). Data Breach Investigations Report 11th edition. https://www.verizon.com/business/en-nl/resources/reports/DBIR_2018_Report.pdf.

Verizon (2019, May 19). 2019 Data Breach Investigations Report. https://www.verizon.com/business/en-gb/resources/reports/2019-data-breach-investigations-report.pdf.

Wang, X., Feng, D., Lai, X., and Yu, H. (2004). Collisions for hash functions md4, md5, haval-128 and ripemd, *Cryptology EPrint Archive.*

Wang, X. and Xu, J. (2021). Deterrence and leadership factors: Which are important for information security policy compliance in the hotel industry, *Tourism Management* **84**, 104282.

Warren, S. D. and Brandeis, L. D. (1890). The right to privacy, *Harvard Law Review* **4**, 5, pp. 193–220.

Weber, P. A., Zhang, N., and Wu, H. (2020). A comparative analysis of personal data protection regulations between the eu and china, *Electronic Commerce Research* **20**, pp. 565–587.

Winder, D. (2020). Hackers made tesla cars autonomously accelerate up to 85 in a 35 zone, https://www.forbes.com/sites/daveywinder/2020/02/19/hackers-made-tesla-cars-autonomously-accelerate-up-to-85-in-a-35-zone/.

Wolfgang Goerlich, J. (2017). Encrypt: Protect the business, prevent the threats, https://www.rsaconference.com/library/presentation/usa/2017/encrypt-protect-the-business-prevent-the-threats.

World Economic Forum (2023). Jobs of tomorrow: Large language models and jobs, https://www.weforum.org/publications/jobs-of-tomorrow-large-language-models-and-jobs/.

Yaga, D., Mell, P., Roby, N., and Scarfone, K. (2018). Blockchain technology overview, doi:https://doi.org/10.6028/NIST.IR.8202.

Yao, A. C. (1982). Protocols for secure computations, in *23rd Annual Symposium on Foundations of Computer Science (SFCS 1982)* (IEEE), pp. 160–164.

Index

Q

R

S

www.ingramcontent.com/pod-product-compliance
Lightning Source LLC
LaVergne TN
LVHW021055220926
848484LV00021B/929